Advances in Genetics, Volume 74

Serial Editors

Stephen F. Goodwin
University of Oxford, Oxford, UK

Theodore Friedmann
University of California at San Diego, School of Medicine, USA

Jay C. Dunlap
Dartmouth Medical School, Hanover, NH, USA

Volume 74

The Genetics of Circadian Rhythms

Edited by

Stuart Brody

Founder, Center for Chronobiology
Division of Biological Sciences
University of California
San Diego, La Jolla, California, USA

ELSEVIER

AMSTERDAM • BOSTON • HEIDELBERG • LONDON
NEW YORK • OXFORD • PARIS • SAN DIEGO
SAN FRANCISCO • SINGAPORE • SYDNEY • TOKYO
Academic Press is an imprint of Elsevier

Academic Press is an imprint of Elsevier

525 B Street, Suite 1900, San Diego, CA 92101-4495, USA
225 Wyman Street, Waltham, MA 02451, USA
32 Jamestown Road, London, NW1 7BY, UK
Radarweg 29, POBox 211, 1000 AE Amsterdam, The Netherlands

First edition 2011

ISBN: 978-0-12-387690-4
ISSN: 0065-2660

For information on all Academic Press publications
visit our website at elsevierdirect.com

Printed and bound in USA

11   12   13   10 9 8 7 6 5 4 3 2 1

Contents

Feedback Loop   64
IV. Other Clock-Associated Genes   76
V. FRQ-Less Rhythms   79
VI. Clock-Controlled Genes   89
VII. Outlook   94
References   95

4   **The Genetics of Plant Clocks**   105
    C. Robertson McClung

      I. Introduction   106
     II. Genome-Wide Characterization of Circadian-Regulated
         Genes and Processes   107
    III. Molecular Organization of Plant Circadian Clocks into
         Interlocked Transcriptional Feedback Loops   110
     IV. Conservation of Clock Genes   122
      V. Circadian Clocks Enhance Fitness   124
     VI. Concluding Remarks   126
         Acknowledgment   126
         References   126

5   **Molecular Genetic Analysis of Circadian Timekeeping
    in *Drosophila***   141
    Paul E. Hardin

      I. Introduction   142
     II. The Circadian Feedback Loops of *Drosophila*   144
    III. Posttranscriptional Regulation of Rhythmic
         Transcription   149
     IV. Light Entrainment of the *Drosophila* Circadian
         Feedback Loops   154
      V. Summary and Conclusions   161
         Acknowledgments   164
         References   165

6   **Genetics of Circadian Rhythms in Mammalian Model
    Organisms**   175
    Phillip L. Lowrey and Joseph S. Takahashi

      I. Introduction   176
     II. The Beginning of Mammalian Clock Genetics   178

7 The Genetics of the Human Circadian Clock   231

Luoying Zhang, Chris R. Jones, Louis J. Ptacek, and
Ying-Hui Fu

Contributors

Numbers in parentheses indicate the pages on which the authors' contributions begin.

Deborah Bell-Pedersen (55) Biology Department, 3258 TAMU, Texas A&M University, College Station, Texas, USA

Stuart Brody (1, 55) Founder, Center for Chronobiology, Division of Biological Sciences, University of California, San Diego, La Jolla, California, USA

Jayna L. Ditty (13) Department of Biology, University of St. Thomas, St. Paul, Minnesota, USA

Ying-Hui Fu (231) Department of Neurology, University of California, San Francisco, California, USA

Susan S. Golden (13) Center for Chronobiology and Division of Biological Sciences, University of California, San Diego, California, USA

Paul E. Hardin (141) Department of Biology and Center for Biological Clocks Research, Texas A&M University, College Station, Texas, USA

Chris R. Jones (231) Department of Neurology, University of Utah, Salt Lake City, Utah, USA

Patricia L. Lakin-Thomas (55) Department of Biology, York University, Toronto Ontario, Canada

Phillip L. Lowrey (175) Department of Biology, Rider University, Lawrenceville, New Jersey, USA

Shannon R. Mackey (13) Biology Department, St. Ambrose University, Davenport, Iowa, USA

C. Robertson McClung (105) Department of Biological Sciences, Dartmouth College, Hanover, New Hampshire, USA

Louis J. Ptacek (231) Department of Neurology, University of California, San Francisco, California, USA

Joseph S. Takahashi (175) Department of Neuroscience, Howard Hughes Medical Institute, University of Texas Southwestern Medical Center, Dallas, Texas, USA

Luoying Zhang (231) Department of Neurology, University of California, San Francisco, California, USA

1

Introduction

Stuart Brody

Founder, Center for Chronobiology, Division of Biological Sciences, University of California, San Diego, La Jolla, California, USA

I. INTRODUCTION

Biological oscillations range from the rapid firing of neurons, on the order of milliseconds, to seasonal (or even longer) oscillations that may be measured in months or years. In between are oscillators that have periods of several minutes or hours (glycolysis), circatidal rhythms, daily rhythms (24 h), and circalunar rhythms. This chapter and the subsequent chapters will focus just on the daily clocks, that is, the circadian rhythms.

The generally agreed-on definition of a circadian rhythm has several parts:

(1) an endogenous and self-sustaining oscillator (ESSO);
(2) a period that is approximately, but not exactly, 24 h (circadian);
(3) capable of being shifted or entrained by environmental signals, primarily light;

Advances in Genetics, Vol. 74
0065-2660/11 $35.00
DOI: 10.1016/B978-0-12-387690-4.00001-5

(4) a period that is temperature compensated, that is, when an organism, or a cell, is grown at a different temperature, the period remains roughly constant. This property is a hallmark of a circadian rhythm, as it is usually not found in other oscillators. This property is not called temperature independent, as the phase of the rhythm is still sensitive to abrupt temperature pulses.

The details about these general properties are given in the subsequent chapters.

The study of circadian rhythms can be traced back prior to the rise of modern molecular biology. The application of genetic techniques to study the clock had a fairly clear starting point, as shown in Section II below. The study of circadian rhythms was, to a large extent, descriptive in nature until recently. Now, it has become a very vibrant experimental area. It may very well return to a more holistic approach once this exciting reductionist phase is over.

The use of mutants in the "clockworks" has enabled us to sharpen the criteria for clock parameters, state variables, hands versus gears, etc. It has allowed us to distinguish major players, that is, key genes and proteins, from the ancillary ones as this volume points out. It has allowed us to (1) describe the circuitry of the clockworks in terms of positive and negative feedback loops; (2) attempt to provide the reasons for constructing clocks out of multiple oscillators; (3) understand the roles of input and output mechanisms, their connections to an oscillator, and the path from the oscillator; (4) understand the complexities of coupled oscillators and networks of cells containing independent oscillators; (5) understand the architecture of the clockworks, that is, how the feedback loops are interconnected; and (6) see what the central themes and motifs are in a variety of clock systems.

The molecular circadian oscillator is now described in terms of many components interlocked to each other via positive and negative feedback loops. Circuit diagrams for clocks in various species show many similarities, even though the key clock genes and proteins have very different sequences. It is interesting to point out that such a biological oscillator with positive and negative feedback loops was proposed many years ago (Monod and Jacob, 1961), but somehow was overlooked by the "clock community."

II. AN UNOFFICIAL, ABBREVIATED/ANNOTATED VERSION OF THE HISTORY OF THE GENETICS OF CIRCADIAN RHYTHMS

(1) The origins of these genetic studies may not be completely known, but a good place to start is Bünning's papers of 1932 and 1935. He analyzed different isolates of the plant, Phaseolus, based on rhythms of their leaf

movement in constant darkness, employing QTL (quantitative trait loci) methods. He came to the conclusion that there were many genes that made small contributions to the free-running rhythm. This chapter discouraged people from looking for single-gene "clock" mutants for many years.

(2) However, Konopka and Benzer (1971) isolated just such single-gene mutants in Drosophila, naming them per^s (period-short), per^l (period-long), and per^0 (arrhythmic). Their success dramatically changed the field.

(3) Clock mutants in microorganisms were then found in *Neurospora crassa* by Feldman and Waser (1971) and in Chlamydomonas by Bruce (1972). The mutant approach allowed manipulative studies that could not be performed by other methods (Feldman and Hoyle, 1973).

(4) A parallel line of genetic studies was initiated by Brody and Martins (1973), summarized by Lakin-Thomas *et al.* (1990), employing a large number of existing Neurospora mutants. The aim of these studies was to determine if mutations affecting biosynthetic pathways or mitochondria or developmental processes also had clock effects.

(5) Molecular techniques then allowed cloning and sequencing of the Drosophila *per* gene (Bargiello and Young, 1984; Reddy *et al.*, 1984) and the Neurospora *frq* gene (McClung *et al.*, 1989). Both these genes coded for unusually long proteins, 1200 amino acids for *per*, and almost 1000 for *frq*. There was no sequence similarity between them, surprisingly enough, except for long stretches of Gly and Ser/Threonine repeats. The function of these regions is still unclear although they may be domain "spacer" regions or phosphorylation sites, etc. The presence of the Gly-Ser repeats led to some good-natured teasing about cloning some homologue to the insect genes for silk! There was also some similarity to a mammalian proteoglycan, an extracellular component.

(6) The first mammalian clock mutant described, the τ mutant, by Ralph and Menaker (1988) ushered in another new chapter in clock genetics. Numerous studies employing this mutant have followed, and it encouraged people to look for mutants in mice (see Chapter 6).

(7) More advances came when Young and his lab group employed a P-1 mediated transposon insertion to isolate another player in the clock field, that is, mutants in the *tim* gene (Sehgal *et al.*, 1994) of Drosophila. In Neurospora, the white-collar (*wc-1* and *wc-2*) mutants, whose phenotypes had been described previously, were found to have "clock" effects (Crosthwaite *et al.*, 1997). In both cases, after these mutants were found, primitive circuit diagrams of feedback loops could then be constructed.

(8) The mouse *Clock* mutant (Vitaterna *et al.*, 1994) was the first circadian clock mutant found in mice and was found after a chemical mutagenesis screen that laid the groundwork for subsequent large-scale searches.

(9) In plants, the isolation of mutants in the *toc-1* gene in Arabidopsis by the Kay lab (Millar *et al.*, 1995) employing a luciferase-based screen was noteworthy for several reasons: (1) it was the first single-gene clock mutant in plants; (2) it employed a clever screen based on a new technology; and (3) it allowed for the unraveling of the circuitry of the clock in another species for comparative studies.

(10) Studies on bacterial clocks were given a significant impetus when a cluster encoding three clock genes was discovered (Ishiura *et al.*, 1998) in cyanobacteria. These genes, *kaiA*, B, and C, were unlike the Drosophila or Neurospora clock genes and have led to a series of interesting discoveries and models (see Chapter 2 for details).

(11) In further genetic analysis, additional mutants were found, double clock mutants were constructed, and dominant/recessive relationships were tested. An example of this for Neurospora was reported by Gardner and Feldman (1980). Most mutants were recessive or codominant, depending upon whether the organism was haploid or diploid or which part of the clock system these mutants affected.

(12) Clock gene interactions have been examined in many organisms. One of the first such studies was of the allelic *frq* series of *N. crassa* mutants and their interactions with the *cel* mutant and other mutants (Lakin-Thomas and Brody, 1985). The authors came to the conclusion that interactions could be classified in three ways: multiplicative in terms of periods; epistatic (one mutant negated the effects of the other); or interactive, where the resulting period suggested some type of gene product interaction. This general scheme could also be applied to inhibitor/mutant interactions as well. Mutant interactions in Drosophila (Rothenfluh *et al.*, 2000) followed the same rules.

(13) More clock mutants were later obtained in all of the clock systems. (see Chapters 2–5 for details). Many mutants had effects on light-resetting behavior, others thought to be in "output" processes, turned out to have some effects on period, and others were employed as critical tests of the transcription–translation loop (TTL) model.

(14) A growing body of evidence, particularly in Neurospora, pointed out that even in clock null mutants, such as frq^{10} or *wc-1* KO, or *wc-2* KO, a visible rhythm could still be seen. This led to the postulate of a second oscillator, downstream of the FRQ/WCC oscillator (see Chapter 3 for details).

(15) In a study identifying the first human clock mutant, the well-known familial advanced sleep phase syndrome (FASPS) was traced down (Toh, 2001) to mutations in a human gene *PER2*, homologous to the *per* gene of Drosophila and mice. This discovery had an important effect on the field since it indicated that the circuitry in humans might be similar to that in the model organisms (see Chapters 5 and 6).

(16) More clock genes have been described in animals as detailed in a thorough compilation of clock and clock-associated genes identified over the past decade (Zhang and Kay, 2010).

Because I have been involved with the clock field since 1972 (Brody and Harris, 1973; Delmer and Brody, 1975), I have observed how the field has changed since then. In the early to mid-1970(s), the techniques employed were simple Mendelian genetics, some enzyme assays, some elementary modeling, some cytology, and lots of descriptive biology. Now, one can make a partial list of the genetic/molecular findings or techniques employed in the clock field: the selection for mutants, as opposed to just finding them; studies on alternative splicing; studies that employ antisense; genomics; modifiers and suppressors; site-specific *in vitro* mutagenesis; RNAi; overexpression/gene dosage; mutants in promoter genes; dissection of regulatory regions ("clock boxes") by deletion analysis; conditional promoters; yeast-2 hybrid "bait" methods; microarrays; QTL methods; ChIP analysis, proteomics, metabolomics, etc.

III. COMMON THEMES

A partial summary of the clock properties in many different organisms is compiled in Table 1.1. One can see many differences and similarities across the systems. But there are other themes common to many clock systems that emerge from the chapters in this volume such as

(1) the centrality of TTL (in eukaryotes);
(2) the interlocking of positive and negative feedback loops;
(3) the hierarchal nature of these loops, ranging from a master/slave relationship to a parallel versus series relationship. The nature of the cross talk between the loops still remains to elucidated;
(4) feedback from the "output" pathways back to the oscillator;
(5) feedback from the core oscillator back to input processes, a type of gating;
(6) the property of temperature compensation, relatively unique to circadian oscillators as opposed to other biological oscillators;
(7) the qualities of the system that allow for robustness and for entrainment;
(8) the widespread clock control of gene expression, ranging from 10% of the genome to almost 100%, depending upon species and sensitivity of the methods employed;
(9) the wide variety of phase times where clock-controlled genes show their maximum expression;
(10) the incorporation of environmental sensors such as light-responsive elements into the clock loops of some organisms;

Table 1.1. A Partial Summary and Comparisons of Important Clock Properties

	Synechococcus	Neurospora	Drosophila	Arabidopsis	Hamsters/mice	Humans
Phosphorylation of clock proteins	✓	✓	✓	✓	✓	✓
Dephosphorylation of clock proteins	✓	✓	✓	✓	✓	✓
Nuclear localization	–	✓	✓	✓	✓	?
Clock protein multimerization	✓	✓	✓	✓	✓	?
Clock protein heteromultimerization	✓	✓	✓	✓	✓	?
Light effects	On clock protein levels or states			✓	Via SCN	?
Multiple loops	?	✓	✓	✓	✓	?
Positive/negative feedback loops	✓	✓	✓	✓	✓	?
Transcription/translation effects	✓	✓	✓	✓	✓	?
Clock boxes (promoter regions)	No	✓	✓	✓	✓	?
Temperature compensation	✓	✓	✓	✓	✓	
% Clock-controlled genes (Est.)	30–100	10–20	10	30	10	
Clock "fitness" experiments	✓			✓	✓	

Citations/details are given in the individual chapters.

(11)   the similarity of architecture of the clock loops, even though many of the individual components are not structurally similar;

(12)   the establishment of the criteria for distinguishing "hands" versus "gears," etc.;

(13)   the employment of genetic techniques (forward, reverse, *in vitro*, cloning, etc.) to dissect clock systems;

(14)   the emergence of some of the details behind clock-controlled chromatin remodeling;

(15)   the emergence of the role of "energy metabolism," whether it be redox control or energy charge or other key parameters in clock control;

(16)   the role of proteolysis in specific regulation of clock components.

IV. SOME REFLECTIONS

1. The nature of clock mutations

Most clock mutations appear to be haplo-insufficient or incompletely dominant (also named as semidominant), while others are recessive, and a few are dominant. Based on classical biochemical–genetic studies, the explanation for dominance is as follows: It can be postulated that if the loss of 50% of the dosage leads to a phenotype, then the original diploid state was "poised" on the edge of a rate-limiting step, and there was no excess capacity in that step. This would suggest then that many of the components of the clock mechanism are "poised" in the linear range of a "rate-limiting" step with no excess capacity. Clearly, other explanations for dominant mutations are possible, such as the interference of defective monomers in multimeric complexes.

The method of screening for mutants clearly plays a role in determining the types of mutants found. In some instances, only dominant-negative or incomplete dominant mutants may be found. In other instances, lethals in essential metabolic steps would not show up in screens unless special techniques are employed.

One of the surprising things about the genetics of circadian rhythms is the lack of a battery of temperature-sensitive (T^s) mutants, so elegantly employed for studying the yeast cell cycle, for instance. Such mutants are very useful for looking for mutations in genes that would otherwise be lethal to the cell or organism under one condition but could have a clock phenotype under another condition. They are also more useful than straight on/off "null" mutants, as one can vary the phenotype gradually with temperature. It should also be pointed out that certain specialized methods such as "tilling," targeting-induced local lesions in genomes (Stemple, 2004), have not yet been widely employed (or published) in the clock field.

2. A word about quantitation

It is interesting that the circadian rhythm field describes oscillators in terms of period (τ). Other fields of science and engineering measure oscillations as frequencies ($1/\tau$) or rates. In addition, all biochemical reactions are expressed not in hours, but as rates, that is, micromoles per minute. Therefore, to fully understand the biochemical nature of the clock mechanism, the clock process will have to be expressed as rates, with rate constants, K_m, K_i values, etc., since there does not appear to be any biochemical process that measures time as a "1-day period." This concept is particularly useful when considering temperature compensation (see below). Rate analysis can also be of some use when discussing the relative importance of a given step affected by a mutation in the clock mechanism. The idea of a clock coefficient, first put forth in 1990 (Lakin-Thomas et al., 1990), can be employed. This quotient is the percentage change in the clock rate due to a mutation divided by the percentage change in a reaction rate. For instance, if only a small (2%) change in the clock rate was produced by a large (90%) change in the reaction rate, then this reaction may be of minor importance from the kinetic stand point. It still may be critical to clock functioning, and to viability, but not as central as other players. The clock coefficient concept can also be employed for analyzing the effect of gene dosage.

3. Temperature compensation

Temperature compensation is very characteristic of circadian oscillators, although temperature-compensated ultradian oscillators have been described as well (Lloyd et al., 2003). Temperature compensation is found even in warm-blooded animals (Reyes et al., 2008; Ruby and Heller, 1996) where large temperature variations do not occur. It may have had a selective advantage over time to allow entrainment to a 24-h LD cycle. If the period varied too much with temperature, then it might be out of the range for proper entrainment. The nature of the mechanism of temperature compensation has been debated for a long time, and a whole journal issue was once devoted to it (Ruoff, 1997). An amplitude model was proposed some time ago (Lakin-Thomas et al., 1991) positing that as the temperature increased, the amplitude also increased and that compensated somehow for the increase in reaction rates. It was not clear at that time what the mechanism for an increase in amplitude might be and how this could compensate for the rate increase. Subsequently, some possible mechanisms have been suggested by studies of mutants with an altered T-C, such as in Synechococcus (Murakami et al., 2008; Nakajima et al., 2005; Terauchi et al., 2007), Neurospora (Mehra et al., 2009), plants (Portolés and Mas, 2010; Salome et al., 2010), mammalian fibroblasts (Tsuchiya et al., 2003), and zebrafish (Lahiri et al., 2005).

4. Clock proteins

Some of the key players in the clock mechanism are very large proteins, that is, FRQ ~ 1000 amino acids long, PER ~ 1200, etc. These proteins are much larger than most cellular proteins, exclusive of structural proteins such as myosin, etc. This leads to the question of, what do these large regions of the clock proteins do? Some of the domains have already been identified, such as self-dimerization domains, phosphorylation sites, sites for interaction with other proteins, nuclear localization signals, etc. But there are still large sections of these proteins that are uncharacterized. Perhaps these regions are involved with or somehow connected to intracellular biochemical regulation or output pathway regulation?

5. Forward

Each of the systems/model organisms has a unique role in our deciphering of the clockworks. The bacterial system (Chapter 2) appears to be excellent for obtaining detailed information about protein interactions and structure at the 3-D level and for expressing these interactions quantitatively (K_m, K_i, etc.). In addition, there will be interesting new information about the control of input and output pathways and their relation to cell division and other processes.

The Neurospora system (Chapter 3) seems very well suited to examining the connection between the clock and biochemical regulation, both in terms of control over key output pathways (allosteric reactions) and their input back to the clock. The Neurospora system also allows the study of synchronization within a common cytoplasm. Interesting facts/hypotheses should arise as to how nuclei communicate phase information to each other and how this is then expressed in a patch of the growing front of the colony.

The study of a model plant (Arabidopsis) clock illustrated in Chapter 4 may lead to many unexpected findings about the "photic" life of this organism and its connection to the clockworks. The existence of so many photoreceptors in plants and the seemingly redundant sets of clock genes pose challenges, but also rewards.

The Drosophila system (Chapter 5) seems well suited for understanding the complexities of a nervous system/clock relationship that does not involve the SCN and that has many different types of cells in the brain involved with the clock. A different set of findings may emerge in this organism vis-a-vis the question of central versus peripheral oscillators and their response to stimuli. In addition, it will be fruitful to study the integration of different environmental signals, such as temperature changes and the LD cycle in this organism.

The use of hamsters and mice for clock studies (Chapter 6) harnesses the power of genetics/anatomy/physiology, etc. to make rodents excellent model systems. There may be a considerable body of evidence from these studies as to hierarchical control, organization of the SCN, and the relationship between clock functioning and disease. However, one still needs a diurnal mammalian genetic model that better approximates the human rhythms than laboratory mice or hamsters. The study of clocks in humans (Chapter 7) of course has many important implications for everyday life and for disease.

So each system will contribute according to its special characteristics and will provide us with some ideas about the diversity and architecture of biological oscillators.

References

Bargiello, T. A., and Young, M. W. (1984). Molecular genetics of a biological clock in *Drosophila*. *Proc. Natl. Acad. Sci. USA* **81**, 2142–2146.

Brody, S., and Harris, S. (1973). Circadian rhythms in neurospora: Spatial differences in pyridine nucleotide levels. *Science* **180**, 498–500.

Brody, S., and Martins, S. (1973). Effects of morphological and auxotrophic mutations on the circadian rhythms of *Neurospora crassa*. *Genetics* **74**(Suppl.), s31.

Bruce, V. G. (1972). Mutants of the biological clock in *Chlamydomonas reinhardtii*. *Genetics* **70**, 537–548.

Bünning, E. (1932). Über die Erblichkeit der Tagesperiodizitat bei den Phaseolus-Blättern. *Jahrb. Wiss Bot.* **77**, 283–320.

Bünning, E. (1935). Zur Kenntnis der erblichen Tagesperioidzität bei den Primarblatter von Phaseolus multiflorus. *Jahrb. Wiss. Bot.* **81**, 411–415.

Crosthwaite, S. C., Dunlap, J. C., and Loros, J. J. (1997). Neurospora *wc-1 and wc-2* transcription, photoresponses, and the origins of circadian rhythmicity. *Science* **276**, 763–769.

Delmer, D. P., and Brody, S. (1975). Circadian Rhythms in *Neurospora crassa*: Oscillation in the level of an adenine nucleotide. *J. Bacteriol.* **121**, 548–553.

Feldman, J. F., and Hoyle, M. N. (1973). Isolation of circadian clock mutants of *Neurospora crassa*. *Genetics* **75**, 606–613.

Feldman, J. F., and Waser, N. (1971). New mutations affecting circadian rhythmicity in Neurospora. In "Biochronometry" (M. Menaker, ed.), pp. 652–656. National Academy of Sciences, Washington, DC.

Gardner, G. F., and Feldman, J. F. (1980). The *frq* locus in *Neurospora crassa*: A key element in circadian clock organization. *Genetics* **96**, 877–886.

Ishiura, M., Kutsuna, S., Aoki, S., Iwasaki, H., Andersson, C. R., Tanabe, A., Golden, S. S., Johnson, C. H., and Kondo, T. (1998). Expression of a gene cluster *kaiABC* as a circadian feedback process in Cyanobacteria. *Science* **281**, 1519–1523.

Konopka, R. J., and Benzer, S. (1971). Clock mutants of *Drosophila melanogaster*. *Proc. Natl. Acad. Sci. USA* **68**, 2112–2116.

Lahiri, K., Vallone, D., Gondi, S. B., Santoriello, C., Dickmeis, T., and Foulkes, N. S. (2005). Temperature regulates transcription in the zebrafish circadian clock. *PLoS Biol.* **3**, e351.

Lakin-Thomas, P. L., and Brody, S. (1985). Circadian rhythms in *Neurospora crassa*: Interactions between clock mutations. *Genetics* **109**, 49–66.

Lakin-Thomas, P. L., Cote, G. G., and Brody, S. (1990). Circadian rhythms in *Neurospora crassa*: Biochemistry and genetics. *Crit. Rev. Microbiol.* **17**, 365–416.

Lakin-Thomas, P. L., Brody, S., and Coté, G. (1991). Amplitude model for the effects of mutations on period and phase resetting of the Neurospora circadian oscillator. *J. Biol. Rhythms* **6**, 281–297.

Lloyd, D., Lemar, K. M., Salgado, E. J., Gould, T. M., and Murray, D. B. (2003). Respiratory oscillations in yeast: Mitochondrial reactive oxygen species, apoptosis and time; a hypothesis. *FEMS Yeast Res.* **3**, 333–339.

McClung, C. R., Fox, B. A., and Dunlap, J. C. (1989). The Neurospora clock gene frequency shares a sequence element with the Drosophila clock gene period. *Nature* **339**, 558–562.

Mehra, A., Shi, M., Baker, C. L., Colot, H. V., Loros, J. J., and Dunlap, J. C. (2009). A role for casein Kinase 2 in the mechanism underlying circadian temperature compensation. *Cell* **137**, 749–760.

Millar, A. J., Carré, I. A., Strayer, C. A., Chua, N.-H., and Kay, S. A. (1995). Circadian clock mutants in Arabidopsis identified by luciferase imaging. *Science* **267**, 1161–1163.

Monod, J., and Jacob, F. (1961). General conclusions: Teleonomic mechanisms in cellular metabolism, growth, and differentiation. Cellular Regulatory Mechanisms. Cold Spring Harbor Symposia on Quantitative Biology XXVI. The Biological Laboratory, Cold Spring Harbor, New York, p. 399.

Murakami, R., Miyake, A., Iwase, R., Hayashi, F., Uzumaki, T., and Ishiura, M. (2008). ATPase activity and its temperature compensation of the cyanobacterial clock protein KaiC. *Genes Cells* **13**, 387–395.

Nakajima, M., Imai, K., Ito, H., Nishiwaki, T., Murayama, Y., Iwasaki, H., Oyama, T., and Kondo, T. (2005). Reconstitution of circadian oscillation of cyanobacterial KaiC phosphorylation in vitro. *Science* **308**, 414–415.

Portolés, S., and Mas, P. (2010). The functional Interplay between protein kinase CK2 and CCA1 transcriptional activity is essential for clock temperature compensation in Arabidopsis. *PLoS Genet.* **6**(11), e1001201.

Ralph, M. R., and Menaker, M. (1988). A mutation of the circadian system in golden hamsters. *Science* **241**, 1225–1227.

Reddy, P., Zehring, W. A., Wheeler, D. A., Pirrotta, V., Hadfield, C., Hall, J. C., and Rosbash, M. (1984). Molecular analysis of the *period* locus in *Drosophila melanogaster* and identification of a transcript involved in biological rhythms. *Cell* **38**, 701–710.

Reyes, B. A., Pendergast, J. S., and Yamazaki, S. (2008). Mammalian peripheral circadian oscillators are temperature compensated. *J. Biol. Rhythms* **23**, 95–98.

Rothenfluh, A., Abodeely, M., Price, J. L., and Young, M. W. (2000). Isolation and analysis of six timeless alleles that cause short-or long-period circadian rhythms in Drosophila. *Genetics* **156**, 665–675.

Ruby, N. F., and Heller, H. C. (1996). Temperature Sensitivity of the suprachiasmatic nucleus of ground squirrels and rats in vitro. *J. Biol. Rhythms* **11**, 126–136.

Ruoff, P. (1997). Temperature compensation of circadian and ultradian rhythms. *Chronobiol. Int.* **14**(5), 445–536.

Salome, P. A., Weigel, D., and Robertson McClung, C. (2010). The role of the Arabidopsis morning loop components CCA1, SHY, PRR7, and PRR9 in temperature compensation. *Plant Cell* **22**, 3650–3661.

Sehgal, A. J., Price, J. L., Man, B., and Young, M. (1994). Loss of circadian behavior rhythms and *per* oscillations in the Drosophila mutant *timeless*. *Science* **263**, 1603–1606.

Stemple, D. L. (2004). TILLING-a high throughput harvest for functional genomics. *Nat. Rev. Genet.* **5**, 145–150.

Terauchi, K., Kitayama, Y., Nishiwaki, T., Miwa, K., Murayama, Y., Oyama, T., and Kondo, T. (2007). The ATPase activity of KaiC determines the basic timing for circadian clock of cyanobacteria. *Proc. Natl. Acad. Sci. USA* **104**, 16377–16381.

Toh, K. L. (2001). An hper2 phosphorylation site mutation in familial advanced sleep phase syndrome. *Science* **291**, 1040–1043.

Tsuchiya, Y., Akashi, M., and Nishida, E. (2003). Temperature compensation and temperature resetting of circadian rhythms in mammalian cultured fibroblasts. *Genes Cells* **8,** 713–720.

Vitaterna, M. H., King, D. P. P., Chang, A.-M., Kornhauser, J. M., Lowrey, P., McDonald, J. D., Dove, W. F., Pinto, L. H., Turek, F. W., and Takahashi, J. S. (1994). Mutagenesis and mapping of a mouse gene *Clock*, essential for circadian behavior. *Science* **264,** 719–725.

Zhang, E., and Kay, S. A. (2010). Clocks not winding down: Unraveling circadian networks. *Nat. Rev. Mol. Cell Biol.* **11,** 764–775.

2

The Itty-Bitty Time Machine: Genetics of the Cyanobacterial Circadian Clock

Shannon R. Mackey,* Susan S. Golden,† and Jayna L. Ditty‡

*Biology Department, St. Ambrose University, Davenport, Iowa, USA
†Center for Chronobiology and Division of Biological Sciences, University of California, San Diego, California, USA
‡Department of Biology, University of St. Thomas, St. Paul, Minnesota, USA

I. Introduction
II. The Kai-Based Oscillator: Early Studies
   A. Identification of the *kai* genes
   B. Basic Kai protein properties
   C. Kai protein structure and homotypic and dynamic heterotypic interactions
III. KaiC: Breakthroughs into Oscillator Timing and Clock Synchronization
   A. Which came first? Feedback loops and the *in vitro* oscillator
   B. Ordered KaiC autokinase and autophosphatase activities drive the circadian oscillator
   C. Stabilization of the KaiC-P cycle by monomer shuffling
   D. KaiC ATPase activity as a basis for oscillator timing and cell-division control
IV. Input Pathways: Light-Dependent Cellular Metabolism Synchronizes the Clock with Local Time
   A. Pex protein synchronizes phase with LD cycles
   B. CikA protein bridges input and output pathways
   C. LdpA regulates continuous entrainment of the clock
V. Output Pathways: Multiple Independent Pathways Merge to Coordinate Cellular and Physiological Processes
   A. Global clock-controlled regulation of gene expression

Advances in Genetics, Vol. 74
Copyright 2011, Elsevier Inc. All rights reserved.

0065-2660/11 $35.00
DOI: 10.1016/B978-0-12-387690-4.00002-7

ABSTRACT

The cyanobacterium *Synechococcus elongatus* PCC 7942 has been used as the prokaryotic model system for the study of circadian rhythms for the past two decades. Its genetic malleability has been instrumental in the discovery of key input, oscillator, and output components and has also provided monumental insights into the mechanism by which proteins function to maintain and dictate 24-h time. In addition, basic research into the prokaryotic system has led to interesting advances in eukaryotic circadian mechanisms. Undoubtedly, continued genetic and mutational analyses of this single-celled cyanobacterium will aid in teasing out the intricacies of the Kai-based circadian clock to advance our understanding of this system as well as other more "complex" systems. © 2011, Elsevier Inc.

ABBREVIATIONS

CT	circadian time
DD	constant dark
FRP	free-running period
HPK	histidine protein kinase
KaiC-P	KaiC phosphorylation
LD	light/dark
LL	constant light
PsR	*pseudo*-receiver
RR	response regulator

Definitions

Aschoff's rule	A phenomenon originally described by Jürgen Aschoff in which the circadian period of diurnal organisms decreases with increasing light intensity (nocturnal organisms display the opposite response).
Continuous entrainment	The modulation of the internal period as a result of gradual changes in the environment, including the changes in light intensity that arise from the rotation of the Earth on its axis.
Discrete entrainment	The modification of the relative phase of the rhythm by an abrupt change in an environmental cue, such as a brief dark pulse during the maintenance of the organism in constant light conditions.
Q_{10} value	The ratio of the rate of a given process at one temperature to the rate at a temperature 10 °C lower; Q_{10} values of rhythms maintained by a circadian clock are near 1.0.
Subjective day/night	The time during constant conditions that the organism would ordinarily be in the light or dark, respectively, of a light/dark (LD) cycle.
Two-component signal transduction system	A regulatory system common in prokaryotic organisms that consists of two proteins—a histidine protein kinase that autophosphorylates in response to some stimulus, and the response regulator protein to which it subsequently transfers it phosphate to elicit a response to the initial stimulus.

I. INTRODUCTION

Most organisms are subjected to daily fluctuations in light and temperature as a result of the full rotation of the Earth on its axis approximately every 24 h. Endogenous circadian biological clocks evolved that allow for the anticipation of these daily variations to control rhythmic gene expression, which in turn regulates metabolic and behavioral processes to provide a selective advantage (DeCoursey, 1961; DeCoursey et al., 2000; Johnson, 2005; Michael et al., 2003; Ouyang et al., 1998; Woelfle et al., 2004). For decades after their description, circadian clocks were believed to exist only in eukaryotes, as prokaryotes were thought to lack the complexity that could support a clock, or the lifespan that would benefit from one

(Edmunds, 1983; Kippert, 1987). However, the existence of two incompatible biochemical processes—oxygenic photosynthesis and oxygen-sensitive nitrogen fixation—that occur in some unicellular cyanobacteria led to investigation of the mechanism used to separate these dissonant reactions (Mitsui *et al.*, 1986). Ensuing research uncovered that these and other alternating rhythms in cyanobacteria display the three hallmark circadian characteristics: persistence under constant conditions, entrainment and phase resetting, and temperature compensation (Chen *et al.*, 1991; Grobbelaar and Huang, 1992; Grobbelaar *et al.*, 1986; Huang *et al.*, 1990; Mitsui *et al.*, 1986; Schneegurt *et al.*, 1994; Sweeney and Borgese, 1989).

While several different cyanobacterial strains were identified as having bona fide circadian mechanisms, ultimately *Synechococcus elongatus* PCC 7942 was chosen as the model system due to its genetic malleability (Golden, 1988; Golden *et al.*, 1987). This bacterium offered genetic advantages in that it has a small (2.7 Mb) and a fully sequenced genome (US Department of Energy Joint Genome Institute, www.jgi.doe.gov; Holtman *et al.*, 2005) is naturally transformable (Golden and Sherman, 1984), conjugates with *Escherichia coli* (Elhai and Wolk, 1988), and has a suite of vectors available for cloning *S. elongatus* genes (Clerico *et al.*, 2007). In addition, an easily observable circadian "behavior" was genetically designed by fusing the promoter of the photosynthesis gene *psbAI* (P*psbAI*) or any other *S. elongatus* PCC 7942 promoter to the *luxAB* bioluminescence genes from *Vibrio harveyi* (Kondo *et al.*, 1993; Liu *et al.*, 1995). The resulting bioluminescence obeys all three rules that deem a process under circadian control and can be monitored automatically in high-throughput assays; these reporter strains were paramount to the rapid discovery and characterization of the genes involved in the cyanobacterial circadian mechanism. As such, the *S. elongatus* PCC 7942 model now serves as a leader for understanding a biological clock system and its connection with metabolism, cell division, and other fundamental cellular processes.

In the less than two decades since the development of a tractable prokaryotic clock model system (Kondo *et al.*, 1993), we have made considerable headway in our understanding of the circadian mechanism in cyanobacteria. What has become increasingly apparent is that the prokaryotic circadian clock is quite complex, despite some historical prejudices regarding the lack of complexity in bacterial systems, and basic research into the Kai clock system has led to interesting insights into eukaryotic circadian mechanisms. Our purpose for this review is to summarize the most current information regarding input, oscillator, and output pathways in *S. elongatus* by highlighting the importance of genetic and mutational analyses that helped elucidate the players, biochemical modifications, and protein interactions that drive the circadian oscillator and the rhythms it controls in this single-celled, yet highly complex organism.

II. THE KAI-BASED OSCILLATOR: EARLY STUDIES

A. Identification of the *kai* genes

Since the initial physiological characterization of circadian rhythms in cyanobacteria, genetic investigations into the underlying circadian clock have played a significant role in the discovery of genes that encode key clock proteins. Due to the photoautotrophic nature of *S. elongatus*, maintaining the cells in constant dark (DD) conditions to monitor their circadian activity is not possible. Instead, circadian rhythms of cyanobacterial cells are measured in constant light (LL); the times in LL in which the organism would ordinarily be in the light or the dark of a light/dark (LD) cycle are considered subjective day and subjective night, respectively. To determine the pervasiveness of the clock system on gene expression, a promoterless *luxAB* transposon was inserted throughout the genome randomly to report promoter activity via light production. This study showed that of the over 800 colonies that produced detectable bioluminescence, all displayed the same near 24-h periodicity of rhythmic promoter activity. Two distinct classes of rhythms were detected as displaying expression patterns that are 12 h out of phase from one another: class 1 peaked at subjective dusk, while class 2 peaked at subjective dawn (Liu *et al.*, 1995). The use of the *luxAB* luciferase reporter system provided a practical method to screen tens of thousands of *S. elongatus* colonies to identify mutants defective in maintaining circadian time. Chemical mutagenesis generated a variety of circadian phenotypes that included arrhythmia and altered period lengths, which extended between the shortest at 16 h to the longest at 60 h (Kondo and Ishiura, 1994).

Complementation of the mutants with an *S. elongatus* genomic DNA library showed that each of the circadian defects was rescued by a single locus that contained three adjacent genes (Ishiura *et al.*, 1998). These genes were named *kaiA*, *kaiB*, and *kaiC* from the Japanese kanji for *kaiten*, which translates as a cycle of events suggestive of the turning of the heavens. The *kai* genes are essential for circadian rhythms to persist in *S. elongatus*. Deletion of each of the *kai* genes, singly, by operon, or all three together, renders the clock arrhythmic (Ditty *et al.*, 2005; Ishiura *et al.*, 1998). Importantly, *kai* genes are not essential for cell viability, and *kai* mutants do not display a growth deficiency when grown in pure culture (Ishiura *et al.*, 1998).

The *kai* genes are expressed from two promoters: one promoter drives expression of *kaiA* (P*kaiA*), and a promoter upstream of *kaiB* (P*kaiBC*) expresses a *kaiBC* dicistronic message (Ishiura *et al.*, 1998). Expressing *luxAB* reporter genes from either of these promoters produces class 1 rhythms in bioluminescence that peak about 12 h after release into LL and every 24–25 h thereafter. The cyclic nature of *kai* transcription is mirrored in the accumulation of mRNA, with *kaiA* and *kaiBC* messages showing circadian fluctuations in absolute levels throughout the day (Ishiura *et al.*, 1998). The *kai* genes and their protein products have been shown to possess autoregulatory properties similar to those of eukaryotic circadian systems (Ishiura _et al._, 1998). The presence of KaiA positively regulates P*kaiBC*, as overexpression of KaiA increases the levels of bioluminescence from a P*kaiBC*::*luxAB* reporter. This activation is most likely indirect, as KaiA does not contain a predicted DNA-binding domain. When produced in excess, KaiC represses expression from the *kaiBC* promoter, suggesting a role in negative regulation of its own expression; however, some level of KaiC is necessary for full activity from the *kaiBC* promoter because levels of *kaiBC* expression decrease when *kaiC* is inactivated (Ishiura _et al._, 1998; Iwasaki _et al._, 2002; Nishiwaki _et al._, 2004). Together, the Kai proteins form a negative transcription–translation feedback loop, yet this feedback loop is not required for sustained circadian oscillation as is seen in other eukaryotic systems (Bell-Pedersen _et al._, 2005; Dunlap, 1999).

Interestingly, neither the *kaiA* nor *kaiBC* promoter region contains specific _cis_ elements needed to maintain circadian rhythms _in vivo_. Placing *kaiBC* under control of the heterologous P*trc* promoter from _E. coli_ restores wild-type rhythms of bioluminescence and *kaiBC* mRNA accumulation in a reporter strain that lacks endogenous *kaiBC* genes (Xu _et al._, 2003). Driving either *kaiA* or *kaiBC* expression from a class 2 promoter, such that peak levels of *kai* expression should be 12 h out of phase from that of their wild-type counterpart, can complement the respective *kai* null mutation (Ditty _et al._, 2005). These results provided preliminary evidence to hint that transcriptional–translational feedback is not the key parameter required to maintain rhythmicity in the prokaryotic model.

A short-term overexpression of KaiC can reset and shift the phase of peak expression in subsequent cycles (Ishiura _et al._, 1998; Xu _et al._, 2000). When the levels of *kaiBC* mRNA are on the rise during the subjective day, a pulse of elevated *kaiC* expression causes a phase advance. A short pulse of *kaiC* overexpression during the subjective night, when *kaiBC* levels are declining, causes a phase delay. The value of the phase shift is proportional to the phase difference between the time when the pulse is given and the time when *kaiBC* levels would normally be at their maximum (Ishiura _et al._, 1998). Murayama _et al._ (2008) quantitatively assessed KaiC-mediated transcriptional regulation; *kaiBC* mRNA abundance increased proportionally with the KaiC phosphorylation (KaiC-P) ratio. KaiC-P state was found to be a major determinant in the activation of P*kaiBC*; however, no similar correlation between decreasing P*kaiBC* activity and decreasing KaiC-P ratio was found (Murayama _et al._, 2008).

B. Basic Kai protein properties

Original primary motif searches within the Kai proteins revealed little about how these proteins work together to form a circadian oscillator, as they have no homology to known eukaryotic circadian clock proteins; this turned out to be a fortuitous characteristic in that there were few preconceived notions about their biochemical functions. It was foreshadowed that phosphorylation of KaiC would be integral to its activity as bioinformatic analysis of the KaiC primary amino acid sequence revealed ATP-binding domains that predicted testable phosphorylation events (Ishiura et al., 1998). KaiC is composed of two tandemly duplicated domains (CI and CII) with each domain containing a Walker A P-loop ATP-binding motif, an imperfect Walker B motif, and catalytic carboxylate glutamate residues that are also present in other proteins known to bind ATP (Fig. 2.1). The CI domain also contains two DXXG motifs (where X represents any amino acid residue) that are conserved among members of the GTPase family (Ishiura et al., 1998). Site-directed mutagenesis resulting in a K52H mutant protein of the P-loop 1 ATP-binding motif of CI disrupted the binding of ATP in vitro and led to an inability to complement a kaiC null mutation in vivo. A corresponding K294H mutant in the P-loop 2 of CII had no effect on the ability of KaiC to bind ATP and was able to restore rhythmicity to the kaiC null strain, albeit with a 70-h period (Nishiwaki et al., 2000). Neither mutation affected GTP binding. Mutations in the DXXG motifs also resulted in mutant rhythm phenotypes. The G71A mutant in CI had a lowered amplitude and distorted waveform in PkaiBC expression, and the G114A mutant had a 27-h period and a bimodal waveform. A glutamine residue immediately following the CI DXXG motif, when changed to Q115R, abolished the circadian rhythm (Nishiwaki et al., 2000).

Functions for KaiA and KaiB became apparent when they were incubated with KaiC in the presence of ATP. Although neither KaiA nor KaiB can undergo an autophosphorylation reaction, both alter the rate of KaiC autokinase activity (Iwasaki et al., 2002; Kitayama et al., 2003; Williams et al., 2002). The addition of KaiA to KaiC in vitro increases the rate at which KaiC autophosphorylates by 2.5-fold, while the addition of only KaiB to KaiC does not have any effect. KaiB does, however, abrogate the positive action of KaiA on KaiC when combined together, such that autophosphorylation by KaiC is reduced to about half of the KaiA/KaiC combination. The phosphorylation state of KaiC, as detected by immunoblot from cell extracts, cycles throughout the day, which led to the hypothesis that KaiC-P would be critical for circadian timing (Iwasaki et al., 2002). Mutations in the kai genes or other clock components that affect the ability of KaiC to phosphorylate in a rhythmic fashion alter the functionality of the clock (Iwasaki et al., 2002; Mehra et al., 2006; Xu et al., 2003).

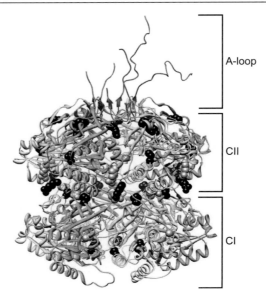

A-loop

CII

CI

Figure 2.1. Period-altering amino acid substitutions mapped to structure of a KaiC hexamer. KaiC monomers consist of two tandemly duplicated domains, CI and CII. These monomers oligomerize into hexameric structures with a thin "waist" region connecting the CI and CII regions. At the C-terminus of each KaiC monomer exists an "A-loop," which determines the steady-state level of KaiC phosphorylation. Amino acid substitutions in the KaiC protein (Ishiura *et al.*, 1998) that result in short-period rhythms (A87V, R215C, and R321Q) are denoted by gray circles, while those substitutions that give rise to period lengths greater than that of wild-type cells (S157C, P236S, R253H, M273I, T409A, and Y442H) are depicted as black circles. Image provided by Yong-Ick Kim, UC-San Diego.

C. Kai protein structure and homotypic and dynamic heterotypic interactions

Numerous structural studies revealed that KaiC, purified from *S. elongatus* PCC 7942 or thermophilic strains, interacts homotypically (Hayashi *et al.*, 2003, 2004, 2006; Kang *et al.*, 2009; Ming *et al.*, 2007; Mori *et al.*, 2002). KaiC, when purified in the presence of ATP, forms a dumb-bell shape consisting of six KaiC monomers and 12 bound ATP molecules (Hayashi *et al.*, 2003; Mori *et al.*, 2002). A thinner region connects the CI and CII structural domains, and a channel exists in the center of the ring structure that narrows near the CI regions of the KaiC monomers (Fig. 2.1). Mutations in each of the Walker A motifs affect the clock phenotype, increase the K_m for ATP, and inhibit hexamerization (Hayashi *et al.*, 2003); however, mutants of the KaiC CI domain Walker A motifs and catalytic carboxylate glutamate residues have greater effect on hexamerization than their CII domain counterparts, suggesting that the CI domain of KaiC is important for hexamerization. The CII domain of KaiC was shown to have phosphorylation activity similar to the wild-type levels, while the CI domain lacks phosphorylation activity. As such, the CII domain, particularly the last 25-aminoacyl residues, was shown to interact with KaiA, while CI alone does not (Hayashi et al., 2004, 2006; Pattanayek et al., 2006).

Structural information obtained for KaiA provided additional hints at its function (Garces et al., 2004; Uzumaki et al., 2004; Vakonakis et al., 2004; Williams et al., 2002). KaiA forms a dimer in solution through interaction of a novel fold in its C-terminus. The C-terminal domain of KaiA is capable of binding a small peptide from the CII domain of KaiC in solution and is alone responsible for the stimulation of KaiC autophosphorylation (Uzumaki et al., 2004; Vakonakis et al., 2004; Williams et al., 2002). The N-terminal domain of KaiA resembles the receiver domains of proteins typically involved in bacterial two-component signal transduction pathways; however, this domain is considered a pseudo-receiver (PsR) in that it does not contain the conserved aspartate residue of bona fide receivers that accepts a phosphoryl group from a partner kinase (Stock et al., 2000). The hypothesized function of the N-terminal domain of KaiA is to act as a conduit for environmental information to the Kai oscillator to regulate the extent by which KaiA stimulates the autophosphorylation of KaiC (Williams et al., 2002; see Section IV.C).

The X-ray crystal structure of KaiB reveals four subunits that are organized into a dimer of two asymmetric dimers that bind to KaiC in the presence of KaiA and ATP (Iwase et al., 2005; Pattanayek et al., 2008). The KaiB C-terminal tail is flexible, negatively charged, and via its interaction with KaiC has been hypothesized to displace ATP and interfere with KaiC subunit exchange by interacting with the CII hexameric barrel (Pattanayek et al., 2008). The same structural studies also suggest that KaiB binding to KaiC sterically inhibits the binding of KaiA.

As suggested by the biochemical relationship among the Kai proteins, KaiA, KaiB, and KaiC physically interact with one another both in vitro and in vivo. Initial yeast two-hybrid assays demonstrated both homotypic and heterotypic interactions among these clock components (Iwasaki et al., 1999). In vitro analysis using GST-Kai protein fusions also showed that KaiA and KaiB only weakly interact, but their interaction is strengthened in the presence of KaiC. Both the CI and CII domains of KaiC interact with the other two Kai proteins, and the interaction between KaiA and KaiB is strengthened more by the CI domain than by full-length KaiC; CII had the least positive effect on KaiA/KaiB binding and further demonstrates the differences in function of the two KaiC domains. Mutational analysis also supports the importance of Kai interactions in maintaining circadian rhythmicity. A mutation in kaiA that causes a 33-h free-running period (FRP) was shown to enhance the interaction of KaiA with KaiB (Iwasaki et al., 1999). Mutations in kaiC that cause enhanced interactions between KaiA and KaiC lengthen the circadian period to 44 h and 60 h, respectively, whereas a KaiC mutant that interacts weakly with KaiA results in a short 16-h period in LL (Taniguchi et al., 2001). These data show that strong interactions between KaiA, KaiB, and KaiC can slow the rate at which the clock proceeds.

   Not only do the Kai proteins interact, but they do so dynamically both in vivo and in vitro. KaiA, KaiB, and KaiC have been shown to interact in a ratio of 1:1:4 by weight (Kitayama et al., 2003) with the largest amounts of KaiA and KaiB proteins interacting with KaiC 16–20 h after being released into LL to form a large complex during the early subjective night (Kitayama et al., 2003). In the cell, KaiB and KaiC protein levels display robust rhythmicity in either LD or LL conditions with peak levels occurring 4–6 h ($\sim$CT 15) after the peak in mRNA. KaiA protein levels are thought to be more consistent throughout the circadian cycle (Ivleva et al., 2005; Xu et al., 2000). The result of the dynamic interaction of the Kai proteins is the alteration of the KaiC-P state at various times of day with the highest levels of KaiC-P appearing to accumulate around CT 12–16 with reduced levels at CT 0–4 (Iwasaki et al., 2002). Kai protein stability is also rhythmic, as KaiC is most stable with a 20-h half-life at CT 16 when accumulation and phosphorylation are at maximum levels; in the early subjective day, KaiC stability is decreased to a 6-h half-life (Imai et al., 2004). It has been suggested that the Kai system may resemble an F1-ATPase system, whereby KaiA rotates inside KaiC through phosphorylation and release of ADP, and KaiB association slows down the rotation antagonizing the function of KaiA on KaiC-P (Wang, 2005).

III. KAIC: BREAKTHROUGHS INTO OSCILLATOR TIMING AND CLOCK SYNCHRONIZATION

What became clear from early studies into the Kai oscillator was that dynamic interactions among KaiA, KaiB, and KaiC lead to KaiC-P events that are central to the timing mechanism, but questions still remained regarding the biochemical basis for timekeeping. Pivotal genetic and mutational studies led to four major findings that have adjusted previous paradigms established for not only the cyanobacterial circadian system but also clock mechanisms in general: first, the canonical transcription–translation feedback loop, central to timing models in eukaryotic circadian oscillators, takes a back seat to strictly posttranslational oscillations in KaiC phosphorylation; second, ordered KaiC autokinase and autophosphatase activity drives the Kai circadian oscillator; third, shuffling of various KaiC monomer phosphoforms between KaiC hexamers provides a mechanism for stability and synchronization; and fourth, the innate and slow-turnover ATPase activity of KaiC dictates timing in the Kai oscillator.

A. Which came first? Feedback loops and the *in vitro* oscillator

Clearly, initial studies into the Kai oscillator demonstrated that the *kai* genes and their protein products possess autoregulatory properties, similar to those of eukaryotic circadian systems (Ishiura et al., 1998). However, placing *kai* genes under the control of heterologous promoters was shown to have little effect on timekeeping (Ditty et al., 2005; Xu et al., 2003), indicating that transcriptional control of clock genes is not central to Kai-based timekeeping. Arguably, the biggest breakthrough toward understanding the mechanism of oscillations came from the realization that transcription and translation are not required at all for the maintenance of circadian rhythmicity in the prokaryotic model. An *in vivo* temperature-compensated cycling of KaiC-P was sustained in the absence of nascent rhythmic accumulation of Kai proteins or *kai* mRNA for 56 h in DD (Tomita et al., 2005). In addition, the presence of transcription and translation inhibitors had no effect on DD KaiC-P rhythms, suggesting a self-sustainable posttranslational rhythm in KaiC-P. The C28a *kaiC4* allele, which directs a 28-h period, resulted in a KaiC-P rhythm extended by 4 h in DD (Tomita et al., 2005). Most convincingly, *in vitro* incubation of KaiA, KaiB, and KaiC with ATP leads to a stable, temperature compensated, near 24-h rhythm in KaiC-P for multiple cycles (Fig. 2.2A), and various KaiC variants that cause period changes *in vivo* had similarly correlated changes in the period of the *in vitro* KaiC-P rhythms (Nakajima et al., 2005).

While *in vitro* oscillation of KaiC-P in the absence of transcriptional–translational feedback shattered the dogma, further investigation into the timing mechanism *in vivo* revealed that transcription–translation-based oscillation of *kai* genes is still important. In a KaiA-overexpressing strain, which forces constitutive KaiC-P, temperature-sensitive and phase resettable rhythms in P*kaiBC* and KaiB and KaiC are detected, albeit with a shorter period and higher trough (Kitayama et al., 2008). Strains producing unphosphorylated S431A/T432A mutant KaiC proteins are arrhythmic at the level of the P*kaiBC* promoter, and S431E/T432E KaiC mutant proteins that are phospholocked generate damped and 48-h period rhythms. In a K294H KaiC mutant strain that lacks autokinase activity, KaiC is unphosphorylated; however, a weak P*kaiBC* oscillation with a long period is evident (Kitayama et al., 2008). Therefore, multiple coupled oscillatory systems are important for precise and robust rhythms, and the importance of the transcription–translation feedback loop plays second chair to the

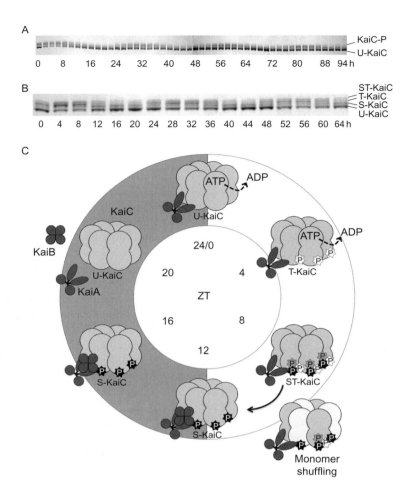

Figure 2.2. *In vitro* KaiC-P oscillation and biochemical model for KaiC-P oscillatory activity. (A) *In vitro* oscillation of KaiC-P for 4 days taken at 2-h time points. U-KaiC, unphosphorylated KaiC; KaiC-P, phosphorylated KaiC. (B) *In vitro* KaiC-P oscillation with resolved phosphoforms. *In vitro* reactions were incubated for 3 days with samples taken at 4-h time points. Four phosphoforms of KaiC exist over the course of a phosphorylation cycle. ST-KaiC, S431 and T432 KaiC double phosphoform; T-KaiC, T432 KaiC phosphoform; S-KaiC, S431 KaiC phosphoform; and U-KaiC, unphosphorylated KaiC. Immunoblot images provided by Yong-Ick Kim, UC-San Diego. (C) Model for ordered phosphorylation and monomer shuffling activity in KaiC. Initially at ZT 0, unphosphorylated KaiC (U-KaiC) associates with KaiA to induce phosphorylation of KaiC (T-KaiC) at T432 (white star) at ZT 4, which is subsequently followed by phosphorylation at S431 (black star) and possibly T426 (gray star) for a fully phosphorylated KaiC (ST-KaiC) by ZT 8. Fully phosphorylated KaiC initiates the dephosphorylation stage and is amenable to monomer shuffling between ZT 8 and 12 to synchronize multiple intracellular KaiC hexamers, leading to the lone S431 phosphoform (S-KaiC) by ZT 12 and association with KaiB. KaiC continues autodephosphorylation through ZT 16, yielding U-KaiC by ZT 20. Note that indicated KaiC phosphoforms at each ZT time represent the most prominent KaiC-P phosphoform relative to the population. P, phosphate.

primary posttranslational KaiC-P oscillation. It is the work of the entire clock system, with its input and output pathways, that helps to coordinate this independent Kai oscillation with the environmental day/night cycle.

B. Ordered KaiC autokinase and autophosphatase activities drive the circadian oscillator

KaiC protein autophosphorylates when incubated with radiolabeled ATP, and crystallographic, mass spectrometric, and mutational analyses demonstrated at least two main, and one additional, phosphorylation sites in KaiC (Nishiwaki et al., 2004; Pattanayek et al., 2004, 2009; Xu et al., 2004, 2009). Kinetic studies divulged an intricate and ordered phosphorylation of KaiC on two adjacent aminoacyl residues in CII, S431, and T432. Four phosphoforms of KaiC exist over the course of a phosphorylation cycle with T432-KaiC-P and S431-KaiC-P phosphoforms predominating during the phosphorylation and dephosphorylation phases, respectively (Fig. 2.2A and B). When KaiA is mixed with unphosphorylated KaiC, T432 is initially phosphorylated followed by S431 to result in a fully phosphorylated KaiC. T432 and S431-KaiC double phosphorylation initiates the switch in KaiC from autokinase to autophosphatase activity, resulting in the dephosphorylation of T432 followed by S431. As expected, KaiA association is only seen during the initial KaiC-P phase. Association of KaiC with KaiA and KaiB results from S431-KaiC-P, and these interactions are important for maintaining the amplitude of KaiC-P (Nishiwaki et al., 2007; Rust et al., 2007). An additional third phosphorylation site at T426 seems to play a role in KaiC-P status. Structural evidence suggests that T426 and S431 face each other across a looped region of KaiC (Xu et al., 2004), and that T426 can be phosphorylated in vitro (Pattanayek et al., 2009). In vivo studies of T426A and T426E mutant strains, which are expected to mimic unphosphorylated and phospholocked states, respectively, abolish rhythmicity (Xu et al., 2009). Various substitutions at T426, S431, and T432 cannot support KaiC-P rhythms, and the T426N mutant KaiC variant specifically hinders the rate of dephosphorylation. When overexpressed with wild-type KaiC, T426N or T426E KaiC variants lengthen the period of the clock to 31 h in vivo, inhibit interactions with KaiA and KaiB, and increase the Q_{10} value, a measure of temperature compensation, to 1.3. While direct phosphorylation of T426 is debatable, this residue could be transiently phosphorylated or be involved in stabilizing phosphorylation at S431 (Pattanayek et al., 2009; Xu et al., 2009).

Additional structural and mutational analyses of KaiC revealed how the KaiA and KaiB proteins act to shift the active state of KaiC from autokinase to autophosphatase (Kim et al., 2008). Aminoacyl residues 488–497 at the C-terminus of every KaiC monomer constitute a regulatory switch called the "A-loop," and whether these residues are buried or exposed determines the steady-state level of phosphorylation (Fig. 2.1). The current model depicts KaiA as stabilizing the exposed loop to direct KaiC phosphorylation, as a KaiC487 mutant that lacks the A-loop and additional tail residues (498–519) structurally mimics the exposed state and is constitutively phosphorylated irrespective of the presence of KaiA. A KaiC497 mutant protein is truncated such that it has the A-loop but not the solvent-exposed tail, which mimics an A-loop buried state resulting in unphosphorylated KaiC, either alone or in combination with KaiA. Truncations partway through the A-loop are hyperphosphorylated, but to a lesser level than that of KaiC487. In addition, ectopic expression of KaiC487 or KaiC497 abolishes rhythmicity in wild-type *S. elongatus* PCC 7942, as measured by the P*kaiBC*::*luxAB* reporter strain, demonstrating dominant-negative effects. KaiA directly binds and stabilizes the exposed state of the A-loop, while KaiB indirectly stabilizes the buried state of the A-loop by hindering KaiC interaction with KaiA. It is hypothesized that A-loop displacement moves bound ATP closer to the sites of phosphorylation (Kim *et al.*, 2008).

In addition to Kai protein interactions, concentrations of the Kai proteins affect the *in vitro* KaiC-P oscillation (Nakajima *et al.*, 2010). In a reaction with 3.5 μM KaiC, a KaiA concentration range from 0.6 to 6.0 μM is necessary for rhythmic KaiC-P and the period and amplitude are continuously influenced by KaiA concentration; outside this concentration range, the KaiC-P rhythm is damped. Although KaiB protein at a concentration of 1.75 μM or higher is necessary for a KaiC-P rhythm, excess KaiB does not affect the period or amplitude. As such, KaiA and KaiB are "parameter-tuning" and "state-switching" regulators of the KaiC-P rhythm, respectively (Nakajima *et al.*, 2010). The ability of KaiA to fine-tune KaiC-P rhythms based on concentration may give future insight to an entrainment mechanism for the clock.

C. Stabilization of the KaiC-P cycle by monomer shuffling

The Kai circadian system functions as a result of quantitative fluctuations in clock components and KaiC-P levels that ultimately dictate cellular activity. Therefore, the question of how such stable oscillations are synchronized among multiple KaiC hexamers in dynamic interaction with all other aspects of the clock drew attention. Studies on the rhythm in single cells revealed that the oscillations in individual cells are extremely resilient to outside perturbations; single *S. elongatus* cells maintain a robust circadian oscillation with a correlation time of several months (Amdaoud *et al.*, 2007; Mihalcescu *et al.*, 2004). Therefore, stability in the clock is innate to the clock machinery of individual cells without the need for cell-to-cell communication.

The stability phenomenon has provided fodder for many mathematical models predicting how synchronization occurs within the Kai oscillator (Clodong *et al.*, 2007; Eguchi *et al.*, 2008; Emberly and Wingreen, 2006;

Kurosawa *et al.*, 2006; Markson and O'Shea, 2009; Mehra *et al.*, 2006; Miyoshi *et al.*, 2007; Mori *et al.*, 2007; Nagai *et al.*, 2010; Rust *et al.*, 2007; Takigawa-Imamura and Mochizuki, 2006; van Zon *et al.*, 2007; Yoda *et al.*, 2007). Stability is based on monomer exchange among KaiC hexamers and is linked to KaiC dephosphorylation. Using native and FLAG-tagged KaiC monomers, the monomers were shown to shuffle among KaiC hexamers, and this shuffling is phosphorylation-dependent as S431A and T432A mutant monomers do not shuffle (Kageyama *et al.*, 2006). KaiA inhibited shuffling and KaiB had no effect, suggesting that monomer shuffling acts as a potential dephosphorylation mechanism by diluting KaiC-P with nascent, unphosphorylated KaiC (Kageyama *et al.*, 2006). This model was tested by monitoring resilience of the *in vitro* KaiC-P rhythm by mixing KaiC hexamers from different phases offset by 4 h. When KaiC hexamers from six different circadian phases are mixed, the overall KaiC-P pattern is immediately rhythmic, suggesting that individual KaiC hexamers synchronize due to monomer shuffling. KaiC-P cycling persists without damping for 10 days *in vitro* as long as sufficient ATP is present in the reaction. Unphosphorylated S431A/T432A KaiC proteins do not affect monomer shuffling while the phospholocked S431D/T432E KaiC proteins increase shuffling by almost twofold and fix KaiC in the dephosphorylation stage, leading to arrhythmia. In addition, KaiC hexamers in the dephosphorylation state are dominant in shuffling as they have been shown to exchange monomers with KaiC proteins in other states. Therefore, shuffling is tightly linked with dephosphorylation and is mediated early in the dephosphorylation phase (ZT 28–32 h; Ito *et al.*, 2007).

D. KaiC ATPase activity as a basis for oscillator timing and cell-division control

While earlier studies on nucleotide binding revealed insights into KaiC function and hexamerization (Hayashi *et al.*, 2003, 2004, 2006; Ishiura *et al.*, 1998; Nishiwaki *et al.*, 2000), investigations into the rate of ATP consumption by KaiC revealed that KaiC consumes ATP at an extremely low level, as each monomer only requires approximately 15 ATP molecules for one circadian cycle (Terauchi *et al.*, 2007). Peak ATPase activity occurs 4 h before phosphorylation of KaiC *in vitro* (Terauchi *et al.*, 2007), and *in vivo* KaiC ATPase is at its peak at CT 12 (Dong *et al.*, 2010).

One major finding regarding KaiC ATPase activity is the linear correlation between ATPase activity and circadian frequency. The ATPase activities of short and long period KaiC mutant proteins are higher and lower than wild-type KaiC, respectively, such that timing of the circadian cycle correlates with the rate of ATP hydrolysis (Terauchi *et al.*, 2007). Per circadian dogma, period must be temperature compensated. Using purified KaiC from *Thermosynechococcus elongatus* BP-1, ATPase activity was shown to be temperature-compensated in the wild-type, but not in a S431A/T432A KaiC-P double mutant, providing a possible mechanism for temperature compensation of the circadian period (Murakami *et al.*, 2008).

Recent studies of the role of KaiC ATPase activity reveal a link between ATPase activity and clock output by providing a checkpoint for cell division (Dong *et al.*, 2010; see Section V.F). An oscillator state that correlates with elevated KaiC-P under normal conditions is responsible for a daily pause in cytokinesis; however, the correlating activity matches that of ATPase activity rather than that of phosphorylation of KaiC. For example, cells expressing the S431D/T432E phospholocked KaiC variant mimic the peak KaiC-P state, but this phosphomimicry does not inhibit cell division. Conversely, the S431A/T432A unphosphorylated KaiC variant blocks cell division. This conundrum was resolved through the recognition that S431A/T432A and other variants that cause cell elongation have elevated ATPase activity, and low ATPase mutants have no effect on cell division (Dong *et al.*, 2010). Thus, although the phosphorylation cycle is the trackable mark of oscillator progression, it only hints at other processes that comprise the oscillator time stamp.

Generating an *in vitro* Kai circadian oscillation in KaiC-P was a monumental breakthrough that allowed prokaryotic chronobiologists to think outside the box for an endogenous timekeeping mechanism (Fig. 2.2C). The combination of ordered KaiC autokinase and autophosphatase activity drives the Kai circadian oscillator, and ATPase activity provides the biochemical basis for timekeeping. Shuffling of various KaiC monomer phosphoforms among KaiC hexamers provides a mechanism for stability and synchronization. A complete clock system is more than an oscillator alone, but rather the sum of all of the clock parts; recent insights into input and output pathways have further shaped the cyanobacterial circadian model system.

IV. INPUT PATHWAYS: LIGHT-DEPENDENT CELLULAR METABOLISM SYNCHRONIZES THE CLOCK WITH LOCAL TIME

As an obligate photoautotroph, *S. elongatus* benefits from the ability to organize its cellular processes in anticipation of "feeding time" that occurs with the rising sun. In contrast to the dedicated photoreceptors of eukaryotic clock systems that transduce light cues to their central oscillator components, no photoreceptors have been identified as essential to the input pathways of the cyanobacterial clock. Multiple genetic screens for phase-resetting mutants as well as directed gene inactivation of each of the seven predicted blue light photoreceptors failed to demonstrate a direct link between these photoreceptors and input to the oscillator (Mackey *et al.*, 2009). Instead, the *S. elongatus* input pathways rely on the metabolic changes within the cell that result from changes in light quantity to synchronize their circadian cycle to that of the environment.

Mutations have been identified in three *S. elongatus* genes—*pex* [*period extender, cikA (circadian input kinase)*, and *ldpA (light-dependent period)*]—that alter the ability of *S. elongatus* to effectively synchronize the phasing of behaviors with repetitive LD cues, reset phase in response to changes in light cues, and adjust period in response to subtle differences in light intensity, respectively. Inactivation of any of these three genes results in a short circadian period as compared to that of wild-type cells (Katayama *et al.*, 2003; Kutsuna *et al.*, 1998, 2007; Schmitz *et al.*, 2000); these data suggest that the *S. elongatus* input pathway (s) is used to delay the internal Kai-based oscillation to better match daily time. In addition to these components, KaiA has been implicated as being a key player in the coordination between environmental stimuli and the timing of the KaiC-P rhythm (Wood *et al.*, 2010). Given the central role that KaiC plays in the circadian system, it is not surprising that a mutation of *kaiC* exists that allows cells to maintain near 24-h time but prevents the ability to synchronize that timing with the solar day (Kiyohara *et al.*, 2005).

A. Pex protein synchronizes phase with LD cycles

The *pex* gene was originally identified through its apparent complementation of a 22-h short-period *kaiC* mutant; further analysis showed that the resulting wild-type rhythm was due to the presence of an ectopic copy of the *pex* gene, which resulted in a 2-h extension of the 22-h mutant period (Kutsuna *et al.*, 1998). The *pex* gene encodes a 148 amino acid protein whose structure has a winged-helix motif similar to DNA-binding domains of the PadR family (Arita *et al.*, 2007). Overexpression of the *pex* gene by an inducible promoter demonstrated a dose-dependent lengthening of the period of the circadian rhythm in both wild-type and *kai* circadian period mutant backgrounds, while inactivation of *pex* results in a 1-h period shortening in LL conditions (Kutsuna *et al.*, 1998). Cells that lack *pex* do not fully synchronize to the entraining LD cycles, such that the resulting phase of the rhythm in LL is advanced as compared to wild type (Takai *et al.*, 2006). This defect in synchronization likely results from a lack of Pex protein accumulation during the dark (subjective night) phase of the cycle, during which wild-type *pex* mRNA and Pex protein levels peak at ZT 16 and ZT 20, respectively, during LD12:12 cycles (Takai *et al.*, 2006).

Pex protein is predicted to function as a dimer and binds preferentially to the upstream promoter region of the *kaiA* gene; individual alanine substitution of arginine residues in the wing region abolishes the DNA-binding activity of Pex (Arita *et al.*, 2007). Cells that lack *pex* display a substantial increase in *kaiA* promoter activity and mRNA accumulation while overexpression of Pex protein reduces *kaiA* expression (Kutsuna *et al.*, 2007). The period length that results from *pex* inactivation or overexpression can be recapitulated by direct manipulation of *kaiA* levels using an inducible promoter to overexpress wild-type or antisense *kaiA* messages, respectively. Pex is predicted to function during the dark phase of an LD12:12 cycle, where its accumulation likely leads to the repression of *kaiA* through direct binding of the *kaiA* promoter region; the resulting decrease in KaiA protein relieves the positive effect of KaiA on KaiC-P and delays the internal oscillation such that it more closely matches that of the environment (Kutsuna *et al.*, 2007).

B. CikA protein bridges input and output pathways

Unlike *pex* mutants that display a subtle phase difference as compared to that of wild type, cells that lack the *cikA* gene are unable to effectively reset the phase of their circadian rhythm to abrupt changes in environmental stimuli, such as pulses of darkness. The *cikA* gene was originally identified as altering expression of a P*psbAII* luciferase reporter by causing a shortened circadian period ($\sim$22 h), altered phase angle, and low-amplitude rhythmicity in LL conditions (Schmitz *et al.*, 2000). Subsequent analyses revealed a cell-division defect in the *cikA* mutant (Miyagishima *et al.*, 2005). The role of CikA in circadian periodicity appears to be nonoverlapping relative to those of the Kai proteins because inactivation of *cikA* in *kai* mutant backgrounds results in additional period shortening and adjustments to the phase angle as compared to the *kai* mutation alone (Schmitz *et al.*, 2000). Overexpression of *cikA in vivo* results in arrhythmia (Mutsuda *et al.*, 2003). CikA protein accumulates in a circadian fashion with lower levels during the subjective day and an increase during the subjective night (Ivleva *et al.*, 2006). In the absence of *kaiC*, CikA levels remain low and constant in LL, but not in LD conditions, which suggests that the clock system can regulate CikA protein levels in the absence of external cues. Thus, the level of CikA protein is modulated both by the environment and by the circadian system.

Bioinformatic analysis of the CikA protein predicted an N-terminal GAF domain, which is typically involved in bilin chromophore attachment in bacterial photoreceptors. The presence of this domain initially suggested a role for CikA in phototransduction, which led to the testing of CikA as a component on the input pathway of the clock. As predicted, inactivation of *cikA* prevents *S. elongatus* cells from resetting the phase of their circadian rhythm by more than 2 h in response to pulses of darkness that would otherwise cause wild-type cells to adjust their phase by up to 8–10 h (Schmitz *et al.*, 2000). This phenotype is not likely attributed to the GAF domain, which lacks the conserved cysteine amino acid residue responsible for bilin adduct formation and phototransduction; attempts to purify CikA with an attached bilin *in vivo* were unsuccessful (Mutsuda *et al.*, 2003). Rather, this GAF domain regulates the phosphorylation state of the CikA central histidine protein kinase (HPK) domain, which is similar to that of sensor protein kinases involved in bacterial two-component regulatory systems (Stock et al., 2000). Removal of the GAF domain results in severely decreased CikA autophosphorylation at its conserved histidine residue (H393) in vitro (Fig. 2.3A; Mutsuda et al., 2003). A CikA variant that lacks the GAF domain is unable to complement a cikA null in vivo (Zhang et al., 2006), which is likely due to the decreased autophosphorylation of CikA, an essential posttranslational modification for normal CikA function.

The C-terminal end of the CikA protein contains a PsR domain that is receiver-like in sequence and structure (Gao et al., 2007), with the exception of the conserved aspartic acid residue that is involved in phosphotransfer from the HPK partner protein of a true two-component system. No phosphorylation event could be detected in experiments designed to allow the transfer of a phosphate group from the HPK of CikA to its PsR domain (Mutsuda et al., 2003). The PsR is involved in multiple roles that provide CikA with its fundamental phenotypic characteristics. It is predicted that the PsR domain physically blocks the H393 residue through direct interaction with the CikA HPK domain (Gao et al., 2007); this interaction would repress autophosphorylation and is consistent with in vitro experiments that show a 10-fold increase in phosphorylated CikA when the PsR domain is removed (Mutsuda et al., 2003). Additionally, the PsR domain is necessary for CikA localization (Fig. 2.3B). A fluorescent-tagged CikA variant that lacks the PsR domain is found throughout the cell and no longer localizes to the pole of the cell like that of the wild-type CikA protein (Fig. 2.3C; Zhang et al., 2006). Overexpression of the CikA–PsR alone in a wild-type background produces a phenotype similar to that of the cikA null, which suggests that the free PsR constructs may be competing with wild-type CikA for binding sites at the pole to allow for proper CikA function (Zhang et al., 2006) or preventing autophosphorylation of HPK domains on full-length CikA protein.

The role of CikA in resetting of the circadian phase appears to derive from its effect on KaiC activity. In response to a 5-h dark pulse that resets the phase of bioluminescent reporter gene expression by 8 h, CikA protein levels increase (Ivleva et al., 2006). Pex protein levels also increase in the dark to repress kaiA expression (Takai et al., 2006). CikA inhibits the ATPase activity of KaiC protein (Fig. 2.3D; Dong et al., 2010), and the Pex-induced decrease in KaiA prevents KaiA-stimulated autophosphorylation of KaiC. Together, these input components lead to a decrease in the population of phosphorylated KaiC molecules (Ivleva et al., 2006). The combined data suggest that resetting of the circadian phase is dependent upon changes in KaiC-P status.

Consistent with a role for KaiC-P in resetting the circadian phase rather than in maintenance of circadian period, two mutants that lack the ability to reset their rhythm do not display robust fluctuations in the ratio of KaiC-P throughout the circadian cycle. Both a cikA null and a cyanobacterial strain

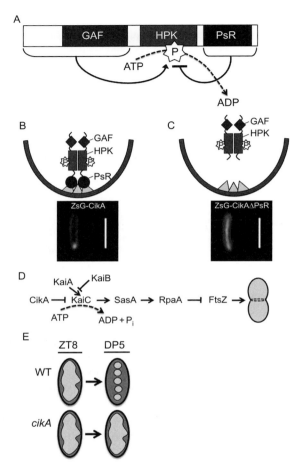

Figure 2.3. Multiple roles for CikA in the *S. elongatus* circadian system. (A) Autophosphorylation of CikA at its histidine protein kinase (HPK) domain is positively influenced by the N-terminal GAF domain but inhibited by the C-terminal *pseudo*-receiver (PsR) domain. (B) The PsR domain (circle) of CikA is proposed to interact with as-yet-to-be-identified proteins (triangles) at the pole of the cyanobacterial cell. Micrograph image of an *S. elongatus* cell harboring a ZsGreen–CikA fusion protein exhibits polar localization. Scale bar = 5 μm. (C) Absence of the CikA–PsR domain results in the delocalization of this CikA variant to exist throughout the cytoplasm of the cell as shown in the cartoon and micrograph image, which visualizes the ZsGreen–CikAΔPsR fusion protein. Scale bar = 5 μm. ZsGreen fusion constructs from Zhang *et al.* (2006); micrograph images courtesy of Julie Bordowitz, UC-San Diego. (D) CikA inhibits the ATPase activity of KaiC, such that in the absence of CikA, the ATPase activity rises above a maximum threshold to activate the SasA/RpaA two-component system of the circadian output pathway. RpaA is predicted to indirectly repress the localization of FtsZ protein that is involved in determining the midline of the cell to allow for division to proceed. (E) At ZT 8, the nucleoid region of the wild type (WT) and *cikA* null is diffuse. After being subjected to a 5-h dark pulse (DP5) beginning at ZT 8, the *cikA* null does not display complete chromosome compaction like that of a wild-type cell. (See Color Insert.)

that harbors a point mutation in the *kaiC* gene, named *kaiC* (*pr1*), are unable to reset the phase of their rhythm in response to external dark pulses, yet produce the rhythmic output of bioluminescence in LL conditions (Kiyohara *et al.*, 2005; Schmitz *et al.*, 2000). Neither of these phase-resetting mutants undergoes complete chromosome compaction when placed in 5-h darkness that would otherwise result in tightly condensed chromosomes in a wild-type cell (Fig. 2.3E; Smith and Williams, 2009). Taken together, these data suggest a relationship among discrete entrainment, KaiC activity, and nucleoid topology, although the pathway that unites them is unknown.

C. LdpA regulates continuous entrainment of the clock

The *ldpA* locus was initially identified in a screen for light input mutants that disrupted the ability to reset to pulses of darkness throughout the circadian cycle. Further analyses demonstrated that the true role of the *ldpA* gene and its protein product is in continuous entrainment of the circadian rhythm through modulation of the period of the internal rhythm in response to changes in light intensity (Katayama *et al.*, 2003). Like other diurnal organisms that possess intrinsic biological timing mechanisms, the circadian period decreases (indicative of a faster clock) as light intensity increases, a phenomenon named Aschoff's rule (Aschoff, 1981); however, mutants that lack *ldpA* maintain the short-period (22–23 h) phenotype associated with high-light intensity regardless of the actual environmental light intensity (Katayama *et al.*, 2003). This light-dependent period mutant phenotype also affects the enhanced susceptibility of the *kaiC22a* mutation to changing light intensity. Inactivation of *ldpA* in the C22a mutant showed no substantial change in period across light intensities that would otherwise display up to a 3-h difference in period; the *ldpA* null displayed the high-light intensity phenotype across the light spectrum (Katayama *et al.*, 2003). These data indicate that *ldpA* is epistatic to *kaiC22a* with regards to light input.

In an *ldpA* mutant background, CikA protein is maintained at its trough level throughout the circadian cycle; this level corresponds to the amount of CikA that is present in high-light conditions in a wild-type strain (Ivleva *et al.*, 2005). In contrast, KaiA protein levels are elevated as compared to wild type, which may lead to the short-period phenotype of the *ldpA* null, as *kaiA* induction is associated with a 1-h period shortening *in vivo* (Kutsuna *et al.*, 2007). Despite this increase in KaiA protein and decrease in CikA protein, both of which would be predicted to result in increased expression from the P*kaiBC* promoter (Ishiura *et al.*, 1998; Taniguchi *et al.*, 2010), there is no noticeable change in the level of bioluminescence from a P*kaiBC*::*luc* reporter in the absence of *ldpA*.

The *ldp*A gene encodes a 352 amino acid protein that carries two redox-active Fe_4S_4 clusters (Ivleva et al., 2005). The presence of these Fe_4S_4 clusters, coupled with the circadian phenotype of the *ldp*A mutant, suggested that LdpA is involved in carrying redox signals from the photosynthetic apparatus to the circadian oscillator to align the circadian rhythm with the changing external environment. Addition of DCMU (3-(3,4-dichlorophenyl)-1,1-dimethylurea), which inhibits electron transport from the photosystem II reaction center to the plastoquinone (PQ) pool, was able to partially rescue the *ldp*A phenotype; in the presence of DCMU, the *ldp*A mutant displayed a small change in period length across light intensities (Katayama et al., 2003). These data suggest that photosynthetic activity contributes to continuous entrainment of the circadian oscillator, but that LdpA is not the sole component in this pathway.

The addition of the quinone analogue DBMIB (2,5-dibromo-3-methyl-6-isopropyl-*p*-benzoquinone) results in a saturation of electrons (overall reduction) of the PQ pool by preventing electron transport from PQ to cytochrome b_6f; introduction of sublethal concentrations of DBMIB leads to the rapid degradation of both the LdpA and CikA proteins (Ivleva et al., 2005, 2006). DBMIB-dependent degradation of CikA occurs more slowly in the absence of LdpA. The PsR domain of CikA binds directly the DBMIB molecule and is necessary and sufficient for DBMIB-dependent degradation (Ivleva et al., 2006); thus, the environmental input that is received by CikA and transduced to the central oscillator reflects that of the metabolic state of the cell. These data provide a molecular mechanism for the phenotypic phenomenon of Aschoff's rule.

In contrast to the eukaryotic clock systems that require compartmentalization of clock components to allow for proper circadian rhythmicity, the cyanobacterial clock proteins of the oscillator, input, and output pathways interact closely with one another to dictate the internal timing mechanisms. Using a 6xHis-tagged LdpA variant, KaiA, CikA, and a component of the output pathway (*Synechococcus* adaptive sensor, SasA) were shown to form a complex *in vivo* (Ivleva et al., 2005). Interaction with KaiA peaked during LL 8–12, despite overall soluble KaiA protein accumulation peaking between LL 16 and 20. CikA copurification with LdpA remained relatively constant across the circadian cycle. The interaction between KaiA and CikA is likely indirect and possibly mediated through LdpA because copurification of LdpA and CikA was uninterrupted in a *kai*A mutant, and the interaction between LdpA and KaiA is maintained in a *cik*A mutant background (Ivleva et al., 2005). The path by which CikA transduces input information to the oscillator does not appear to require LdpA, as an *ldp*A mutant, unlike a *cik*A null, can reset its phase in response to dark pulses. CikA copurifies with KaiA and KaiC in large multimeric complexes *in vivo* that change in their overall molecular weight throughout the circadian cycle (Ivleva et al., 2006); these interactions place CikA in the heart of the clock complex.

The interactions between components of the input pathways and the central oscillator suggest direct transduction of environmental information among protein partners. The current model places KaiA as the conduit for environmental information to the Kai-based oscillator. The N-terminus of KaiA contains a PsR, similar to that of CikA, which lacks the necessary aspartic acid residue necessary for phosphoryl transfer (Williams *et al.*, 2002). This PsR domain binds directly the oxidized, but not reduced, form of the DBMIB quinone analogue (Wood *et al.*, 2010). The bound oxidized DBMIB results in irreversible aggregation of KaiA protein, such that KaiA is no longer able to stimulate the autokinase activity of KaiC *in vitro* (Wood *et al.*, 2010). Although the electron donor for KaiA is not yet known, the copurification of CikA and LdpA with the Kai proteins, the direct binding of DBMIB to the CikA–PsR, and the rapid degradation of CikA and LdpA in response to DBMIB together implicate the possibility of direct electron transfer from the photosynthesis machinery via known input proteins to the redox-sensitive KaiA protein as a mechanism of entrainment of the cyanobacterial clock.

V. OUTPUT PATHWAYS: MULTIPLE INDEPENDENT PATHWAYS MERGE TO COORDINATE CELLULAR AND PHYSIOLOGICAL PROCESSES

The presence of circadian rhythms in a diverse range of organisms—including cyanobacteria, fungi, insects, plant, and mammals—has led most chronobiologists to hypothesize that these internal oscillations beget an adaptive advantage to those organisms through behaviors anticipatory of the changing LD cycle (Dunlap *et al.*, 2004). Microarray analyses demonstrate that 89% of the *S. elongatus* genes that encode proteins involved in photosynthesis display peak expression near subjective dawn, which would allow for their photosynthesis-related products to be most prominent during the daytime (Vijayan *et al.*, 2009). The enhanced fitness of mammals has been demonstrated in the decreased survival rates of clock mutants as compared to those organisms with circadian oscillations that are consonant with the external LD cycles (DeCoursey *et al.*, 2000). The natural variation in FRPs of *Arabidopsis thaliana* plants that live at different latitudes correlates with the day length to which those plants are subjected (Dodd *et al.*, 2005; Michael *et al.*, 2003), which suggests a selective pressure to possess an internal rhythm that matches that of the environment.

Because *S. elongatus* does not fix nitrogen (Herrero *et al.*, 2001) and has no physiological requirement for temporally separating this oxygen-sensitive process from that of photosynthesis as do diazotrophic cyanobacteria, the evolutionary advantage for maintaining internal time was not initially clear. To test the evolutionary advantage that the clock may bestow upon *S. elongatus*, strains that harbor mutations in one of the three genes—*kaiA*, *kaiB*, or *kaiC*—that encode the core oscillator proteins were placed in direct competition with wild-type cells in a variety of environmental conditions (Ouyang *et al.*, 1998; Woelfle *et al.*, 2004). When grown independently as pure cultures, the mutant cultures that produce long period, short period, or arrhythmic phenotypes from bioluminescent reporters grew at a rate indistinguishable from that of wild type in either LL or LD conditions. When two cultures were grown together in a continuously diluted culture, which allows for growth to continue for 30–45 generations, the strain whose FRP most closely matched that of the given LD cycle would out-compete the other strain (Ouyang *et al.*, 1998; Woelfle *et al.*, 2004). If cells are subjected to an LD12:12 cycle, a wild-type strain that produces a FRP of 25 h outcompetes a 23-h mutant, yet this mutant will quickly dominate the population if provided an LD11:11 cycle (Ouyang *et al.*, 1998). This advantage occurs only when the competing cultures are subjected to LD cycles and not in LL conditions, and only when strains are cocultured, which suggests that the internal timekeeping system confers an adaptive advantage to *S. elongatus* only when in a competitive environment. Notably, nature is a competitive environment, and the laboratory setting is likely the only location in which cyanobacteria would be in pure culture and LL conditions.

A. Global clock-controlled regulation of gene expression

Even before the *kai* genes were identified, a screen was conducted to determine the extent by which the circadian system regulates the timing of *S. elongatus* gene expression; the strategy used a "promoter trap" that inserted promoterless *luxAB* throughout the genome to randomly sample promoter activity via bioluminescence (Liu *et al.*, 1995). Two major classes of peak expression patterns emerged that were 12 h out of phase with one another. Class 1 appears to be the "default" phasing for promoters because the majority of promoter activities peak at subjective dusk, including heterologous *E. coli* promoters (P*conII* and P*trc*) that have been introduced into the cyanobacterial chromosome (Tsinoremas *et al.*, 1996; Xu *et al.*, 2003). Taken together, these data pinpoint the circadian system as a global regulator for gene expression in *S. elongatus*.

   The widespread circadian regulation of transcription is not necessarily indicative of rhythmic mRNA and/or protein accumulation. Microarray analyses show that only 30–64% of *S. elongatus* transcripts accumulate in a rhythmic fashion in LL (Ito *et al.*, 2009; Vijayan *et al.*, 2009); similar to those of the promoter trap assay, transcript rhythms fall into the two major categories of peaking near subjective dusk (CT 8–12) or subjective dawn (CT 20–24). Interestingly, not all of the known clock components display rhythmic mRNA levels throughout the day. The *kaiBC* and *cikA* transcripts accumulate with a near 24-h rhythm, yet those of other known clock components, including *sasA*, *ldpA*, and *pex*, do not (Ito *et al.*, 2009). The timing of the peak KaiB, KaiC, and CikA protein levels lag their respective mRNA rhythms by 6–8 h; this fact implies that posttranscriptional and posttranslational modifications play an important role in the functionality of those proteins, especially in the dark where mRNA levels are quickly reduced within 4 h, yet KaiC-P rhythms persist for 3 days in DD (Ito *et al.*, 2009; Tomita *et al.*, 2005).

The inhibitory role that KaiC plays on its own expression is paralleled throughout most of the genome, such that constitutive overexpression of *kaiC* leads to repression of nearly every promoter tested (Ito *et al.*, 2009; Nakahira *et al.*, 2004). For class 1 genes, the level of promoter activity after KaiC-induced repression is at a level consistent with the trough of the activity level in a wild-type background. These promoters are divided into two main types based on the level of expression that remained after the KaiC-induced repression occurred. The "clock-dominated" promoters display high-amplitude rhythms in the presence of the clock, but almost no expression occurred when the clock was disrupted by excess KaiC. Other "clock-modulated" promoters maintained relatively higher levels of arrhythmic expression with the nonfunctional clock. The mRNA levels that result from KaiC overexpression are similar to those seen at subjective dawn with a functional clock, such that class 1 genes produce mRNA levels that mirror the trough level in LL and class 2 genes escalate mRNA production to reach peak levels (Ito *et al.*, 2009).

B. Chromosome topology as a mediator of rhythmic gene expression

Multiple lines of evidence have hinted that the pervasiveness of clock control of gene expression in *S. elongatus* is not necessarily a direct result of the transcription–translation feedback loop of the *kai* genes and their protein products (Ito *et al.*, 2009; Kutsuna *et al.*, 2005; Nakahira *et al.*, 2004; Nakajima *et al.*, 2005; Terauchi *et al.*, 2007; Tomita *et al.*, 2005; Xu *et al.*, 2003). Negative regulation of *kaiBC* by excess KaiC is not due to specific *cis*-acting elements within the promoter. Although a negative regulatory region exists within the *kaiBC* promoter region, removal of this negative element does not prevent the promoter from responding to KaiA or KaiC overexpression. Further, there do not appear to be specific *cis*- or *trans*-acting elements that contribute to the phasing of expression for class 1 or 2 genes during LL conditions (Min and Golden, 2000; Min *et al.*, 2004). Rather, the local DNA topology likely determines the times at which a promoter is accessible to the transcriptional machinery. Early evidence for this "oscilloid model" (Mori and Johnson, 2001a) stems from introduction of *E. coli* promoters (P*fis* and P*tyrT*), whose expression patterns in *E. coli* are dependent upon the surrounding DNA topology. These promoters drive luciferase reporter genes with a near 24-h rhythm and specific phasing to suggest the importance of DNA superhelicity and arrangement within the cell in the maintenance of circadian-regulated gene expression (Min *et al.*, 2004).

When *S. elongatus* cells are sampled over time and stained with DAPI (4′, 6-diamidino-2-phenylindole) dye to visualize the nucleoid, a rhythm in the compaction/decompaction of the chromosome can be observed (Smith and Williams, 2006). This rhythm occurs in a circadian pattern in LD12:12 cycles and continues with a period of approximately 24 h when cells are transferred to LL. The chromosome slowly condenses during the light (or subjective day) with full compaction occurring just before the anticipated light to dark transfer, and decompaction proceeds during the (subjective) night (Smith and Williams, 2006). The predictable changes in DNA topology over time are not limited to the chromosome; endogenous plasmids of *S. elongatus* exhibit the rhythmic fluctuations in superhelicity, and luciferase reporters display similar patterns of bioluminescence when expressed from the endogenous pANS plasmid as they do from the chromosome (Woelfle *et al.*, 2007).

By comparing microarray analysis and the superhelical conformation of the pANS plasmid of *S. elongatus*, each topological state of the DNA can be associated with a particular state of gene expression (Vijayan *et al.*, 2009). The activation or repression of genes results directly from response to the change in superhelicity of the chromosome. These responses were tested by adding novobiocin, a DNA gyrase inhibitor that results in a more relaxed chromosome, to cells at sublethal concentrations. Almost 80% of the tested promoters responded in a predictable manner: those genes that are highly expressed when the plasmid is supercoiled DNA were repressed, while other genes had increased expression upon novobiocin addition to suggest that they are normally repressed when the DNA is supercoiled (Vijayan *et al.*, 2009).

The link between the Kai oscillator and changes in DNA topology likely lies in KaiC (Fig. 2.4). The structure of KaiC resembles the hexameric proteins of the RecA/DnaB family—recombinase and helicase proteins that can bind DNA—and KaiC can interact with forked, double-stranded DNA molecules (Mori *et al.*, 2002). In a *kaiC* null background, the rhythms in chromosome condensation are lost; however, the rhythm is not dependent solely upon the presence of KaiC, as the inactivation of *kaiA* produces a phenotype indistinguishable from that of the *kaiC* mutant (Smith and Williams, 2006). The period of the genome compaction rhythm is dependent upon the timing of the Kai oscillator. The 14-h periodicity of a *kaiC14* mutant is paralleled in the compaction rhythm. An inactivation of *sasA* still allows the chromosome to condense in a rhythmic fashion, but the transfer of information is lost without SasA present, and the result is arrhythmic gene expression (Smith and Williams, 2006). Additionally, unlike its wild-type counterpart, the *kaiC* (*prl*) allele cannot induce repression globally when overexpressed (Kiyohara *et al.*, 2005) and cannot undergo dark-induced chromosome compaction. The *kaiC* (*prl*) mutant

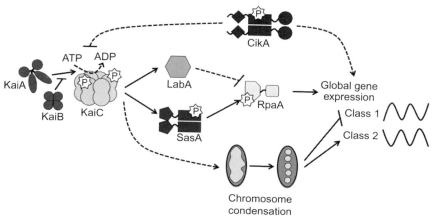

Figure 2.4. Model of *S. elongatus* circadian output pathways. Internal timekeeping is mediated via regulation of KaiC ATPase and phosphorylation activities, where KaiA enhances KaiC autophosphorylation and KaiB averts the KaiA-induced stimulation. KaiC ATPase activity is repressed (indirectly) by CikA in a KaiA-independent pathway. Temporal information from the Kai complex is transduced through at least three protein-based pathways that regulate gene expression. KaiC stimulates SasA autophosphorylation; phosphorylated SasA subsequently transfers its phosphoryl group to RpaA, which leads to RpaA activation. This phosphotransfer event is predicted to result in changes in genome-wide gene expression patterns. LabA is involved in the negative regulation of RpaA function through indirect mechanisms. Residual rhythmic output in the absence of RpaA results from CikA-mediated gene expression pathways. KaiC activity also regulates chromosome compaction; a compact chromosome turns off expression of class 1 genes, while class 2 genes peak at times when compaction is at its highest. The exact mechanism by which KaiC influences chromosome topology is not yet known. Solid lines and dashed lines represent direct and indirect mechanisms, respectively. Arrows represent positive influence, while blunt lines represent inhibition. P, phosphate.

will allow for wild-type rhythms to occur in LL conditions. These data suggest that the *kaiC* (*pr1*) allele is defective in its ability to regulate stimulus-induced chromosome compaction. Pulsed *kaiC* (*pr1*) overexpression at different times throughout the cycle cannot induce phase shifts like that of wild type, which suggests that there is a role for KaiC feedback to input pathways or compaction as an input to the cell (Smith and Williams, 2006, 2009).

C. The SasA–RpaA two-component system as a positive limb of output

The positive limb of the *S. elongatus* output pathway comprises the two-component regulatory system proteins SasA and RpaA (Regulator of phycobiliosome associated). The importance of SasA in the *S. elongatus* circadian timing mechanism was recognized through its direct interaction with KaiC using a yeast two-hybrid screen (Iwasaki et al., 2000). Immunoprecipitation experiments showed that either full-length or the N-terminal domain (residues 1–97) of SasA is capable of interaction with KaiC; either the CI or CII domain of KaiC is sufficient for this binding. No direct interactions between SasA and either KaiA or KaiB were detected in vivo or in vitro (Iwasaki et al., 2000). The sasA gene encodes a 387-amino acid protein that belongs to the HPK family of proteins involved in His-to-Asp two-component signal transduction (Stock et al., 2000). Further bioinformatic analyses revealed the N-terminal domain of SasA has 60% amino acid similarity to the full-length KaiB protein (Iwasaki et al., 2000); however, the tetrameric structure of KaiB (Iwase et al., 2005) differs from the thioredoxin-like fold of the KaiC-interacting N-terminus of SasA (Klewer et al., 2002).

Removal of sasA shortens the circadian period and severely decreases the amplitude of circadian gene expression to near-baseline levels (Iwasaki et al., 2000). The clock is still running and can be reset by light and temperature cues, but the timekeeping mechanism can no longer effectively transduce that information to clock-controlled processes. Rhythmic accumulation of kaiA and kaiBC mRNA does not persist in a sasA mutant background; KaiA protein levels are reduced to approximately 70% compared to those of wild type, while KaiB and KaiC protein levels decrease to levels that are nearly undetectable using standard protocols. Constitutive overexpression of SasA causes arrhythmia as measured by bioluminescent reporters. Short pulses of SasA overexpression shift the phase of the rhythm with advances in the oscillation when sasA levels are decreasing and delay when sasA levels are on the rise (Iwasaki et al., 2000). The direction of the phase shifts that result from short pulses of SasA overexpression are contradictory to those seen when KaiC is overexpressed for short durations. This discrepancy further demonstrates that SasA is important for the functionality of the clock system, but that its function is distinct and separate from that of the Kai complex.

The interaction between the N-terminal sensor domain of SasA and KaiC allows for KaiC to increase the rate by which SasA autophosphorylates, though the reciprocal experiment does not show a SasA-dependent change in KaiC autokinase activity (Smith and Williams, 2006). The autokinase activity of SasA is crucial to its function, as an amino acid substitution at the conserved histidine residue (H162Q) results in a sasA null phenotype (Iwasaki et al., 2000). This KaiC-induced activation of SasA is predicted to be limited during the late subjective night/early subjective day of the circadian cycle in which SasA is part of a large multimeric complex that contains at least the KaiA, KaiB, and KaiC proteins (Kageyama et al., 2003), and likely also CikA (Ivleva et al., 2006) and LdpA (Ivleva et al., 2005). SasA transfers its phosphate group to RpaA (Fig. 2.4), which is predicted to activate the protein to serve as a transcription factor through its DNA-binding domain (Takai et al., 2006); the exact target(s) of RpaA has not yet been elucidated.

RpaA in *S. elongatus* was identified using bioinformatic analyses to identify sequences that encode receiver domains commonly found in RR proteins (Takai *et al.*, 2006). Of 24 identified sequences, only disruption of *rpaA* resulted in the attenuated expression from the *kaiBC* promoter that would be expected of the partner protein of SasA. The phenotype of the *rpaA* null is more severe than that of a *sasA* mutant in that elimination of *rpaA* results in arrhythmia from nearly all promoters regardless of light intensity; however, the residual rhythmicity from a P*psbAI* reporter in the absence of *rpaA* shows that the oscillator continues to function in the absence of this output pathway (Takai *et al.*, 2006). The overall decrease in the level of gene expression in the absence of either *sasA* or *rpaA* suggests that the SasA–RpaA system acts as a positive regulator of the output pathway.

D. LabA and CikA as negative regulators of output

The *labA* (*low-amplitude and bright*) gene was identified because its absence lessened the inhibitory effects of excess KaiC protein on the *kaiBC* promoter (Taniguchi *et al.*, 2007). In an otherwise wild-type background, a *labA* null mutant displays decreased amplitude that results from elevated trough levels from bioluminescence reporters, but maintains a wild-type circadian period. These phenotypes are paralleled in the overall increased levels of *kaiBC* mRNA and KaiC protein with a maintained 24-h rhythmic accumulation and KaiC-P rhythm (Taniguchi *et al.*, 2007). Overexpression of *labA* reduces the overall level of bioluminescence and corresponding KaiC protein levels, but a wild-type period is maintained by the *luxAB* reporter and KaiC-P status. These data support the role of LabA as a negative limb of the output pathway through which KaiC-P acts to repress gene expression.

LabA and SasA are predicted to function in separate pathways that converge at RpaA (Fig. 2.4). The *labA*/*sasA* double mutant displays an intermediate phenotype of overall low-amplitude rhythms, but overall higher bioluminescence levels than that of the *sasA* mutant alone; inactivation of *labA* can attenuate the bioluminescence repression that results from SasA overexpression (Taniguchi *et al.*, 2007). The *rpaA* gene is epistatic to *labA* because inactivation of both *labA* and *rpaA* results in the phenotype that is similar to that of *rpaA* alone (Taniguchi *et al.*, 2007). Thus, RpaA likely receives circadian output information through (at least) two independent pathways: SasA positively influences RpaA function via phosphotransfer, and LabA negatively regulates RpaA function by indirect mechanisms.

The residual rhythmicity of the *labA*/*sasA* double mutant may be the result of the CikA protein, which is most notable for its role in the *S. elongatus* input pathway for discrete entrainment. A transposon insertion in the *cikA* gene was identified in the same screen that identified *labA*; inactivation of *cikA* resulted in decreased repression of the P*kaiBC* promoter by excess KaiC (Taniguchi *et al.*, 2010). Inactivation of *labA*, *sasA*, and *cikA* in the same cell resulted in arrhythmic bioluminescence from luciferase reporters; however, the oscillator itself appears to be intact, as the pattern of KaiC-P over the circadian cycle was similar to that of wild type (Taniguchi *et al.*, 2010).

Genetic analyses demonstrated that LabA and CikA work through independent output pathways to negatively regulate expression of *kaiBC*. The *cikA/labA* double mutant displays the additive phenotype of a short period and low-amplitude rhythm associated with the removal of *cikA*, but with an overall increase in bioluminescence as seen in the *labA* null (Taniguchi *et al.*, 2010). In each single mutant and the double mutant, KaiC protein levels were higher than those of wild type; however, only the *cikA* null showed a change in the phase of the KaiC-P pattern to suggest that CikA-mediated repression of *kaiBC* expression is separate from that of the KaiC-P-mediated feedback in which LabA participates.

E. CpmA and Group 2 sigma factors affect subsets of genes

The *circadian phase modifier* (*cpmA*) gene has also been implicated as a component in output from the *S. elongatus* clock. Although overexpression of the *cpmA* gene does not alter the circadian rhythm, its inactivation results in a drastic change in the relative phase angle of a small subset of cyanobacterial reporters (Katayama *et al.*, 1999). In a *cpmA* background, rhythms from the *kaiBC* promoter are unaffected, but the phase of expression from a P*kaiA::luxAB* reporter is shifted by 10 h, such that its expression peaks almost in antiphase to that of P*kaiBC* (Katayama *et al.*, 1999). Despite this discrepancy in timing from the *kai* promoters, robust rhythms of gene expression continue with wild-type period lengths. This fact suggests that *cpmA* is not part of the oscillator, or a general output pathway, but rather specifically relays temporal information to a subset of genes.

The basic transcriptional machinery is implicated in contributing to circadian output pathways. In bacteria, the RNA polymerase holoenzyme can include different sigma factors, which are the subunits that recognize promoter elements (Gross *et al.*, 1998). The Group 2 sigma factors in cyanobacteria are a family of proteins that have sequence similarity to the housekeeping sigma-70 RpoD1 protein but are not essential for viability of the cells (Tanaka *et al.*, 1992). Inactivation of any of the Group 2 sigma factor genes (*rpoD2*, *rpoD3*, *rpoD4*, or *sigC*) either alone or in pairs alters the rhythm from a P*psbAI::luxAB* reporter by affecting the phase, period, or amplitude of expression (Nair *et al.*, 2002; Tsinoremas *et al.*, 1996). The inactivation of *sigC* lengthens the period of expression from the P*psbAI* promoter but has little effect on the expression patterns from either the P*kaiBC* or P*purF* promoters. The P*kaiBC* promoter is only slightly perturbed as a result of the *rpoD2* single mutant, or the *rpoD3/rpoD4* and *rpoD2/rpoD3* double mutants, which suggests that there is greater buffering of the *kaiBC* promoter compared to others. The current model proposes that RNA polymerase forms holoenzymes with different Group 2 sigma factors throughout the circadian cycle. Durations of activity for some sigma factors likely overlap; redundancy in their roles is indicated by discrepancies in their phenotypes. The connection between the SasA–RpaA pathway and the regulation by the Group 2 sigma factors are still unclear.

F. Cell division

Circadian oscillations in gene expression persist in *S. elongatus* cells with generation times much shorter (as fast as 6 h) than a full circadian cycle (Kondo *et al.*, 1993; Mori *et al.*, 1996). The Kai-based system regulates cell division such that the number of cells in a continuously diluted culture displays a circadian rhythm in LL that matches the phase of its synchronizing LD cycle (Mori *et al.*, 1996). In an arrhythmic *kaiC* mutant, the rhythm in cell division can be driven by external LD cycles, but does not continue in LL (Mori and Johnson, 2001b). The circadian clock gates cell division, such that there are times in the circadian cycle where cell division is forbidden. For *S. elongatus*, this forbidden phase occurs early to mid-subjective night (Mori *et al.*, 1996). The regulation is unidirectional, as the cell-division cycle does not influence the timing of circadian oscillation (Mori and Johnson, 2001b). Cells maintain wild-type rhythms in bioluminescence from transcriptional fusion reporters regardless of doubling time, which varies in response to light intensity, as well as during stationary phase where cells do not undergo cell division, and during chemically induced inhibition of cell division (Mori and Johnson, 2001b).

Because DNA synthesis occurs at a constant rate in free-running cells (Mori *et al.*, 1996), the checkpoint for the cell-division cycle is likely cytokinesis. Coordination of protein components from the three intermingled sectors of the clock—input, oscillator, and output—has been implicated in the transduction pathway leading to the cell-cycle checkpoint. This gating results from elevated ATPase activity of KaiC, rather than overall abundance or phosphorylation state of this central oscillator component (Dong *et al.*, 2010); the two latter properties play important roles in the maintenance of circadian period and global repression of *S. elongatus* promoters.

Of the known clock genes, only the *cikA* and *kaiB* individual mutants produce elongated cells indicative of altered circadian control of cell division (Dong *et al.*, 2010; Miyagishima *et al.*, 2005), although deletion of *cikA* does not eliminate the gating phenomenon. The elongated cell phenotype in the absence of either CikA or KaiB is suppressed when *kaiC*, *sasA*, or *rpaA* is removed from the cell (Dong *et al.*, 2010). The current model predicts that when KaiC ATPase activity rises above a necessary threshold, cell division is not allowed to proceed because the gate is closed (Fig. 2.3D). Conformational changes in KaiC that are predicted to occur when ATPase activity is above the threshold allow for its stimulation of SasA autophosphorylation and subsequent phosphotransfer to RpaA. Phosphorylated RpaA likely inhibits the formation and localization of FtsZ rings that function to denote the midline of the cell where cell division would occur. Support for this model includes mutants with elevated KaiC ATPase activity that display mislocalized FtsZ proteins (Dong *et al.*, 2010).

Interestingly, the forbidden phase of cell division is coincident with the formation of the clock complex and with the transition from a decondensed to condensed nucleoid. It is possible that cell division is tightly regulated by the circadian system in order to protect the clock itself from disruptions that would result in an otherwise desynchronized population. Mathematical models that describe the coupling of the cell-division cycle with that of the circadian rhythm (Yang *et al.*, 2010) help to explain how a population of cyanobacterial cells is able to maintain a similar circadian phase with one another after multiple generations (Mihalcescu *et al.*, 2004).

VI. CONCLUSIONS

The benefits of the genetic and mutational exploitation of the *S. elongatus* PCC 7942 model for a circadian oscillator have been multifold. The ease of working with this single-celled system was monumental to unveiling the nuts and bolts of the clock and continues to provoke new ideas about circadian mechanisms. The idea of a self-sustained oscillation in the absence of a traditional feedback loop (Nakajima *et al.*, 2005, 2010; Tomita *et al.*, 2005) flew in the face of canonical circadian models for timing (Cheng *et al.*, 2003; Cyran *et al.*, 2003; Dunlap, 1999; Harmer *et al.*, 2001; Okamura *et al.*, 2002; Van Gelder *et al.*, 2003) and has subsequently transformed the way, or at least opened the possibilities as to how, eukaryotic-centered chronobiologists think about circadian mechanisms. Recent work in the unicellular picoeukaryotic alga *Ostreococcus tauri* and in human red blood cells has shown that circadian rhythms in peroxiredoxin oxidation and oligomerization, respectively, exist in the absence of transcription and in eukaryotic cells lacking a nucleus (O'Neill *et al.*, 2011; O'Neill and Reddy, 2011). With redox sensing as an essential part of LdpA, CikA, and KaiA function, oscillators in cellular metabolism and redox status are emerging as interesting avenues for further investigation to the understanding of timekeeping mechanisms.

Although a transcriptional-translational feedback loop is not required, one does exist. The theory of, and evidence for, multiple oscillators in circadian systems is not new (Bell-Pedersen *et al.*, 2005; de Paula *et al.*, 2007; Indic *et al.*, 2007). There are multiple lines of evidence for the presence of at least a second oscillator in the *S. elongatus* model. Inactivation of *sigC* causes period changes in some, but not all reporters leading to the idea that there might be more than one oscillator in the *S. elongatus* system that become uncoupled from one another when SigC is missing (Nair *et al.*, 2002). In microarray analyses using a *kaiABC* null strain, 17 genes were found to continue to display rhythmic mRNA accumulation, again suggestive that there is a weak secondary oscillator that controls those transcripts in the absence of the Kai-based oscillator (Vijayan *et al.*, 2009). In addition, genome-wide proteomic studies to determine which clock components (or other proteins) fluctuate their abundance in a circadian manner suggest that only 30–64% of mRNA levels are rhythmic, some of which are known clock components. A more thorough proteomic analysis would allow for a determination of the physiological relevance of these proteins in the circadian system. While there is evidence to predict multiple oscillators, their mechanisms have not yet been described. And, more importantly, the mechanism by which these multiple oscillators communicate with one another to drive precise rhythms has yet to be elucidated.

In addition to the oscillator mechanism itself, there are many questions still remaining with regards to the prokaryotic clock system as a whole. The current data suggest that KaiA is the conduit for relaying input information to the oscillator as a result of direct expression regulation by Pex (Arita *et al.*, 2007) and interactions with CikA (Ivleva *et al.*, 2006) and LdpA (Ivleva *et al.*, 2005), but the story is not complete. While sensing redox state via the quinone pool seems to be the cue, questions still remain regarding how CikA and LdpA relay this environmental information to the central oscillator. The existence of a cognate RR for CikA that relays redox information to the clock directly is currently elusive. In terms of output, how KaiC-P, KaiC ATPase activity, the SasA/RpaA two-component regulatory system, LabA, Group 2 factors, and chromosome compaction coordinate to dictate precise rhythmic activity is not understood. KaiC overexpression has been shown to have differential effects on the promoters that exist on the chromosome and on plasmids. On the chromosome, KaiC overexpression eliminates rhythmic gene expression to the trough levels of all promoters tested and leads to complete chromosome compaction. However, gene expression from the same reporter construct on the pANS plasmid results in elevated, arhythmic expression (Woelfle *et al.*, 2007). How the clock controls DNA topology and gene expression, how it differentially controls gene expression based on chromosome or plasmid status, and how topology changes feedback on the clock itself are also still in question.

We have also learned to be careful of our own biases. Since its discovery and as dictated by its name, CikA is known to function in the input pathway for the Kai oscillator (Ivleva *et al.*, 2006; Schmitz *et al.*, 2000; Zhang *et al.*, 2006). However, recent lines of evidence elucidated by genetic studies have shown that CikA also functions as an inhibitor of KaiC ATPase activity (Dong *et al.*, 2010) and as a component of one of three independent output pathways to negatively regulate expression of *kaiBC* (Taniguchi *et al.*, 2010). Regardless of its multiple personalities, the activity of CikA in input, the oscillator, or output is currently placed in an indirect role; exactly how CikA functions in each of these parts of the oscillator is unclear.

No circadian system is understood to completion; however, insights from classic circadian studies and dogma-shattering breakthroughs have shaped the pathway for understanding the Kai oscillator, illustrating the importance of multiple model systems and basic science. What we have learned thus far is that it is important to keep an eye on the past but an open mind in the future regarding overall clock processes. Undoubtedly, the ease of this single-celled prokaryotic model will aid in teasing out the intricacies of a circadian clock model, which will lead to advances in other systems as well.

Acknowledgments

The authors thank Dr. Yong-Ick Kim for structural and immunoblot images used in Figs. 2.1 and 2.2, respectively, Dr. Julie Bordowitz for micrograph images used in Fig. 2.3, and Dr. Michael Vitalini for careful reading of the chapter. The *S. elongatus* PCC 7942 sequence data were produced by the US Department of Energy Joint Genome Institute (http://www.jgi.doe.gov/) in collaboration with the user community. Research in the lab of SSG was supported by NIH grant GM62419.

References

Amdaoud, M., Vallade, M., Weiss-Schaber, C., and Mihalcescu, I. (2007). Cyanobacterial clock, a stable phase oscillator with negligible intercellular coupling. *Proc. Natl. Acad. Sci. USA* **104,** 7051–7056.

Arita, K., Hashimoto, H., Igari, K., Akaboshi, M., Kutsuna, S., Sato, M., and Shimizu, T. (2007). Structural and biochemical characterization of a cyanobacterium circadian clock-modifier protein. *J. Biol. Chem.* **282,** 1128–1135.

Aschoff, J. (1981). Freerunning and entrained circadian rhythms. Handbook of Behavioral Neurobiology: Biological Rhythms, pp. 81–93. Plenum Press, New York.

Bell-Pedersen, D., Cassone, V. M., Earnest, D. J., Golden, S. S., Hardin, P. E., Thomas, T. L., and Zoran, M. J. (2005). Circadian rhythms from multiple oscillators: Lessons from diverse organisms. *Nat. Rev. Genet.* **6,** 544–556.

Chen, T. H., Chen, T. L., Hung, L. M., and Huang, T. C. (1991). Circadian rhythm in amino acid uptake by *Synechococcus* RF-1. *Plant Physiol.* **97,** 55–59.

Cheng, P., Yang, Y., Wang, L., He, Q., and Liu, Y. (2003). WHITE COLLAR-1, a multifunctional neurospora protein involved in the circadian feedback loops, light sensing, and transcription repression of *wc-2*. *J. Biol. Chem.* **278,** 3801–3808.

Clerico, E. M., Ditty, J. L., and Golden, S. S. (2007). Specialized techniques for site-directed mutagenesis in cyanobacteria. Methods in Molecular Biology, pp. 155–172. Humana Press, Totowa, NJ.

Clodong, S., Duhring, U., Kronk, L., Wilde, A., Axmann, I., Herzel, H., and Kollmann, M. (2007). Functioning and robustness of a bacterial circadian clock. *Mol. Syst. Biol.* **3,** 90.

Cyran, S. A., Buchsbaum, A. M., Reddy, K. L., Lin, M. C., Glossop, N. R., Hardin, P. E., Young, M. W., Storti, R. V., and Blau, J. (2003). vrille, Pdp1, and dClock form a second feedback loop in the *Drosophila* circadian clock. *Cell* **112**, 329–341.

de Paula, R. M., Vitalini, M. W., Gomer, R. H., and Bell-Pedersen, D. (2007). Complexity of the *Neurospora crassa* circadian clock system: Multiple loops and oscillators. *Cold Spring Harb. Symp. Quant. Biol.* **72**, 345–351.

DeCoursey, P. J. (1961). Effect of light on the circadian activity rhythm of the flying squirrel, *Glaucomys volans*. *Z. Vgl. Physiol.* **44**, 331–354.

DeCoursey, P. J., Walker, J. K., and Smith, S. A. (2000). A circadian pacemaker in free-living chipmunks: Essential for survival? *J. Comp. Physiol. A* **186**, 169–180.

Ditty, J. L., Canales, S. R., Anderson, B. E., Williams, S. B., and Golden, S. S. (2005). Stability of the *Synechococcus elongatus* PCC 7942 circadian clock under directed anti-phase expression of the kai genes. *Microbiology* **151**, 2605–2613.

Dodd, A. N., Salathia, N., Hall, A., Kevei, E., Toth, R., Nagy, F., Hibberd, J. M., Millar, A. J., and Webb, A. A. (2005). Plant circadian clocks increase photosynthesis, growth, survival, and competitive advantage. *Science* **309**, 630–633.

Dong, G., Yang, Q., Wang, Q., Kim, Y. I., Wood, T. L., Osteryoung, K. W., van Oudenaarden, A., and Golden, S. S. (2010). Elevated ATPase activity of KaiC applies a circadian checkpoint on cell division in *Synechococcus elongatus*. *Cell* **140**, 529–539.

Dunlap, J. C. (1999). Molecular bases for circadian clocks. *Cell* **96**, 271–290.

Dunlap, J. C., Loros, J. J., and DeCoursey, P. J. (2004). Chronobiology: Biological Timekeeping. Sinauer Associates, Sunderland, MA.

Edmunds, L. N., Jr. (1983). Chronobiology at the cellular and molecular levels: Models and mechanisms for circadian timekeeping. *Am. J. Anat.* **168**, 389–431.

Eguchi, K., Yoda, M., Terada, T. P., and Sasai, M. (2008). Mechanism of robust circadian oscillation of KaiC phosphorylation *in vitro*. *Biophys. J.* **95**, 1773–1784.

Elhai, J., and Wolk, C. P. (1988). Conjugal transfer of DNA to cyanobacteria. *Methods Enzymol.* **167**, 747–754.

Emberly, E., and Wingreen, N. S. (2006). Hourglass model for a protein-based circadian oscillator. *Phys. Rev. Lett.* **96**, 038303.

Gao, T., Zhang, X., Ivleva, N. B., Golden, S. S., and LiWang, A. (2007). NMR structure of the *pseudo*-receiver domain of CikA. *Protein Sci.* **16**, 465–475.

Garces, R. G., Wu, N., Gillon, W., and Pai, E. F. (2004). *Anabaena* circadian clock proteins KaiA and KaiB reveal a potential common binding site to their partner KaiC. *EMBO J.* **23**, 1688–1698.

Golden, S. S. (1988). Mutagenesis of cyanobacteria by classical and gene-transfer-based methods. *Methods Enzymol.* **167**, 714–727.

Golden, S. S., and Sherman, L. A. (1984). Optimal conditions for genetic transformation of the cyanobacterium *Anacystis nidulans* R2. *J. Bacteriol.* **158**, 36–42.

Golden, S. S., Brusslan, J., and Haselkorn, R. (1987). Genetic engineering of the cyanobacterial chromosome. *Methods Enzymol.* **153**, 215–231.

Grobbelaar, N., and Huang, T.-C. (1992). Effect of oxygen and temperature on the induction of a circadian nitrogenase activity rhythm in *Synechococcus* RF-1. *Plant Physiol.* **140**, 391–394.

Grobbelaar, N., Huang, T.-C., Lin, H. Y., and Chow, T. J. (1986). Dinitrogen fixing endogenous rhythm is *Synechococcus* RF-1. *FEMS Microbiol. Lett.* **37**, 173–177.

Gross, C. A., Chan, C., Dombroski, A., Gruber, T., Sharp, M., Tupy, J., and Young, B. (1998). The functional and regulatory roles of sigma factors in transcription. *Cold Spring Harb. Symp. Quant. Biol.* **63**, 141–155.

Harmer, S. L., Panda, S., and Kay, S. A. (2001). Molecular bases of circadian rhythms. *Annu. Rev. Cell Dev. Biol.* **17**, 215–253.

Hayashi, F., Suzuki, H., Iwase, R., Uzumaki, T., Miyake, A., Shen, J. R., Imada, K., Furukawa, Y., Yonekura, K., Namba, K., and Ishiura, M. (2003). ATP-induced hexameric ring structure of the cyanobacterial circadian clock protein KaiC. *Genes Cells* **8,** 287–296.

Hayashi, F., Ito, H., Fujita, M., Iwase, R., Uzumaki, T., and Ishiura, M. (2004). Stoichiometric interactions between cyanobacterial clock proteins KaiA and KaiC. *Biochem. Biophys. Res. Commun.* **316,** 195–202.

Hayashi, F., Iwase, R., Uzumaki, T., and Ishiura, M. (2006). Hexamerization by the N-terminal domain and intersubunit phosphorylation by the C-terminal domain of cyanobacterial circadian clock protein KaiC. *Biochem. Biophys. Res. Commun.* **348,** 864–872.

Herrero, A., Muro-Pastor, A. M., and Flores, E. (2001). Nitrogen control in cyanobacteria. *J. Bacteriol.* **183,** 411–425.

Holtman, C. K., Chen, Y., Sandoval, P., Gonzales, A., Nalty, M. S., Thomas, T. L., Youderian, P., and Golden, S. S. (2005). High-throughput functional analysis of the *Synechococcus elongatus* PCC 7942 genome. *DNA Res.* **12,** 103–115.

Huang, T. C., Tu, J., Chow, T. J., and Chen, T. H. (1990). Circadian rhythm of the prokaryote *Synechococcus* sp. RF-1. *Plant Physiol.* **92,** 531–533.

Imai, K., Nishiwaki, T., Kondo, T., and Iwasaki, H. (2004). Circadian rhythms in the synthesis and degradation of a master clock protein KaiC in cyanobacteria. *J. Biol. Chem.* **279,** 36534–36539.

Indic, P., Schwartz, W. J., Herzog, E. D., Foley, N. C., and Antle, M. C. (2007). Modeling the behavior of coupled cellular circadian oscillators in the suprachiasmatic nucleus. *J. Biol. Rhythms* **22,** 211–219.

Ishiura, M., Kutsuna, S., Aoki, S., Iwasaki, H., Andersson, C. R., Tanabe, A., Golden, S. S., Johnson, C. H., and Kondo, T. (1998). Expression of a gene cluster *kaiABC* as a circadian feedback process in cyanobacteria. *Science* **281,** 1519–1523.

Ito, H., Kageyama, H., Mutsuda, M., Nakajima, M., Oyama, T., and Kondo, T. (2007). Autonomous synchronization of the circadian KaiC phosphorylation rhythm. *Nat. Struct. Mol. Biol.* **14,** 1084–1088.

Ito, H., Mutsuda, M., Murayama, Y., Tomita, J., Hosokawa, N., Terauchi, K., Sugita, C., Sugita, M., Kondo, T., and Iwasaki, H. (2009). Cyanobacterial daily life with Kai-based circadian and diurnal genome-wide transcriptional control in *Synechococcus elongatus. Proc. Natl. Acad. Sci. USA* **106,** 14168–14173.

Ivleva, N. B., Bramlett, M. R., Lindahl, P. A., and Golden, S. S. (2005). LdpA: A component of the circadian clock senses redox state of the cell. *EMBO J.* **24,** 1202–1210.

Ivleva, N. B., Gao, T., LiWang, A. C., and Golden, S. S. (2006). Quinone sensing by the circadian input kinase of the cyanobacterial circadian clock. *Proc. Natl. Acad. Sci. USA* **103,** 17468–17473.

Iwasaki, H., Taniguchi, Y., Ishiura, M., and Kondo, T. (1999). Physical interactions among circadian clock proteins, KaiA, KaiB and KaiC in cyanobacteria. *EMBO J.* **18,** 1137–1145.

Iwasaki, H., Williams, S. B., Kitayama, Y., Ishiura, M., Golden, S. S., and Kondo, T. (2000). A KaiC-interacting sensory histidine kinase, SasA, necessary to sustain robust circadian oscillation in cyanobacteria. *Cell* **101,** 223–233.

Iwasaki, H., Nishiwaki, T., Kitayama, Y., Nakajima, M., and Kondo, T. (2002). KaiA-stimulated KaiC phosphorylation in circadian timing loops in cyanobacteria. *Proc. Natl. Acad. Sci. USA* **99,** 15788–15793.

Iwase, R., Imada, K., Hayashi, F., Uzumaki, T., Morishita, M., Onai, K., Furukawa, Y., Namba, K., and Ishiura, M. (2005). Functionally important substructures of circadian clock protein KaiB in a unique tetramer complex. *J. Biol. Chem.* **280,** 43141–43149.

Johnson, C. H. (2005). Testing the adaptive value of circadian systems. *Methods Enzymol.* **393,** 818–837.

Kageyama, H., Kondo, T., and Iwasaki, H. (2003). Circadian formation of clock protein complexes by KaiA, KaiB, KaiC, and SasA in cyanobacteria. *J. Biol. Chem.* **278,** 2388–2395.

Kageyama, H., Nishiwaki, T., Nakajima, M., Iwasaki, H., Oyama, T., and Kondo, T. (2006). Cyanobacterial circadian pacemaker: Kai protein complex dynamics in the KaiC phosphorylation cycle *in vitro*. *Mol. Cell* **23**, 161–171.

Kang, H. J., Kubota, K., Ming, H., Miyazono, K., and Tanokura, M. (2009). Crystal structure of KaiC-like protein PH0186 from hyperthermophilic archaea *Pyrococcus horikoshii* OT3. *Proteins* **75**, 1035–1039.

Katayama, M., Tsinoremas, N. F., Kondo, T., and Golden, S. S. (1999). *cpmA*, a gene involved in an output pathway of the cyanobacterial circadian system. *J. Bacteriol.* **181**, 3516–3524.

Katayama, M., Kondo, T., Xiong, J., and Golden, S. S. (2003). *ldpA* encodes an iron-sulfur protein involved in light-dependent modulation of the circadian period in the cyanobacterium *Synechococcus elongatus* PCC 7942. *J. Bacteriol.* **185**, 1415–1422.

Kim, Y. I., Dong, G., Carruthers, C. W., Jr., Golden, S. S., and LiWang, A. (2008). The day/night switch in KaiC, a central oscillator component of the circadian clock of cyanobacteria. *Proc. Natl. Acad. Sci. USA* **105**, 12825–12830.

Kippert, F. (1987). Endocytobiotic coordination, intracellular calcium signaling, and the origin of endogenous rhythms. *Ann. N. Y. Acad. Sci.* **503**, 476–495.

Kitayama, Y., Iwasaki, H., Nishiwaki, T., and Kondo, T. (2003). KaiB functions as an attenuator of KaiC phosphorylation in the cyanobacterial circadian clock system. *EMBO J.* **22**, 2127–2134.

Kitayama, Y., Nishiwaki, T., Terauchi, K., and Kondo, T. (2008). Dual KaiC-based oscillations constitute the circadian system of cyanobacteria. *Genes Dev.* **22**, 1513–1521.

Kiyohara, Y. B., Katayama, M., and Kondo, T. (2005). A novel mutation in *kaiC* affects resetting of the cyanobacterial circadian clock. *J. Bacteriol.* **187**, 2559–2564.

Klewer, D. A., Williams, S. B., Golden, S. S., and LiWang, A. C. (2002). Sequence-specific resonance assignments of the N-terminal, 105-residue KaiC-interacting domain of SasA, a protein necessary for a robust circadian rhythm in *Synechococcus elongatus*. *J. Biomol. NMR* **24**, 77–78.

Kondo, T., and Ishiura, M. (1994). Circadian rhythms of cyanobacteria: Monitoring the biological clocks of individual colonies by bioluminescence. *J. Bacteriol.* **176**, 1881–1885.

Kondo, T., Strayer, C. A., Kulkarni, R. D., Taylor, W., Ishiura, M., Golden, S. S., and Johnson, C. H. (1993). Circadian rhythms in prokaryotes: Luciferase as a reporter of circadian gene expression in cyanobacteria. *Proc. Natl. Acad. Sci. USA* **90**, 5672–5676.

Kurosawa, G., Aihara, K., and Iwasa, Y. (2006). A model for the circadian rhythm of cyanobacteria that maintains oscillation without gene expression. *Biophys. J.* **91**, 2015–2023.

Kutsuna, S., Kondo, T., Aoki, S., and Ishiura, M. (1998). A period-extender gene, *pex*, that extends the period of the circadian clock in the cyanobacterium *Synechococcus* sp. strain PCC 7942. *J. Bacteriol.* **180**, 2167–2174.

Kutsuna, S., Nakahira, Y., Katayama, M., Ishiura, M., and Kondo, T. (2005). Transcriptional regulation of the circadian clock operon *kaiBC* by upstream regions in cyanobacteria. *Mol. Microbiol.* **57**, 1474–1484.

Kutsuna, S., Kondo, T., Ikegami, H., Uzumaki, T., Katayama, M., and Ishiura, M. (2007). The circadian clock-related gene *pex* regulates a negative *cis* element in the *kaiA* promoter region. *J. Bacteriol.* **189**, 7690–7696.

Liu, Y., Golden, S. S., Kondo, T., Ishiura, M., and Johnson, C. H. (1995). Bacterial luciferase as a reporter of circadian gene expression in cyanobacteria. *J. Bacteriol.* **177**, 2080–2086.

Mackey, S. R., Ditty, J. L., Zeidner, G., Chen, Y., and Golden, S. S. (2009). Mechanisms for entraining the cyanobacterial circadian clock system with the environment. *In* "Bacterial Circadian Programs" (J. L. Ditty, S. R. Mackey, and C. H. Johnson, eds.), pp. 141–156. Springer-Verlag, Berlin.

Markson, J. S., and O'Shea, E. K. (2009). The molecular clockwork of a protein-based circadian oscillator. *FEBS Lett.* **583**, 3938–3947.

Mehra, A., Hong, C. I., Shi, M., Loros, J. J., Dunlap, J. C., and Ruoff, P. (2006). Circadian rhythmicity by autocatalysis. *PLoS Comput. Biol.* **2**, e96.

Michael, T. P., Salome, P. A., Yu, H. J., Spencer, T. R., Sharp, E. L., McPeek, M. A., Alonso, J. M., Ecker, J. R., and McClung, C. R. (2003). Enhanced fitness conferred by naturally occurring variation in the circadian clock. *Science* **302,** 1049–1053.

Mihalcescu, I., Hsing, W., and Leibler, S. (2004). Resilient circadian oscillator revealed in individual cyanobacteria. *Nature* **430,** 81–85.

Min, H., and Golden, S. S. (2000). A new circadian class 2 gene, *opcA*, whose product is important for reductant production at night in *Synechococcus elongatus* PCC 7942. *J. Bacteriol.* **182,** 6214–6221.

Min, H., Liu, Y., Johnson, C. H., and Golden, S. S. (2004). Phase determination of circadian gene expression in *Synechococcus elongatus* PCC 7942. *J. Biol. Rhythms* **19,** 103–112.

Ming, H., Miyazono, K., and Tanokura, M. (2007). Cloning, expression, purification, crystallization and preliminary crystallographic analysis of selenomethionine-labelled KaiC-like protein PH0186 from *Pyrococcus horikoshii* OT3. *Acta Crystallogr. Sect. F Struct. Biol. Cryst. Commun.* **63,** 327–329.

Mitsui, A., Kumazawa, S., Takahashi, A., Ikemoto, H., and Arai, T. (1986). Strategy by which nitrogen-fixing unicellular cyanobacteria grow photoautotrophically. *Nature* **323,** 720–722.

Miyagishima, S. Y., Wolk, C. P., and Osteryoung, K. W. (2005). Identification of cyanobacterial cell division genes by comparative and mutational analyses. *Mol. Microbiol.* **56,** 126–143.

Miyoshi, F., Nakayama, Y., Kaizu, K., Iwasaki, H., and Tomita, M. (2007). A mathematical model for the Kai-protein-based chemical oscillator and clock gene expression rhythms in cyanobacteria. *J. Biol. Rhythms* **22,** 69–80.

Mori, T., and Johnson, C. H. (2001a). Circadian programming in cyanobacteria. *Semin. Cell Dev. Biol.* **12,** 271–278.

Mori, T., and Johnson, C. H. (2001b). Independence of circadian timing from cell division in cyanobacteria. *J. Bacteriol.* **183,** 2439–2444.

Mori, T., Binder, B., and Johnson, C. H. (1996). Circadian gating of cell division in cyanobacteria growing with average doubling times of less than 24 hours. *Proc. Natl. Acad. Sci. USA* **93,** 10183–10188.

Mori, T., Saveliev, S. V., Xu, Y., Stafford, W. F., Cox, M. M., Inman, R. B., and Johnson, C. H. (2002). Circadian clock protein KaiC forms ATP-dependent hexameric rings and binds DNA. *Proc. Natl. Acad. Sci. USA* **99,** 17203–17208.

Mori, T., Williams, D. R., Byrne, M. O., Qin, X., Egli, M., McHaourab, H. S., Stewart, P. L., and Johnson, C. H. (2007). Elucidating the ticking of an *in vitro* circadian clockwork. *PLoS Biol.* **5,** e93.

Murakami, R., Miyake, A., Iwase, R., Hayashi, F., Uzumaki, T., and Ishiura, M. (2008). ATPase activity and its temperature compensation of the cyanobacterial clock protein KaiC. *Genes Cells* **13,** 387–395.

Murayama, Y., Oyama, T., and Kondo, T. (2008). Regulation of circadian clock gene expression by phosphorylation states of KaiC in cyanobacteria. *J. Bacteriol.* **190,** 1691–1698.

Mutsuda, M., Michel, K. P., Zhang, X., Montgomery, B. L., and Golden, S. S. (2003). Biochemical properties of CikA, an unusual phytochrome-like histidine protein kinase that resets the circadian clock in *Synechococcus elongatus* PCC 7942. *J. Biol. Chem.* **278,** 19102–19110.

Nagai, T., Terada, T. P., and Sasai, M. (2010). Synchronization of circadian oscillation of phosphorylation level of KaiC *in vitro*. *Biophys. J.* **98,** 2469–2477.

Nair, U., Ditty, J. L., Min, H., and Golden, S. S. (2002). Roles for sigma factors in global circadian regulation of the cyanobacterial genome. *J. Bacteriol.* **184,** 3530–3538.

Nakahira, Y., Katayama, M., Miyashita, H., Kutsuna, S., Iwasaki, H., Oyama, T., and Kondo, T. (2004). Global gene repression by KaiC as a master process of prokaryotic circadian system. *Proc. Natl. Acad. Sci. USA* **101,** 881–885.

Nakajima, M., Imai, K., Ito, H., Nishiwaki, T., Murayama, Y., Iwasaki, H., Oyama, T., and Kondo, T. (2005). Reconstitution of circadian oscillation of cyanobacterial KaiC phosphorylation *in vitro*. *Science* **308,** 414–415.

Nakajima, M., Ito, H., and Kondo, T. (2010). *In vitro* regulation of circadian phosphorylation rhythm of cyanobacterial clock protein KaiC by KaiA and KaiB. *FEBS Lett.* **584**, 898–902.

Nishiwaki, T., Iwasaki, H., Ishiura, M., and Kondo, T. (2000). Nucleotide binding and autophosphorylation of the clock protein KaiC as a circadian timing process of cyanobacteria. *Proc. Natl. Acad. Sci. USA* **97**, 495–499.

Nishiwaki, T., Satomi, Y., Nakajima, M., Lee, C., Kiyohara, R., Kageyama, H., Kitayama, Y., Temamoto, M., Yamaguchi, A., Hijikata, A., Go, M., Iwasaki, H., *et al.* (2004). Role of KaiC phosphorylation in the circadian clock system of *Synechococcus elongatus* PCC 7942. *Proc. Natl. Acad. Sci. USA* **101**, 13927–13932.

Nishiwaki, T., Satomi, Y., Kitayama, Y., Terauchi, K., Kiyohara, R., Takao, T., and Kondo, T. (2007). A sequential program of dual phosphorylation of KaiC as a basis for circadian rhythm in cyanobacteria. *EMBO J.* **26**, 4029–4037.

Okamura, H., Yamaguchi, S., and Yagita, K. (2002). Molecular machinery of the circadian clock in mammals. *Cell Tissue Res.* **309**, 47–56.

O'Neill, J. S., and Reddy, A. B. (2011). Circadian clocks in human red blood cells. *Nature* **469**, 498–503.

O'Neill, J. S., van Ooijen, G., Dixon, L. E., Troein, C., Corellou, F., Bouget, F. Y., Reddy, A. B., and Millar, A. J. (2011). Circadian rhythms persist without transcription in a eukaryote. *Nature* **469**, 554–558.

Ouyang, Y., Andersson, C. R., Kondo, T., Golden, S. S., and Johnson, C. H. (1998). Resonating circadian clocks enhance fitness in cyanobacteria. *Proc. Natl. Acad. Sci. USA* **95**, 8660–8664.

Pattanayek, R., Wang, J., Mori, T., Xu, Y., Johnson, C. H., and Egli, M. (2004). Visualizing a circadian clock protein: Crystal structure of KaiC and functional insights. *Mol. Cell* **15**, 375–388.

Pattanayek, R., Williams, D. R., Pattanayek, S., Xu, Y., Mori, T., Johnson, C. H., Stewart, P. L., and Egli, M. (2006). Analysis of KaiA-KaiC protein interactions in the cyanobacterial circadian clock using hybrid structural methods. *EMBO J.* **25**, 2017–2028.

Pattanayek, R., Williams, D. R., Pattanayek, S., Mori, T., Johnson, C. H., Stewart, P. L., and Egli, M. (2008). Structural model of the circadian clock KaiB-KaiC complex and mechanism for modulation of KaiC phosphorylation. *EMBO J.* **27**, 1767–1778.

Pattanayek, R., Mori, T., Xu, Y., Pattanayek, S., Johnson, C. H., and Egli, M. (2009). Structures of KaiC circadian clock mutant proteins: A new phosphorylation site at T426 and mechanisms of kinase, ATPase and phosphatase. *PLoS One* **4**, e7529.

Rust, M. J., Markson, J. S., Lane, W. S., Fisher, D. S., and O'Shea, E. K. (2007). Ordered phosphorylation governs oscillation of a three-protein circadian clock. *Science* **318**, 809–812.

Schmitz, O., Katayama, M., Williams, S. B., Kondo, T., and Golden, S. S. (2000). CikA, a bacteriophytochrome that resets the cyanobacterial circadian clock. *Science* **289**, 765–768.

Schneegurt, M. A., Sherman, D. M., Nayar, S., and Sherman, L. A. (1994). Oscillating behavior of carbohydrate granule formation and dinitrogen fixation in the cyanobacterium *Cyanothece* sp. strain ATCC 51142. *J. Bacteriol.* **176**, 1586–1597.

Smith, R. M., and Williams, S. B. (2006). Circadian rhythms in gene transcription imparted by chromosome compaction in the cyanobacterium *Synechococcus elongatus*. *Proc. Natl. Acad. Sci. USA* **103**, 8564–8569.

Smith, R. M., and Williams, S. B. (2009). Chromosome compaction: Output and phase. *In* "Bacterial Circadian Programs" (J. L. Ditty, S. R. Mackey, and C. H. Johnson, eds.), pp. 169–182. Springer-Verlag, Berlin.

Stock, A. M., Robinson, V. L., and Goudreau, P. N. (2000). Two-component signal transduction. *Annu. Rev. Biochem.* **69**, 183–215.

Sweeney, B. M., and Borgese, M. B. (1989). A circadian rhythm in cell division in a prokaryote, the cyanobacterium *Synechococcus* WH7803. *J. Phycol.* **25**, 183–186.

Takai, N., Ikeuchi, S., Manabe, K., and Kutsuna, S. (2006). Expression of the circadian clock-related gene *pex* in cyanobacteria increases in darkness and is required to delay the clock. *J. Biol. Rhythms* **21,** 235–244.

Takigawa-Imamura, H., and Mochizuki, A. (2006). Predicting regulation of the phosphorylation cycle of KaiC clock protein using mathematical analysis. *J. Biol. Rhythms* **21,** 405–416.

Tanaka, K., Masuda, S., and Takahashi, H. (1992). Multiple *rpoD*-related genes of cyanobacteria. *Biosci. Biotechnol. Biochem.* **56,** 1113–1117.

Taniguchi, Y., Yamaguchi, A., Hijikata, A., Iwasaki, H., Kamagata, K., Ishiura, M., Go, M., and Kondo, T. (2001). Two KaiA-binding domains of cyanobacterial circadian clock protein KaiC. *FEBS Lett.* **496,** 86–90.

Taniguchi, Y., Katayama, M., Ito, R., Takai, N., Kondo, T., and Oyama, T. (2007). *labA*: A novel gene required for negative feedback regulation of the cyanobacterial circadian clock protein KaiC. *Genes Dev.* **21,** 60–70.

Taniguchi, Y., Takai, N., Katayama, M., Kondo, T., and Oyama, T. (2010). Three major output pathways from the KaiABC-based oscillator cooperate to generate robust circadian *kaiBC* expression in cyanobacteria. *Proc. Natl. Acad. Sci. USA* **107,** 3263–3268.

Terauchi, K., Kitayama, Y., Nishiwaki, T., Miwa, K., Murayama, Y., Oyama, T., and Kondo, T. (2007). ATPase activity of KaiC determines the basic timing for circadian clock of cyanobacteria. *Proc. Natl. Acad. Sci. USA* **104,** 16377–16381.

Tomita, J., Nakajima, M., Kondo, T., and Iwasaki, H. (2005). No transcription-translation feedback in circadian rhythm of KaiC phosphorylation. *Science* **307,** 251–254.

Tsinoremas, N. F., Ishiura, M., Kondo, T., Andersson, C. R., Tanaka, K., Takahashi, H., Johnson, C. H., and Golden, S. S. (1996). A sigma factor that modifies the circadian expression of a subset of genes in cyanobacteria. *EMBO J.* **15,** 2488–2495.

Uzumaki, T., Fujita, M., Nakatsu, T., Hayashi, F., Shibata, H., Itoh, N., Kato, H., and Ishiura, M. (2004). Crystal structure of the C-terminal clock-oscillator domain of the cyanobacterial KaiA protein. *Nat. Struct. Mol. Biol.* **11,** 623–631.

Vakonakis, I., Klewer, D. A., Williams, S. B., Golden, S. S., and LiWang, A. C. (2004). Structure of the N-terminal domain of the circadian clock-associated histidine kinase SasA. *J. Mol. Biol.* **342,** 9–17.

Van Gelder, R. N., Herzog, E. D., Schwartz, W. J., and Taghert, P. H. (2003). Circadian rhythms: In the loop at last. *Science* **300,** 1534–1535.

van Zon, J. S., Lubensky, D. K., Altena, P. R., and ten Wolde, P. R. (2007). An allosteric model of circadian KaiC phosphorylation. *Proc. Natl. Acad. Sci. USA* **104,** 7420–7425.

Vijayan, V., Zuzow, R., and O'Shea, E. K. (2009). Oscillations in supercoiling drive circadian gene expression in cyanobacteria. *Proc. Natl. Acad. Sci. USA* **106,** 22564–22568.

Wang, J. (2005). Recent cyanobacterial Kai protein structures suggest a rotary clock. *Structure* **13,** 735–741.

Williams, S. B., Vakonakis, I., Golden, S. S., and LiWang, A. C. (2002). Structure and function from the circadian clock protein KaiA of *Synechococcus elongatus*: A potential clock input mechanism. *Proc. Natl. Acad. Sci. USA* **99,** 15357–15362.

Woelfle, M. A., Ouyang, Y., Phanvijhitsiri, K., and Johnson, C. H. (2004). The adaptive value of circadian clocks: An experimental assessment in cyanobacteria. *Curr. Biol.* **14,** 1481–1486.

Woelfle, M. A., Xu, Y., Qin, X., and Johnson, C. H. (2007). Circadian rhythms of superhelical status of DNA in cyanobacteria. *Proc. Natl. Acad. Sci. USA* **104,** 18819–18824.

Wood, T. L., Bridwell-Rabb, J., Kim, Y. I., Gao, T., Chang, Y. G., LiWang, A., Barondeau, D. P., and Golden, S. S. (2010). The KaiA protein of the cyanobacterial circadian oscillator is modulated by a redox-active cofactor. *Proc. Natl. Acad. Sci. USA* **107,** 5804–5809.

Xu, Y., Mori, T., and Johnson, C. H. (2000). Circadian clock-protein expression in cyanobacteria: Rhythms and phase setting. *EMBO J.* **19,** 3349–3357.

Xu, Y., Mori, T., and Johnson, C. H. (2003). Cyanobacterial circadian clockwork: Roles of KaiA, KaiB and the kaiBC promoter in regulating KaiC. *EMBO J.* **22,** 2117–2126.

Xu, Y., Mori, T., Pattanayek, R., Pattanayek, S., Egli, M., and Johnson, C. H. (2004). Identification of key phosphorylation sites in the circadian clock protein KaiC by crystallographic and mutagenetic analyses. *Proc. Natl. Acad. Sci. USA* **101,** 13933–13938.

Xu, Y., Mori, T., Qin, X., Yan, H., Egli, M., and Johnson, C. H. (2009). Intramolecular regulation of phosphorylation status of the circadian clock protein KaiC. *PLoS One* **4,** e7509.

Yang, Q., Pando, B. F., Dong, G., Golden, S. S., and van Oudenaarden, A. (2010). Circadian gating of the cell cycle revealed in single cyanobacterial cells. *Science* **327,** 1522–1526.

Yoda, M., Eguchi, K., Terada, T. P., and Sasai, M. (2007). Monomer-shuffling and allosteric transition in KaiC circadian oscillation. *PLoS One* **2,** e408.

Zhang, X., Dong, G., and Golden, S. S. (2006). The *pseudo*-receiver domain of CikA regulates the cyanobacterial circadian input pathway. *Mol. Microbiol.* **60,** 658–668.

3

The Genetics of Circadian Rhythms in *Neurospora*

**Patricia L. Lakin-Thomas,* Deborah Bell-Pedersen,†
and Stuart Brody‡**

*Department of Biology, York University, Toronto Ontario, Canada
†Biology Department, 3258 TAMU, Texas A&M University, College Station,
Texas, USA
‡Founder, Center for Chronobiology, Division of Biological Sciences, University
of California, San Diego, La Jolla, California, USA

Advances in Genetics, Vol. 74
Copyright 2011, Elsevier Inc. All rights reserved.

0065-2660/11 $35.00
DOI: 10.1016/B978-0-12-387690-4.00003-9

ABSTRACT

This chapter describes our current understanding of the genetics of the *Neurospora* clock and summarizes the important findings in this area in the past decade. *Neurospora* is the most intensively studied clock system, and the reasons for this are listed. A discussion of the genetic interactions between clock mutants is included, highlighting the utility of dissecting complex mechanisms by genetic means. The molecular details of the *Neurospora* circadian clock mechanism are described, as well as the mutations that affect the key clock proteins, FRQ, WC-1, and WC-2, with an emphasis on the roles of protein phosphorylation. Studies on additional genes affecting clock properties are described and place these genes into two categories: those that affect the FRQ/WCC oscillator and those that do not. A discussion of temperature compensation and the mutants affecting this property is included. A section is devoted to the observations pertinent to the existence of other oscillators in this organism with respect to their properties, their effects, and their preliminary characterization. The output of the clock and the control of clock-controlled genes are discussed, emphasizing the phasing of these genes and the layers of control. In conclusion, the authors provide an outlook summarizing their suggestions for areas that would be fruitful for further exploration. © 2011, Elsevier Inc.

I. INTRODUCTION

A. The *Neurospora* clock system

Neurospora has been one of the primary organisms for circadian research for many years, and work on the *Neurospora* circadian system has often been reviewed (Brunner and Kaldi, 2008; Dunlap and Loros, 2006; Heintzen and Liu, 2007; Lakin-Thomas and Brody, 2004; Liu and Bell-Pedersen, 2006; Vitalini et al., 2010). One output of the circadian system, which is easily assayed under constant environmental conditions (usually constant darkness and constant temperature), is the daily rhythm of conidiation (asexual spore formation) on agar growth medium: the conidiation rhythm is seen as a pattern of areas of thick conidiating growth (bands) alternating with thinner areas of growth (interbands). These areas are formed as the endogenous clock directs the growth front to produce primarily apical extensions of the filamentous mycelium for approximately 11 h (forming an "interband") followed by approximately 11 h of extension of the apical filaments plus the production of aerial hyphae at right angles to the surface of the medium on which the spores develop (forming a "band" region). This "fossil record" of the developmental state of the growth front can be seen on agar medium in petri plates (Fig. 3.1) or in long glass cylindrical tubes ("growth tubes" or "race tubes"). This conidiation rhythm has been the standard assay for the state of the circadian system in labs working with *Neurospora*. All known clock genes were identified and/or characterized by observing changes in the conidiation rhythm. Molecular rhythms can also be observed in the *Neurospora* system and are most often assayed in specialized cultures in which mycelial disks are submerged in liquid medium (Loros et al., 1989; Nakashima, 1981). Conidiation is suppressed to some extent by these culture conditions, but the underlying clock mechanisms continue functioning. Molecular rhythms can also be assayed on solid agar medium in conditions more physiologically similar to those the organism would encounter in nature (Gooch et al., 2008; Ramsdale and Lakin-Thomas, 2000; Schneider et al., 2009).

　　This review focuses on what genetics has told us about the mechanism of the underlying circadian oscillator that drives the observed rhythms, with an emphasis on findings within the past 10 years. We will also consider one aspect of the output of the oscillator: the downstream effects of the clock on the rhythmic control of the expression of "clock-controlled genes," or ccgs. Circadian systems must also have input pathways to synchronize the clock with the

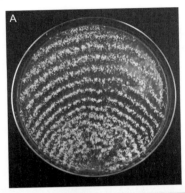

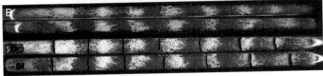

Figure 3.1. The developmental rhythm of conidiation. (A) A 15-cm petri plate was inoculated at the lower edge with the *bd*; *vvd*^{P} *pan-2* strain of *Neurospora* and allowed to grow in constant light. Yellow bands of conidiation formed at approximately 11-h intervals. This strain is unusual in that it continues to form conidiation bands in constant light with a short period. (B) Two 30-cm race tubes were inoculated at the left with the *csp-1*, *bd* strain, exposed to light for 24 h and transferred to constant darkness for the remainder of growth. Growth was from left to right. The upper two images are photographs of the tops of the tubes; the lower two images are scans of the bottoms of the tubes. The black marks indicate the positions of the growth fronts of the colonies at 24-h intervals. The period of these cultures was approximately 21.5 h. (For interpretation of the references to color in this figure legend, the reader is referred to the Web version of this chapter.)

external environment, but a consideration of input pathway genetics is beyond the scope of this review; the reader is directed to recent reviews (Chen *et al.*, 2010; Liu, 2003; Price-Lloyd *et al.*, 2005).

B. *Neurospora* as a model organism for circadian research

The circadian system of *Neurospora* is arguably the most intensively studied of all clocks and has many advantages as an experimental system. As a microorganism that is easily grown in culture, *Neurospora* has an advantage over other organisms in that large amounts of material can be synchronously harvested, making biochemical analysis easier. As a haploid organism, *Neurospora* has advantages for genetic analysis in the ease of isolation of new mutations and the expression of mutant phenotypes in the haploid without the complications of dominance relationships between alleles in a diploid. This has resulted in a huge collection of mutant strains, cataloged by (Perkins *et al.*, 2001), now available online (http://www.fgsc.net/2000compendium/NewCompend.html), and continuously updated (http://bmbpcu36.leeds.ac.uk/~gen6ar/newgenelist/genes/gene_list. htm). The genome has been sequenced (Galagan *et al.*, 2003), transformation methods are routine, and most molecular genetics methods are available for *Neurospora* including the use of an inducible promoter for dosage control (Campbell *et al.*, 1994; Geever *et al.*, 1989) and RNAi for gene silencing (Ziv and Yarden, 2010). Modern imaging techniques can be applied to *Neurospora*, including the use of luciferase (Gooch *et al.*, 2008), GFP (Freitag *et al.*, 2004), and mCherry (Castro-Longoria *et al.*, 2010). Knockout mutants can be easily constructed (Colot *et al.*, 2006), and knockouts are available for most identified genes from the Fungal Genetics Stock Center (McCluskey, 2003).

Neurospora has an 80-year history as a lab organism (Perkins and Davis, 2000), and the first report of circadian rhythmicity in *Neurospora* was published in 1959 (Pittendrigh *et al.*, 1959). Classic circadian experiments describing the response of the *Neurospora* clock to pulses and steps of light and temperature were published in the 1960s and 1970s (Francis and Sargent, 1979; Sargent and Briggs, 1967). The primary output, the conidiation developmental pathway, has been extensively studied (Correa and Bell-Pedersen, 2002; Springer, 1993), and the characterization of ccgs at the molecular level is underway (see Section VI). The development of a transcription/translation feedback model to describe the mechanism of the *Neurospora* circadian oscillator (see Section III) has contributed to the development of similar models for other organisms and the finding of analogous genes in other clock systems (see Chapters 2, 4, 5, 6 and 7 in this volume). *Neurospora* was also one of the pioneer lab organisms for biochemical investigations (Davis, 2000), and there is a wealth of information about metabolic pathways, organelles such as mitochondria, signaling pathways, etc.; as the attention of circadian researchers shifts away from transcription/translation mechanisms and toward metabolic pathways (Bass and Takahashi, 2010; Harrisingh and Nitabach, 2008; Hastings *et al.*, 2008), *Neurospora* will again prove its worth as a model organism.

II. EARLY GENETIC ANALYSIS

A. Identification of clock mutants

Neurospora was the second organism, after *Drosophila* (Konopka and Benzer, 1971), in which mutations affecting the circadian clock were isolated (Feldman and Hoyle, 1973). These were identified in brute force screenings for period changes as observed in the conidiation rhythm on solid agar medium.

The first locus identified in these screens, *frequency* (*frq*), also turned out to be of extraordinary interest. Alleles with both shorter periods (*frq¹*, *frq²*) and longer periods (*frq³*, *frq⁷*) were found at the same locus (see Table 3.1 for periods). These were found to be point mutations, introducing single amino acid changes (Aronson et al., 1994a), and were shown to be codominant in heterokaryons. Null alleles, either a truncated protein (*frq⁹*) (Loros and Feldman, 1986) or an engineered gene deletion (*frq¹⁰*) (Aronson et al., 1994a), were found to be "conditionally rhythmic" (see Section V) and recessive in heterokaryons. A great deal of subsequent work has been carried out on the biochemistry and genetics of the *frq* gene and is summarized below (Section III).

A number of other clock-affecting loci were identified in screens for period mutants, including five *prd* (period) loci (*prd-1*, *prd-2*, *prd-3*, *prd-4*, and *prd-6*) and *chr* (chrono). Additional period-affecting mutations were identified by screening existing libraries of mutants, including several amino acid auxotrophs, drug resistance mutants, and polymerase mutants. Comprehensive lists of mutations and their effects on period can be found in earlier reviews (Lakin-Thomas et al., 1990; Loros and Dunlap, 2001; Morgan et al., 2001). With the few exceptions noted in later sections, the molecular bases for the effects of most of these mutations on the period of the circadian rhythm have not been determined.

Two mutations affecting lipid synthesis are of particular interest: *cel* (*fas*) and *chol-1*. The *cel* (chain elongation) mutation impairs fatty acid elongation, and this strain requires exogenous saturated fatty acids for normal growth. When supplemented with unsaturated fatty acids, the period of the conidiation rhythm lengthens and temperature compensation is impaired. The *chol-1* (choline) mutation impairs synthesis of the lipid phosphatidylcholine, and this

Table 3.1. Interactions Involving *cel* and *chol-1* Mutations in *Neurospora*

	wt[a]	cel[a]	wt[b]	chol-1[b]
frq⁺	20	42	21	63
frq¹	14	34[*,**]	16	55[**]
frq²	17	38[*,**]	18	57
frq³	23	41[*,**]	23	70
frq⁷	28	39[*,**]	29	63[*]
prd-1	24	22[*,**]		
oli^r	19	19[*,**]		

Notes: Periods are rounded to the nearest whole hours. wt, wild type at other loci.
[*]Significantly different from both additive and multiplicative models.
[**]Significantly different from model of epistasis of *cel* or *chol-1* over *frq*.
[a]Grown on unsaturated fatty acid supplement; data from Lakin-Thomas and Brody (1985).
[b]Grown on medium without choline supplementation; data from Lakin-Thomas (1998).

strain requires choline for normal growth. On low-choline medium, the period of the rhythm lengthens in inverse proportion to the choline concentration in the medium, and temperature compensation is impaired. The *cel* and *chol-1* strains will be discussed further in Sections II.B and V.

One large class of period-affecting mutants has a common target in mitochondrial functions and similar period-shortening effects. This class includes oli^r (oligomycin resistance), several cytochrome mutants, and both nuclear- and mitochondrially inherited genes (Lakin-Thomas *et al.*, 1990). The short period of these mitochondrial mutants can be phenocopied by treatment of cultures with inhibitors of mitochondrial function such as chloramphenicol (Brody, 1992). A common effect of several of these mitochondrial mutations was found to be an increase in mitochondrial mass per cell volume (Brody, 1992), but the relationship between this effect and the circadian clock mechanism has not been uncovered.

In the absence of detailed molecular information about the roles of period-affecting mutations, there are several approaches to formulating hypotheses about their effects on the circadian clock. One type of hypothesis focuses on molecular mechanisms and changes in rates: short-period mutations may increase the rate of turnover of a negative-acting clock component, or increase the rate of synthesis of a positive-acting component; long-period mutations may have the opposite effects. Another possibility is a change in affinity of proteins in a regulatory complex: short-period mutations may increase the affinity of two proteins or of a protein for a promoter element, thus leading to precocious activation, and long-period mutations may decrease such affinities. A complementary approach is to consider the amplitude of the oscillator: an increase or decrease in amplitude may cause an increase or decrease in period. This change in amplitude may be brought about by any of the molecular mechanisms listed above. Evidence for changes in amplitude in clock mutants has been discussed in the framework of limit cycle theory (Lakin-Thomas *et al.*, 1991; Shaw and Brody, 2000).

B. Interactions between clock mutants

With quantitative phenotypes such as circadian periods, it is possible to use genetics to ask questions about interactions between gene products in the absence of biochemical evidence. Constructing double- and triple-mutant strains combining several clock mutations in *Neurospora* has provided some insights into potential gene product interactions. The simplest interpretation of additive or multiplicative interactions (each mutation either adds/subtracts a fixed number of hours to the period or multiples the period by a fixed factor) is that each gene product acts independently on the final phenotype (the period). Epistatic, synergistic, or intermediate effects indicate the possibility of interactions between the gene products, and at least a common pathway for effects on the period (Lakin-Thomas and Brody, 1985; Morgan *et al.*, 2001). It should be noted that most studies of genetic interactions have focused on period measurements and very little work has been reported on other clock properties such as phase resetting or temperature compensation.

The *cel* mutation described in Section II.A requires saturated fatty acids for normal growth, and when supplemented with unsaturated fatty acids, the period of the conidiation rhythm lengthens (Mattern *et al.*, 1982). A series of double mutant strains was constructed between *cel* and the series of *frq* alleles (Lakin-Thomas and Brody, 1985), and the data are presented in Table 3.1. A systematic interaction between *cel* and the *frq* alleles was found, with greater divergence from the predicted additive or multiplicative periods found with the longer-period *frq* alleles (Lakin-Thomas and Brody, 1985). The prediction of complete epistasis of *cel* over *frq* (in which all *cel frq* double mutants would have the same period as *cel frq*$^+$) also fails, but with the short-period *frq* alleles diverging the most from this prediction (Lakin-Thomas and Brody, 1985). In contrast, both the *prd-1* and the *oli*^{r} mutations were found to be epistatic to *cel*, blocking the period-lengthening effects of unsaturated fatty acids (Lakin-Thomas and Brody, 1985) (Table 3.1). These data suggest that the gene products of *frq*, *prd-1*, and *oli*^{r} all interact with the unknown mechanism that produces long periods in *cel*.

The *chol-1* mutant described above requires choline supplementation for normal growth and rhythmicity (Lakin-Thomas, 1996, 1998). Double mutants between *chol-1* and the series of *frq* alleles were constructed, and the data are presented in Table 3.1. As with *cel*, the prediction of complete epistasis of *chol-1* over *frq* (in which all *chol-1 frq* double mutants would have the same period as *chol-1 frq*$^+$) also fails, again with the short-period *frq* alleles diverging the most from this prediction (Lakin-Thomas, 1998). The *chol-1 frq*^{1} strain is significantly different from *chol-1 frq*$^+$ (Lakin-Thomas, 1998); an additive prediction fits this double mutant best. The *chol-1 frq*^{2} and *chol-1 frq*^{3} strains are not significantly different from additive, multiplicative, or epistatic predictions (Table 3.1). The *chol-1 frq*^{7} strain fits an epistatic prediction but not additive or multiplicative models (Table 3.1). These results indicate a complex series of interactions between the *frq* alleles and the period-lengthening effect of *chol-1*.

A large number of multiple mutant strains have been constructed using the series of *prd* mutants, *frq* alleles, and the *chr* mutant (Morgan and Feldman, 2001; Morgan *et al.*, 2001), and the periods of some of these strains are presented in Table 3.2. These authors found significant synergistic interactions among a group of mutations including *prd-2*, *prd-3*, *prd-6*, and *frq*^{7}, suggesting these gene products mutually interact (Table 3.2A and B). Synergistic effects on temperature compensation were also found in several double mutant combinations (Morgan and Feldman, 2001). Epistasis between *prd-1*

Table 3.2. Periods (in Hours) of *Neurospora* Double and Triple Clock Mutants from the Data of Morgan and Feldman (2001)

	A. *prd frq and chr frq double mutants*				
	frq^+	frq^1	frq^2	frq^3	frq^7
prd^+	22	17	19	24	29
prd-1	26	19	23	28	35
prd-2	26	19	22	28	38*
prd-3	25	19	23	30*	41*
prd-4	18	14	16	20	24
prd-6	18	15	15	20	21*
chr	24	17	21	26	33

	B. *prd prd and chr prd double mutants*				
	chr^+	chr	prd-2	prd-3	prd-4	prd-6
prd^+	22	24				
prd-1	26	28	26*	30	21	24
prd-2	26	29		33*	21	18
prd-3	25	27			20	18*
prd-4	18	19				16
prd-6	18	19				

C. *Triple mutants*

prd-1; chr; frq^7	36
prd-3; prd-1; frq^7	49
prd-3; prd-2; frq^7	48
prd-2 prd-6; frq^7	21
prd-3; prd-2 prd-6	18

Note: Data from L. Morgan (personal communication).

*Significantly different from both additive and multiplicative models.

and *prd-2* places them in the same pathway (Table 3.2B). It is interesting to note that not all alleles at the *frq* locus produced evidence of interactions; in the case of *prd* double mutants, only the double mutants carrying the frq^7 allele, and in one case, frq^3, differed significantly from the additive or multiplicative models (Table 3.2A). The *chr* mutant did not produce significant interactions in double mutants (Table 3.2A), suggesting that the *chr* gene product affects the period through a pathway independent of *frq*. Triple mutants with various combinations of mutations have also been constructed (Table 3.2C), and although the results are more difficult to interpret, the periods are roughly as would be predicted from the interactions revealed by the data in Table 3.2A and B.

III. THE GENETICS AND BIOCHEMISTRY OF THE FRQ/WCC FEEDBACK LOOP

A. Basic mechanism of the feedback loop

The phenotypes of *frq* mutations described above identified the *frq* gene as potentially a central component of the mechanism of the circadian clock in *Neurospora*, and cloning of the gene (McClung *et al.*, 1989) allowed biochemical and molecular analysis of its function to begin. When negative feedback of *frq* transcription by its protein product FRQ was discovered (Aronson *et al.*, 1994b), following the example of the *per* gene of *Drosophila* (Hardin *et al.*, 1990), a transcription/translation negative feedback model was developed for the circadian clock mechanism of *Neurospora* with the FRQ protein and *frq* RNA as the core components (Aronson *et al.*, 1994b). Additional clock components were later identified, the most important being the two white-collar proteins WC-1 and WC-2 (Crosthwaite *et al.*, 1997). These were incorporated into a model that can be called the FRQ–white-collar complex (FRQ/WCC) model. In a simplified version of the current model for the *Neurospora* clock mechanism (for a detailed description of the model, see Vitalini *et al.*, 2010), transcription of the *frq* gene is activated by a complex of the two white-collar proteins, WC-1 and WC-2, together called the white-collar complex (WCC; Froehlich *et al.*, 2002, 2003). The level of *frq* mRNA increases when WCC is active, FRQ protein is translated, and FRQ protein inhibits the activity of WCC, thereby negatively regulating its own transcription and leading to a fall in *frq* mRNA levels (He *et al.*, 2006; Schafmeier *et al.*, 2006). FRQ protein is degraded (Liu *et al.*, 2000), inhibition of transcription is relieved, and the cycle repeats as *frq* mRNA accumulates again. In a positive feedback loop, FRQ protein also acts to increase the levels of WCC, at posttranscriptional and/or posttranslational steps (Cheng *et al.*, 2001b, 2003; Lee *et al.*, 2000; Merrow *et al.*, 2001; Schafmeier *et al.*, 2006). It is the rhythmic activity of the WCC that drives the observed rhythms by regulating transcription of downstream genes (see Section VI). Interestingly, the levels of the *wc* transcripts do not cycle over the course of the day; however, WC-1 protein levels but not WC-2 protein do cycle (He *et al.*, 2002; Lee *et al.*, 2000; Merrow *et al.*, 2001). Experiments in which *wc-1* or *wc-2* were overexpressed from an inducible promoter demonstrate that, while cycling levels of WC-1 protein are not required for overt rhythmicity, the positive feedback loops are important to stabilize and maintain proper amplitude of the rhythm (Cheng *et al.*, 2001b).

B. Genetic analysis of FRQ and WCC

The earliest work on clock genes in *Neurospora* identified multiple alleles at the *frq* locus that altered the period, and subsequently, null alleles were found to be arrhythmic under standard growth conditions, or "conditionally rhythmic" (described in Section II.A). The two white-collar genes, *wc-1* and *wc-2*, had been previously identified as essential for blue light photoresponses and were subsequently identified as clock components through the arrhythmic phenotypes of null mutants (Crosthwaite *et al.*, 1997). Recent genetic studies have added to our understanding of the functions of these genes.

1. FRQ protein domains

No enzymatic activity has been attributed to the FRQ protein. Its function appears to be to help recruit enzymes involved in the negative feedback loop, including kinases, phosphatases, and a helicase. The full length FRQ protein is 989 amino acids long, with a coiled-coil domain near the N-terminus (Cheng *et al.*, 2001a), and a nuclear localization signal located downstream of the coiled-coil domain. FRQ dimerizes through its coiled-coil domain, and dimerization is required for binding to the WCC. Two alternatively translated forms of FRQ, large FRQ (l-FRQ) and small FRQ (s-FRQ), that differ by 99 amino acids at the N-terminus are translated from two in frame AUGs (AUG1 and AUG3) (Garceau *et al.*, 1997; Liu *et al.*, 1997) within different transcripts that result from alternative splicing events (Colot *et al.*, 2005; Diernfellner *et al.*, 2005, 2007). Low levels of FRQ are present in the nucleus, and nuclear localization is required for FRQ's role in the circadian oscillator (Luo *et al.*, 1998). Despite the presence of the nuclear localization signal, most FRQ protein is cytoplasmic, suggesting that there is an active process to export FRQ from the nucleus. Using FRQ deletion mutants, the C-terminus of FRQ that includes the FRQ/FRQ-interacting RNA helicase (FRH) interaction domain was found to play an important role for cytoplasmic localization of FRQ (Cha *et al.*, 2011). Other domains in FRQ include a casein kinase I (CK1) interaction domain located in the middle of the protein (Guo *et al.*, 2010) and two PEST domains that function in FRQ turnover (Görl *et al.*, 2001). The first PEST domain is just downstream of the CK1 interaction domain, and the second PEST domain is located at the carboxy-terminus of the protein. Between the two PEST domains is a binding site for FRH (Cheng *et al.*, 2005; Guo *et al.*, 2010). All FRQ protein interacts with FRH, and the FRQ/FRH complex is required for interaction of FRQ with the WCC and the inactivation of WCC in the negative feedback loop. The FRQ/FRH complex is also required for FRQ stability and for the positive activity of FRQ in promoting the accumulation of the WC proteins (Guo *et al.*, 2010; Shi *et al.*, 2010).

2. The WCC proteins

WC-1 and WC-2 are PAS domain-containing GATA-type zinc finger transcription factors, found primarily in the nucleus of cells (Ballario *et al.*, 1996; Lee *et al.*, 2000; Linden and Macino, 1997; Schwerdtfeger and Linden, 2000). The WC-1

protein has three PAS domains (A, B, and C), with the N-terminal PAS domain
being a specialized LOV (light, oxygen-, voltage-sensing) domain that functions
as a blue light sensory module (Briggs, 2007; Christie *et al.*, 1999; Liu *et al.*, 2003).
The WC-2 protein has a single PAS domain. WC-1 and WC-2 bind to each
other to form the WCC through the PASC domain of WC-1 and the PAS
domain of WC-2 (Cheng *et al.*, 2002, 2003; Talora *et al.*, 1999).

3. *wc-1* simple sequence repeats

Wild-type *Neurospora* isolates collected from a variety of locations and environments can provide information about natural genetic and phenotypic variation.
A survey of circadian phenotypes among 143 accessions found a correlation
between period of the conidiation rhythm and polymorphisms in simple
sequence repeat (SSR) regions in the *wc-1* gene (Michael *et al.*, 2007).
A correlation between period and latitude of collection was also found. These
results suggest that the variation in *wc-1* SSRs may be a target of selection for
fitness by the local environment (Michael *et al.*, 2007).

C. Role of protein phosphorylation in the FRQ/WCC feedback loop

Current work on the FRQ/WCC feedback loop focuses on filling in details of this
model. It has become increasingly clear that both the mechanism of negative
feedback, whereby FRQ protein inhibits the transcriptional activity of WCC,
and the mechanism of positive feedback, whereby FRQ promotes accumulation
of WCC, depend on posttranslational mechanisms, specifically the phosphorylation and dephosphorylation of the three proteins FRQ, WC-1, and WC-2. The
activity and subcellular localization of these three major components of the
FRQ/WCC are all modulated by phosphorylation. Kinase and phosphatase
genes that have been identified as playing roles in regulating rhythmicity in
Neurospora are listed in Table 3.3.

1. FRQ phosphorylation

The FRQ protein is phosphorylated at multiple sites, and a robust rhythm of
phosphorylation can be seen on Western blots (Garceau *et al.*, 1997). Newly
synthesized FRQ protein is hypophosphorylated and undergoes progressive phosphorylation as it matures, until the hyperphosphorylated protein is degraded by
the ubiquitin/proteasome pathway (He and Liu, 2005a). Recent work has identified numerous phosphorylation sites on the FRQ protein using MS techniques
(Baker *et al.*, 2009; Tang *et al.*, 2009). At least 43 (Tang *et al.*, 2009) or 75 (Baker
et al., 2009) *in vivo* sites were identified, and the majority are progressively

Table 3.3. Recently Identified Clock-Associated Genes in *Neurospora*

Gene name	Rhythm phenotypes of mutants	Function/Identity	References
camk-1	Small effects on period, phase, and light-induced phase shifting of the conidiation rhythm	Calcium/calmodulin-dependent protein kinase	Yang *et al.* (2001)
chr (*ckb-1*)	Long period, low-amplitude rhythms of conidiation, altered temperature compensation, long-period rhythms of FRQ protein, *frq*, and *ccg-1* RNA	Regulatory subunit of casein kinase 2 (CKII)	Mehra *et al.* (2009), Yang *et al.* (2003)
ck-1a	Long-period rhythms of conidiation, FRQ protein, and *frq* RNA	Casein kinase 1a	He *et al.* (2006)
csn-2	Loss of FRQ protein rhythm, lengthened period of conidiation rhythm	Subunit of COP9 signalosome, involved in protein degradation	He *et al.* (2005a)
csn-1 *csn-4* *csn-5* *csn-6* *csn-7*	Irregular and long-period conidiation rhythms	Subunits of COP9 signalosome, involved in protein degradation	Wang *et al.* (2010)
csp-1	Slightly shorter period of conidiation rhythm	Light-induced transcription factor, conidial separation	Schneider *et al.* (2009)
csp-2	Slightly longer period of conidiation rhythm	Conidial separation	Brody *et al.* (2010)
csw-1	Abnormal conidiation rhythm, loss of FRQ protein, and *frq* RNA rhythms	Chromatin remodeling enzyme	Belden *et al.* (2007b)
frh	Loss of conidiation rhythm and rhythms of FRQ protein, *frq* RNA, and *ccg-2* RNA	RNA helicase, component of exosome, regulates RNA processing	Cheng *et al.* (2005), Guo *et al.* (2009, 2010), Shi *et al.* (2010)
fwd1	Loss of conidiation rhythm and rhythms of FRQ protein, *frq* RNA, and *ccg-1* RNA	Component of SCF-type ubiquitin ligase complex, involved in protein degradation	He *et al.* (2003)

(*Continues*)

Table 3.3. (*Continued*)

Gene name	Rhythm phenotypes of mutants	Function/Identity	References
mcb	Loss of rhythms of conidiation, FRQ protein, and *ccg-1* RNA	Regulatory subunit of protein kinase A	Huang *et al.* (2007)
pkac-1	Loss of rhythms of conidiation, FRQ protein, and *ccg-1* RNA	Catalytic subunit of protein kinase A	Huang *et al.* (2007)
ppp-1	Short period and advanced phase of conidiation rhythm	Catalytic subunit of protein phosphatase 1	Yang *et al.* (2004)
pp4	Short period of conidiation and FRQ protein rhythms	Protein phosphatase 4	Cha *et al.* (2008)
pp4 ppp-1	Arrhythmic for conidiation and FRQ protein rhythms	Double mutant for protein phosphatases 1 and 4	Cha *et al.* (2008)
prd-1	Long period of conidiation rhythm in *frq*$^+$, loss of fatty acid effect on FRQ-less rhythm in *cel*, long period of FRQ-less rhythm with geraniol, loss of FRQ-less rhythm in *chol-1*, severe effect on heat-entrainable FRQ-less rhythm	Unknown	Lakin-Thomas and Brody (1985), Li and Lakin-Thomas (2010), Lombardi *et al.* (2007)
prd-2	Long period of conidiation rhythm in *frq*$^+$, long period of FRQ-less rhythm with geraniol, loss of FRQ-less rhythm in *chol-1*, severe effect on heat-entrainable FRQ-less rhythm	Unknown	Li and Lakin-Thomas (2010), Lombardi *et al.* (2007)
prd-3 (*cka*)	Long period of conidiation rhythm, altered temperature compensation, loss of FRQ protein rhythm, and RNA rhythms of *frq*, *ccg-1*, and *ccg-2*	Catalytic subunit of casein kinase 2 (CKII)	Mehra *et al.* (2009), Yang *et al.* (2002)

Table 3.3. (*Continued*)

Gene name	Rhythm phenotypes of mutants	Function/Identity	References
prd-4	Short period of conidiation rhythm	Checkpoint kinase 2 (Chk2), links DNA damage to cell cycle arrest	Pregueiro et al. (2006)
prd-6	Short period of FRQ-less rhythm with geraniol	Unknown	Lombardi et al. (2007)
ras^{bd} (formerly bd)	Increased conidiation output, rhythm resistant to CO_2	ras-1, small G-protein involved in signaling pathways	Belden et al. (2007a)
rco-1	Very long period	Transcription factor, regulates conidiation and photoadaptation	Brody et al. (2010), Olmedo et al. (2010a), Yamashiro et al. (1996)
rgb-1	Low-amplitude, long-period rhythms of FRQ protein and ccg-1 RNA	Regulatory subunit of protein phosphatase 2A	Yang et al. (2004)
rrp44	Long period of FRQ protein and frq RNA molecular rhythms	Component of exosome, regulates RNA processing	Guo et al. (2009)
sod-1	Phenocopies ras^{bd}, slightly shorter period, double mutant with frq^{10} is rhythmic	Superoxide dismutase, removes reactive oxygen species	Belden et al. (2007a), Yoshida et al. (2008)
ult	Short-period conidiation rhythm, double mutant with frq^{10} or wc-2 is rhythmic, shortens the period of the FLO in the presence of geraniol	Unknown	Lombardi et al. (2007)
UV90	Loss of FRQ-less rhythm in chol-1, severe effect on heat-entrainable FRQ-less rhythm, damping of amplitude of FRQ/WCC oscillator	Unknown	Li et al. (2011)
vvd	Short-period conidiation rhythms in constant light in frq$^+$ and frq^{10}	Photoreceptor involved in downregulation of light responses	Schneider et al. (2009)

phosphorylated during maturation of the FRQ protein. Systematic mutagenesis of these sites did not identify individual residues essential for rhythmicity but did identify regions of the FRQ protein in which phosphorylation affects the period: mutations in the central region of FRQ lengthen the period, and mutations in the C-terminal region shorten the period; degradation rates of FRQ protein are correlated with period (Baker *et al.*, 2009; Tang *et al.*, 2009).

Several kinases have been identified as phosphorylating FRQ, including calcium/calmodulin-dependent kinase (CAMK-1) (Yang *et al.*, 2001), casein kinase 1a (CK-1a) (Görl *et al.*, 2001; He *et al.*, 2006), casein kinase II (CKII) (Yang *et al.*, 2002), and protein kinase A (PKA) (Huang *et al.*, 2007). CK-1a and CKII were shown to phosphorylate the majority of sites identified by MS analysis (Tang *et al.*, 2009). Knockout of the *camk-1* gene produces slight effects on period, phase, and light-induced phase shifting of the conidiation rhythm (Yang *et al.*, 2001). CK-1a physically associates with FRQ protein (Görl *et al.*, 2001; He *et al.*, 2006), and disruption of that association by mutation of the binding domain on FRQ results in hypophosphorylation of FRQ and arrhythmic conidiation (He *et al.*, 2006). A knock-in mutation of *ck-1a* that disrupts the kinase domain also results in hypophosphorylation of FRQ and long-period rhythms of conidiation, FRQ protein levels and phosphorylation, and *frq* RNA levels (He *et al.*, 2006). Disruption of *cka*, the gene coding for the catalytic subunit of CKII, abolishes rhythms of FRQ protein levels and phosphorylation and RNA levels of *frq* and two ccgs: *ccg-1* and *ccg-2* (Yang *et al.*, 2002). Disruption of *ckb1*, the gene coding for the regulatory subunit CKB-1 of CKII, produces mostly arrhythmic conidiation patterns with a few long periods, and long-period rhythms of FRQ protein and *frq* and *ccg-1* RNA (Yang *et al.*, 2003). With either a knockout of the gene encoding the major PKA catalytic subunit *pkac-1* or a mutation in the gene encoding the PKA regulatory subunit *mcb*, rhythms of conidiation, rhythms of FRQ levels and phosphorylation, and rhythms of *ccg-1* RNA are abolished (Huang *et al.*, 2007).

Three protein phosphatases, PP1, PP2A, and PP4, have been implicated in regulating the phosphorylation status of FRQ (Cha *et al.*, 2008; Yang *et al.*, 2004). The PP1 and PP2A proteins can dephosphorylate FRQ *in vitro* (Yang *et al.*, 2004). A partially functional mutation of *ppp-1*, the gene that codes for the catalytic subunit of PP1, shortens the period and advances the phase of the conidiation rhythm (Yang *et al.*, 2004). Disruption of *rgb-1*, the gene for a regulatory subunit of PP2A, produces low-amplitude, long-period rhythms of FRQ protein and *ccg-1* RNA (Yang *et al.*, 2004). Deletion of the *pp4* gene shortens the period of the conidiation rhythm and the FRQ protein rhythm (Cha *et al.*, 2008). FRQ protein is hyperphosphorylated and is at low levels in this mutant, and its degradation rate is faster than wild type (Cha *et al.*, 2008). A double mutant strain carrying both the *pp4* deletion and the partially functional *ppp-1* mutation is arrhythmic for conidiation and for FRQ protein levels, and also displays hyperphosphorylated FRQ and a rate of FRQ degradation even faster than the *pp4* deletion strain (Cha *et al.*, 2008), indicating that both PP1 and PP4 normally dephosphorylate FRQ and stabilize it.

There may be several related functions for this phosphorylation of FRQ. Degradation of hyperphosphorylated FRQ protein is essential for the operation of the feedback loop to allow reactivation of WCC. A major role for FRQ phosphorylation is in determining the kinetics of FRQ degradation, and this, in turn, influences the period of the circadian oscillator (Baker *et al.*, 2009; Liu *et al.*, 2000; Ruoff *et al.*, 2005; Tang *et al.*, 2009).

A second role for FRQ phosphorylation may be modulating the interaction between FRQ and WCC. In the *cka* mutant that is defective in CKII, FRQ protein is hypophosphorylated and more FRQ was found to associate with WCC than in the wild type (Yang *et al.*, 2002). In a time series analysis of proteins interacting with FRQ, WCC was found to associate with FRQ preferentially during the times in the circadian cycle when FRQ is hypophosphorylated (Baker *et al.*, 2009). Lastly, mutation analysis of FRQ phosphorylation sites demonstrated that FRQ phosphorylation inhibits FRQ binding to the WCC and FRQ/CK-1a interactions (Cha *et al.*, 2011).

Phosphorylation of FRQ may also modulate the rate of nucleocytoplasmic shuttling of FRQ (Diernfellner *et al.*, 2009). Negative feedback requires hypophosphorylated nuclear FRQ (Schafmeier *et al.*, 2006), and the slow accumulation of cytoplasmic FRQ as it becomes phosphorylated may contribute to the slow kinetics of the circadian cycle; this is supported by the observation that the mutant FRQ^7 protein remains in the nucleus longer than the wild-type protein, and this correlates with the slower phosphorylation kinetics of FRQ^7 (Diernfellner *et al.*, 2009). However, recent data using FRQ kinase mutants and FRQ phosphorylation site mutations indicate that FRQ phosphorylation does not play a major role in FRQ localization in the cell (Cha *et al.*, 2011).

2. WCC phosphorylation

Both the WC-1 and the WC-2 proteins are phosphorylated *in vivo* in the dark, and their phosphorylation increases in response to light exposure. This light-dependent phosphorylation may play a role in the function of WCC as the blue light receptor in *Neurospora*, but this function of WCC can be separated from its function in the FRQ/WCC feedback loop of the circadian system. We will be concerned in this review only with the clock-related function of WCC.

FRQ/FRH regulates the activity of WCC by regulating its phosphorylation status, inhibiting WCC activity as its phosphorylation increases (He *et al.*, 2006; Schafmeier *et al.*, 2005). Five phosphorylation sites on the WC-1 protein downstream of the DNA-binding zinc finger domain have been identified as light independent, and mutation of these sites produces short-period or arrhythmic conidiation (He _et al_., 2005b). Additional WC-1 phosphorylation sites have been identified, as well as one site on WC-2 (Sancar _et al_., 2009). The dark phosphorylation of both WC-1 and WC-2 depends on FRQ/FRH, and hypophosphorylated WCC is more transcriptionally active than the hyperphosphorylated forms (He _et al_., 2006; Schafmeier _et al_., 2005), supporting the conclusion that FRQ inactivates WCC in the negative feedback loop by promoting WCC phosphorylation.

Both CK-1a and CKII phosphorylate WC-1 and WC-2 _in vivo_ in a FRQ-dependent manner (He _et al_., 2006). Mutations of FRQ that abolish the interaction between FRQ and CK-1a result in hypophosphorylation of WC proteins, as do both a kinase-defective _ck-1a_ knock-in mutant and a _cka_ mutant defective in CKII (He _et al_., 2006). The transcriptional activity of WCC is increased by mutations in either CK-1a or CKII (He _et al_., 2006). PKA also phosphorylates WC-1, but independently of FRQ (Huang _et al_., 2007). WC-1 protein is at low levels and is hypophosphorylated in the _pkac-1_ knockout, suggesting that PKA stabilizes WC-1 and primes it for phosphorylation by other kinases (Huang _et al_., 2007).

The phosphatases PP2A and PP4 also play roles in regulating WCC phosphorylation status. WCC is a substrate for PP2A, and the _rgb-1_ mutant, defective in a regulatory subunit of PP2A, increases phosphorylation of WC-1 and WC-2 (Schafmeier _et al_., 2005). Both WC-1 and WC-2 are also hyperphosphorylated when the _pp4_ gene is deleted, and the nuclear enrichment of WCC is lost (Cha _et al_., 2008).

3. Positive feedback and phosphorylation

The mechanism of positive feedback whereby FRQ promotes accumulation of WCC is not completely known. In one simple model (Schafmeier _et al_., 2008), positive feedback depends only on posttranslational regulation of WCC. Degradation of WCC may be triggered by DNA binding, so that FRQ can cause accumulation of WCC by inhibiting its binding to DNA through phosphorylation. In this way, the same mechanism (phosphorylation of WCC) that causes negative feedback (inhibition of DNA binding) may also cause positive feedback (accumulation of WCC). However, the system is more complex as FRQ-dependent transcriptional regulation of _wc-2_ has been demonstrated (Cheng _et al_., 2001b), and FRQ regulates WC-1 levels independent of WC-2 (Cheng _et al_., 2001b, 2002; Lee _et al_., 2000).

4. Nuclear localization and phosphorylation

As a transcription factor, WCC acts on its DNA targets in the nucleus, and therefore, the subcellular localizations and movements of WCC and FRQ are essential to the kinetics of the FRQ/WCC feedback loop. As indicated above, the phosphorylation and dephosphorylation cycle of FRQ is probably not a major player in this process. Instead, an unknown mechanism to actively export FRQ from the nucleus is thought to exist (Cha *et al.*, 2011). Using the fluorescent protein mCherryNC fused to FRQ, the accumulation of FRQ in the nucleus can be visualized across a circadian cycle; a major peak of nuclear accumulation is found at CT 5 and a much smaller peak at CT 19, suggesting two phases of FRQ nuclear transport (Castro-Longoria *et al.*, 2010). WCC is normally enriched in the nucleus, but this enrichment is lost when the *pp4* gene, which codes for protein phosphatase 4, is deleted (Cha *et al.*, 2008). There is conflicting evidence as to whether the *rgb-1* mutation, which inactivates a subunit of PP2A, affects the nuclear localization of WCC (Cha *et al.*, 2008; Schafmeier *et al.*, 2008). The movement of WCC to the cytosol depends on the presence of a functional FRQ protein (Cha *et al.*, 2008; Hong *et al.*, 2008). FRAP analysis of GFP-tagged WC proteins showed rapid shuttling in and out of the nucleus on a time scale of minutes, modulated by FRQ-induced phosphorylation and PP2A phosphatase-mediated dephosphorylation (Schafmeier *et al.*, 2008). These results suggest that FRQ promotes phosphorylation of WCC, which is counteracted by PP2A and PP4, and phosphorylated WCC preferentially accumulates in the cytosol. FRQ also physically interacts with WCC to promote clearance of the complex from the nucleus, and this may also contribute to negative feedback (Cha *et al.*, 2008; Hong *et al.*, 2008).

D. Identification of additional genes regulating the FRQ/WCC feedback loop

Included in this category are genes with clock-affecting phenotypes, either period effects or disruption of rhythmicity, whose functions in the FRQ/WCC feedback loop are known. See Table 3.3 for a summary of the genes described below.

1. *frq* antisense RNA

Two antisense transcripts from the *frq* locus have been identified, and their roles in regulating *frq* expression have been investigated (Kramer *et al.*, 2003). Antisense *frq* RNA is rhythmically produced, but in antiphase to sense RNA, and its transcription is light induced. This light induction requires the WCC, which directly binds to the promoter of the antisense RNA following a short light pulse (Smith *et al.*, 2010). When antisense *frq* expression is abolished, there is a delay in the expression of sense *frq* and a similar delay in the onset of the conidiation rhythm. The most dramatic effect is an increased response to phase resetting by light of the conidiation rhythm in the strains without antisense *frq* (Kramer *et al.*, 2003). Antisense *frq* therefore seems to oppose the function of sense *frq* in the light response and may play a role in moderating the level of *frq* expression (Crosthwaite, 2004).

2. *frh*

A search for additional clock components that associate with FRQ using immunoprecipitation of FRQ protein revealed a protein that was given the name FRH. FRH forms a complex with the entire pool of FRQ protein and mediates the interaction of FRQ with WCC (Cheng *et al.*, 2005; Guo *et al.*, 2010). The *frh* gene is essential in *Neurospora*, but downregulation abolishes circadian conidiation rhythms as well as molecular rhythms in FRQ protein, *frq* RNA, and *ccg-2* RNA, identifying FRH as an essential component of the FRQ/WCC feedback loop (Cheng *et al.*, 2005). A genetic screen for mutations affecting negative feedback identified a mutant allele, frh^{R806H}, that disrupts the interaction between FRQ/FRH and WCC and abolishes conidiation rhythmicity and molecular rhythms without affecting growth and viability (Shi *et al.*, 2010). WC-1 phosphorylation and consequent stabilization appear to require the interaction of both FRQ and FRH with WCC: the frh^{R806H} mutant disrupts those interactions, and both WC-1 and WC-2 are found at low levels in hypophosphorylated forms in this mutant (Shi *et al.*, 2010).

FRQ/FRH associates with the exosome that regulates RNA processing and promotes decay of *frq* RNA (Guo *et al.*, 2009). FRQ/FRH may therefore play two roles in the FRQ/WCC feedback loop: inhibiting WCC activity and promoting decay of *frq* RNA. Downregulating a component of the exosome, RRP44, increases the stability of *frq* RNA and lengthens the period of the molecular rhythms of FRQ protein and *frq* RNA, demonstrating the role of the exosome in regulating *frq* RNA decay (Guo *et al.*, 2009). The exosome may also regulate rhythmicity of some ccgs by regulating their RNA stability: *rrp44* RNA is rhythmic, and downregulating *rrp44* affects the rhythmic expression of two *ccgs* (Guo *et al.*, 2009). FRH has recently been shown to play a role in stability of FRQ protein as well (Guo *et al.*, 2010): FRQ is degraded rapidly in the absence of FRQ–FRH interaction, by a pathway that is independent of FWD-1 (see Section III.D.3).

3. *fwd1* and *csn*

The degradation of FRQ protein is an essential process in the functioning of the FRQ/WCC feedback loop. In *frq* mutants with different periods, the rate of degradation correlates with the period (Ruoff *et al.*, 2005). Protein degradation in eukaryotes is often carried out by the ubiquitin–proteasome pathway, and this pathway has been shown to mediate FRQ degradation in the circadian feedback loop. FRQ protein is ubiquitylated, and the FWD1 protein was identified as the substrate-recruiting subunit of the SCF ubiquitin ligase complex responsible for FRQ degradation (He *et al.*, 2003). Disruption of the *fwd1* gene disrupts rhythms of FRQ protein, *frq* RNA, *ccg-1* RNA, and conidiation rhythms (He *et al.*, 2003). The COP9 signalosome (CSN) is a multisubunit complex that, among other functions, regulates the stability and activity of ubiquitin ligases. Disruption of the *csn-2* gene that encodes a CSN subunit impairs degradation of FRQ, abolishes FRQ protein oscillations, and produces conidiation rhythms with very long periods (He *et al.*, 2005a). A series of knockouts of other CSN subunits (*csn-1*, *-2*, *-4*, *-5*, *-6*, and *-7*) produces similar phentoypes of long-period and irregular conidiation rhythms (Wang *et al.*, 2010). It should be noted that there is also evidence for another pathway for FRQ degradation independent of FWD-1, although the details of this pathway are not yet known (Guo *et al.*, 2010).

4. *csw-1*

The WCC is the transcriptional activator that is responsible for activating *frq* expression, but little is known about events at the promoter. Chromatin remodeling and histone modifications are likely to be involved as they are in the activation of many genes. A survey of knockouts of putative chromatin-remodeling genes in *Neurospora* identified one gene, *csw-1* (clockswitch), with a clock-associated phenotype (Belden *et al.*, 2007b). The knockout produces abnormal, sporadic conidiation patterns and a loss of rhythmicity of *frq* RNA and FRQ protein expression. The CSW protein localizes to the *frq* promoter and affects chromatin structure and WCC association with the *frq* promoter (Belden *et al.*, 2007b).

E. Temperature compensation

One of the basic properties of circadian oscillators is the property of temperature compensation: the period of the rhythm changes very little at different constant environmental temperatures. The wild-type *Neurospora* clock is well compensated in the physiological temperature range of 16–30 °C and poorly compensated between 30 and 36 °C (Gardner and Feldman, 1981; Gooch *et al.*, 2008; Sargent *et al.*, 1966). Not all period-affecting mutations in *Neurospora* have been assayed for their effects on temperature compensation; but of those that have, many have been found to have some effect on this property, either impairing compensation such that the period becomes longer at low temperature or in a few cases causing "overcompensation" in which the period becomes somewhat shorter at lower temperatures (Gardner and Feldman, 1981). Effects on temperature compensation can be slight, such as for *prd-4* (Gardner and Feldman, 1981), or may be greater, as for *cel* (Mattern *et al.*, 1982). Impaired temperature compensation may be uniform across a broad temperature range, as for *frq⁷* (Gardner and Feldman, 1981), or may affect only a limited range of temperatures, as for *chol-1* (Lakin-Thomas, 1998). The observation that many different gene products influence temperature compensation with a range of different effects may indicate that temperature compensation is a property of the system as a whole, rather than a special function of a particular temperature-independent regulatory mechanism. It also seems likely that some components of the circadian system may have evolved to play larger roles in maintaining temperature compensation.

The *frq* gene plays an important role in temperature compensation of the period, as shown by the impaired compensation in the null *frq* mutants *frq⁹* (Loros and Feldman, 1986) and *frq¹⁰* (Aronson *et al.*, 1994a) and altered temperature compensation in period-affecting *frq* mutants (Gardner and Feldman, 1981). In *frq* mutants, FRQ protein stability is related to temperature compensation (Ruoff *et al.*, 2005). Casein kinase 2 (CK2 or CKII, see Section III.C.1) has recently been identified as a regulator of temperature compensation through its phosphorylation of FRQ protein (Mehra *et al.*, 2009). Two clock-affecting genes identified in early genetic screens as period-affecting mutants, *chr* (chrono) and *prd-3* (period-3), also display unusual temperature compensation: *chr* extends temperature compensation beyond the wild-type range, and *prd-3* is overcompensated (Gardner and Feldman, 1981). These two genes have now been identified as coding for subunits of CK2 (Mehra *et al.*, 2009): *chr* (*ckb-1*) encodes subunit CKB-1 and *prd-3* (*cka*) encodes subunit CK2α. The mutations were demonstrated to be hypomorphs that result in reduced FRQ phosphorylation and could be phenocopied by mutations in putative CK2 phosphorylation sites on FRQ. Manipulating the dosage of the CKB-1 protein alters temperature compensation and affects FRQ stability as predicted (Mehra *et al.*, 2009). Importantly, not every kinase that phosphorylates FRQ affects temperature compensation. For example, manipulation of the levels of CK1a alters the period of the developmental rhythm, but the clock is still temperature compensated (Mehra *et al.*, 2009). Together, these data suggest a specific role for CK2 in regulating the pace of the clock at different temperatures via temperature-dependent phosphorylation of FRQ altering FRQ stability.

IV. OTHER CLOCK-ASSOCIATED GENES

Included in this category are genes whose identity is known that affect rhythmicity but do not appear to directly affect the functioning of the FRQ/WCC feedback loop, and genes that affect the period of the conidiation rhythm but whose identity and function are as yet unknown. While the significance and underlying mechanisms for many of these observations are unknown, they provide fertile ground for future investigation. See Table 3.3 for a summary of the genes described below.

A. *csp-1*

This gene codes for a light-inducible transcription factor that affects the conidiation process by preventing conidial separation (Lambreghts *et al.*, 2009; Smith *et al.*, 2010). Strains carrying the *csp-1* mutation show periods about 1 h shorter than wild type (Schneider *et al.*, 2009).

B. *csp-2*

Mutations in *csp-2* also prevent conidial separation, but unlike *csp-1*, this mutation lengthens the period by about 1.5 h (Brody *et al.*, 2010). The *csp-2* gene has been found by mapping and complementation to be allelic with *ghh*, the *Neurospora* homolog of the animal epidermal integrity gene Grainyhead; the period of the *ghh* mutant is about 3 h longer than wild type (A. Paré and B. McGinnis, personal communication).

C. *prd-4*

The *prd-4* (period) gene was identified in the early screens for mutations affecting the period of the conidiation rhythm. The original *prd-4* mutation shortens the period by about 3 h (Gardner and Feldman, 1981), but a null mutation (Pregueiro *et al.*, 2006) displays a wild-type period and normal rhythmicity, indicating that *prd-4* is not essential to the clock mechanism. Cloning of the gene (Pregueiro *et al.*, 2006) provided an identity and function. *Prd-4* is an ortholog of checkpoint kinase 2 (Chk2) that activates a cell cycle checkpoint in response to DNA damage in eukaryotes. DNA-damaging agents reset the clock in *Neurospora*, and this resetting requires PRD-4 (Pregueiro *et al.*, 2006). PRD-4 promotes phosphorylation of FRQ protein in response to DNA damage, thereby resetting the clock. The *prd-4* mutant is semidominant, suggesting a gain of function. Consistent with that, FRQ protein is phosphorylated earlier than normal during the circadian cycle in the *prd-4* mutant, accounting for the decreased period (Pregueiro *et al.*, 2006).

D. *ras^{bd}* (formerly *bd*)

Almost all strains used in laboratories that study the circadian rhythm of conidiation in *Neurospora* carry the *bd* (band) mutation to make the rhythm easier to assay. The effect of the *bd* mutation was originally described as making the conidiation rhythm resistant to inhibition by high levels of CO_2 that accumulate in closed culture vessels such as race tubes (Sargent and Kaltenborn, 1972). Cloning of the *bd* gene (Belden *et al.*, 2007a) revealed that it is an allele of *ras-1*, the small G-protein involved in many signaling pathways in eukaryotes, and the mutation has been renamed ras^{bd}. The addition of a source of reactive oxygen species (ROS) such as menadione to the growth medium can phenocopy the effects of ras^{bd}, suggesting a signaling role for ROS in the conidiation output pathway (Belden *et al.*, 2007a).

E. *rco-1*

This gene is a regulator of conidiation and photoadaptation (Olmedo *et al.*, 2010a; Yamashiro *et al.*, 1996). In the ras^{bd} background, it lengthens the conidiation rhythm to 34–55 h, depending on the culture conditions (Brody *et al.*, 2010).

F. *Sod-1*

This gene codes for superoxide dismutase, an enzyme that removes ROS. Mutation or deletion of this gene phenocopies the ras^{bd} mutation in that a robust conidiation rhythm with a slightly shorter period can be seen under conditions where the wild type appears arrhythmic (Belden *et al.*, 2007a; Yoshida *et al.*, 2008). The addition of ROS to the growth medium also phenocopies ras^{bd} (Belden *et al.*, 2007a), while the removal of ROS reverses the banding phenotype of *sod-1* (Yoshida *et al.*, 2008), implicating ROS in the conidiation output pathway.

G. *ult*

This newly isolated mutation (*ult*, ultradian) shortens the conidiation period to about 12 h (see also Section V) and is the shortest clock mutant yet reported (Lombardi *et al.*, 2007). It is dominant in heterokaryons and is not yet mapped and cloned. It produces a pattern of alternating thin and thick bands of conidiation (Lombardi *et al.*, 2007). The addition of caffeine to the growth medium doubles the period of the *ult* mutant, suggesting a role for the cAMP pathway in the *ult* phenotype (Brody *et al.*, 2010).

H. *vvd*

The *vvd* (vivid) gene has been previously identified as a component of the light input pathway in *Neurospora*. The gene is rapidly transcribed in response to light, and the protein functions to downregulate responses to light (Heintzen *et al.*, 2001; Schwerdtfeger and Linden, 2003; Shrode *et al.*, 2001). Wild-type cultures of *Neurospora* do not express the conidiation rhythm in constant light (LL), but *vvd* mutants express short-period (∼11 h) conidiation rhythms in LL (Schneider *et al.*, 2009).

With increasing light intensity, temperature compensation is gradually lost and the period decreases to as little as 6–7 h. Upon transfer to the dark, the period gradually lengthens until a final steady-state period of 22 h is reached. The expression of rhythmicity depends on the culture: rhythms are seen in petri plates but not in race tubes (Schneider *et al.*, 2009). A mutation in the *csp-1* gene lengthens the period and reduces the sensitivity of the period to light effects. Cultures rhythmic in LL can be phase shifted by dark pulses. The level of FRQ protein is high and constant in *vvd* cultures when they are rhythmic in LL, indicating that a rhythm in FRQ is not essential for the conidiation rhythm; this result is consistent with the finding that the *frq* gene is not required for the *vvd* rhythm in LL (Schneider *et al.*, 2009) (see Section V).

I. Quantitative trait loci

A method for identifying new clock-affecting genes is to exploit the genetic variation in natural populations. Free-running period and entrained phase were used to map quantitative trait loci (QTL) in three populations of progeny derived from crosses of isolates collected in different geographic locations (Kim *et al.*, 2007). Thirty QTL were found that did not map near previously identified clock genes, and these regions are candidates for the locations of new clock-affecting genes (Kim *et al.*, 2007).

J. Screening *Neurospora* knockout library for defects in circadian rhythms

Currently, almost all *Neurospora*'s predicted 10,000 open reading frames (ORFs) have been deleted, providing new tools for assaying the effects of specific gene knockouts on circadian rhythmicity. Screens are underway in several labs to assay alterations in the developmental rhythms in the knockout strain collection, as well as assaying rhythmic reporter genes for alterations in rhythmicity in the mutant strains. These approaches promise to provide a global view of the influence of the genome on circadian clock function.

V. FRQ-LESS RHYTHMS

A. "Conditional rhythmicity" of *frq* and *wc* null mutants

Although the vast majority of research carried out on the *Neurospora* circadian system is focused on the FRQ/WCC feedback loop, it has been known for many years that rhythmic conidiation can be seen in the absence of a functional *frq* gene (Loros and Feldman, 1986). More examples of these FRQ-less rhythms have been reported over the years, and there is now a large collection of conditions and genetic backgrounds in which FRQ-less rhythmicity can be demonstrated. Although null mutants of *frq*, *wc-1*, or *wc-2* are often described as "arrhythmic," it might be more accurate to call them "conditionally rhythmic." Rhythmicity can be induced in these null mutants by conditions such as introducing additional mutations, by adding chemicals to the growth medium, or by simply changing the geometry of the culture vessels.

Rhythmicity in the absence of FRQ/WCC function must be driven by an oscillator, and there has been much discussion over the years as to the relationship between the FRQ/WCC and the FRQ-less oscillator(s) (FLOs). The possibility that FRQ functions to transduce environmental information as part of an input pathway to another oscillator was discussed as early as 1997 (Lakin-Thomas *et al.*, 1997). In 1998, Roenneberg and Merrow proposed the "zeitnehmer" model in which FRQ/WCC functions as a rhythmic input pathway to an oscillator (Roenneberg and Merrow, 1998). Evidence for multiple oscillators was discussed with reference to the restoration of rhythmicity to *frq* null strains by Lakin-Thomas and Brody in 2000 (Lakin-Thomas and Brody, 2000). Subsequently, the term FLOs was coined to describe the oscillator(s) driving FRQ-less rhythms (Iwasaki and Dunlap, 2000). While several labs have obtained a significant amount of data describing the FRQ-less oscillations, we still do not know the identity of the components of the FLOs. Until we do, we can only speculate on the role and organization of the FLOs in the circadian system.

B. FRQ-less conidiation rhythms and their properties

Most reports of FRQ-less rhythms have assayed rhythms of conidiation, and these are described below in approximate chronological order. In some cases, a mutation in the genetic background is required to observe the rhythm, and those genes that have been reported in the past 10 years are included in Table 3.3.

1. Long race tubes

FRQ-less conidiation rhythms were first reported in a *frq*[9] null mutant (Loros and Feldman, 1986). Rhythms take several days to appear, so extra-long race tubes were used to gather more data, and not all cultures are rhythmic. The period is more variable than wild type and is affected by the nutritional composition of the medium. Temperature compensation of the period is defective. The rhythm can be damped out by constant light and entrained to cycles of light pulses. When the knockout frq^{10} became available, it was found to produce similar rhythms (Aronson *et al.*, 1994a).

2. *cel* and *chol-1*

The two lipid-defective mutations *cel* and *chol-1* described in Section II can reveal FRQ-less rhythms under appropriate conditions. When double mutants of *cel* and either frq^{10} or *wc-2* are grown at low temperature or supplemented with unsaturated fatty acids, long-period conidiation rhythms are robustly produced (Lakin-Thomas and Brody, 2000). Double mutants of the choline-requiring mutant *chol-1* with *frq*, *wc-1*, or *wc-2* null mutants produce robust conidiation rhythms when choline supplementation is limited (Lakin-Thomas and Brody, 2000). At intermediate choline concentrations, the periods are in the circadian range and lengthen as choline is decreased. The period of the *chol-1* frq^{10} double mutant is temperature compensated between 22 and 28 °C and lengthens below 22 °C. None of the double mutant *cel* or *chol-1* strains are damped by constant light or entrain to light/dark cycles (Lakin-Thomas and Brody, 2000). Recent studies examining *frq*-driven luciferase rhythms in the *chol-1* mutant strain revealed that under conditions in which the strain shows a long-period developmental rhythm, the *frq*:luc rhythms cycled with a 22-h period (Shi *et al.*, 2007). Although it has been suggested that these data indicate independent functioning of two oscillators (Shi *et al.*, 2007), the possibility of coupling between two oscillators in a frequency demultiplication relationship has not been fully explored.

3. Temperature entrainment

The circadian oscillator in *Neurospora* can be entrained to cycles of high and low temperature, and it was reported that such cycles could also entrain the conidiation rhythm in a *frq* null strain (Merrow *et al.*, 1999; Roenneberg *et al.*, 2005), suggesting that the same oscillator functions in both wild-type and *frq* null strains. This interpretation was challenged when a similar data set was analyzed with different methods and the authors concluded that the conidiation peaks represented "driven" or "masking" responses, not the output of an oscillator (Pregueiro *et al.*, 2005). These experiments used symmetric temperature cycles, with equal periods of warm and cool temperature, which makes it difficult to separate masking effects from entrained peaks. An improved method using short pulses of high temperature demonstrated definitively that the conidiation peaks behave as if they are produced by an entrained oscillator (Lakin-Thomas, 2006a). This improved method uncovered both temperature-driven masking effects and temperature entrainment for both frq^{10} and a *wc-1* null mutant (Lakin-Thomas, 2006a).

4. Farnesol and geraniol

The addition of the chemicals farnesol or geraniol to the growth medium produces reliable rhythms in *frq*, *wc-1*, and *wc-2* null mutants with periods in the circadian range or somewhat longer (Granshaw *et al.*, 2003). It takes several days before conidiation bands appear. Temperature compensation of the period is defective, and the carbon source in the growth medium affects the appearance of banding. Phase resetting by temperature pulses produces a strong type 0 phase response curve in both the frq^{10} and the frq^{+} strains grown on geraniol. However, frq^{10} did not entrain to light/dark cycles, unlike frq^{+} (Granshaw *et al.*, 2003).

5. *ult*

This short-period mutant produces robust 12 h rhythms in frq^{10} double mutants. In a *wc-2* null mutant background, *ult* produces rhythms with short periods of about 16.5 h (Lombardi *et al.*, 2007).

6. *sod-1*

The frq^{10}; *sod-1* double mutant strain, displays a conidiation rhythm in constant darkness, although the triple-mutant frq^{10}; *bd*; *sod-1* is arrhythmic (Yoshida *et al.*, 2008). Light/dark cycles induce a conidiation rhythm that behaves as if it is produced by masking effects of light and not by entrainment of an oscillator (Yoshida *et al.*, 2008).

7. *vvd*

The short-period conidiation rhythms expressed in constant light (LL) by the *vvd* mutant are still expressed in the *vvd*; frq^{10} double mutant with no significant change in period, although the rhythms are not as robust as in *vvd* (Schneider *et al.*, 2009). Introduction of the *wc-1* knockout abolishes the LL rhythms in *vvd*; introducing *wc-2* reduced the clarity of the *vvd* rhythms (Schneider *et al.*, 2009). This suggests that *wc-1* is required as the light receptor for the expression of rhythmicity in *vvd*; frq^{10} in LL.

8. Menadione

This chemical generates ROS and has been shown to phenocopy the ras^{bd} mutation (see Section IV). When added to the growth medium, menadione induces conidiation rhythms in null mutants frq^{10}, *wc-1*, and *wc-2*, in both DD and

LL (Brody *et al.*, 2010). The periods of frq^{10} and wc-1 are short (14–15 h), and wc-2 is long (25 h). In 24-h LD cycles, frq^{10} entrains and wc-1 produces two bands per cycle, while wc-2 fails to entrain. The addition of caffeine to the medium doubles the period of frq^{10}. A mutation in the csp-1 gene shortens the period of the null frq^{9} mutant by about 1 h. The period of wc-2 is temperature compensated in the range 16–28 °C, and frq^{10} is compensated in the range 22–28 °C on menadione (Brody *et al.*, 2010).

C. FRQ-less molecular rhythms and their properties

Several molecular species have been shown to be rhythmic in FRQ-less strains. Because these rhythms have been assayed under different growth conditions in different laboratories, it is not yet clear whether they are related to each other, and how they are related to FRQ-less conidiation rhythms.

1. DAG

The levels of the neutral lipid diacylglycerol (DAG) are rhythmic in wild-type *Neurospora* with a period that reflects the conidiation rhythm (Ramsdale and Lakin-Thomas, 2000). In a *frq* null strain grown on solid agar, the DAG rhythm has a period of about 12 h, similar to short-period conidiation rhythms seen in these cultures (Ramsdale and Lakin-Thomas, 2000).

2. Nitrate reductase

In ammonia-free medium, a rhythm in nitrate reductase activity can be measured in *frq* null and *wc*-1 null strains in both DD and LL (Christensen *et al.*, 2004). This rhythm may be an indicator of rhythmic metabolism, or may be an autonomous oscillator constructed from the negative feedback of glutamine on nitrate reductase expression (Christensen *et al.*, 2004).

3. ccg-16

The mRNA levels of a ccg, *ccg-16*, are rhythmic in DD in a *frq* knockout strain but are not seen in null mutants of either *wc*-1 or *wc*-2 (Correa *et al.*, 2003). This FRQ-less rhythm is also seen in LL when FRQ protein levels are constantly elevated and is more responsive to temperature than to light cues for synchronization (de Paula *et al.*, 2006). The period is temperature compensated between 22 and 27 °C (de Paula *et al.*, 2006). A knockout of *ccg-16* does not affect the conidiation rhythm in a frq^{+} background and does not affect the rhythm of WC-1 protein in a *frq* null background (see Section V.C.4), thereby defining *ccg-16* as an output of the FLO and not a component of either the FRQ/WCC oscillator or the FLO (de Paula *et al.*, 2006).

4. WC-1

A low-amplitude rhythm in WC-1 protein levels has been reported in *frq*[+] strains (de Paula *et al.*, 2006; Lee *et al.*, 2000). This rhythm can also be seen in LL in wild-type strains, and in both DD and LL in a *frq* null strain, although the average levels are lower than in wild type (de Paula *et al.*, 2006).

D. Mutations affecting FRQ-less rhythms

If additional mutations are introduced into *frq* or *wc* null strains and these double mutant strains are assayed under conditions in which FRQ-less rhythms can be seen, the effects of the additional mutations on the FLO(s) can be assayed. Several mutations have been found to affect FRQ-less rhythms, and they are summarized in Table 3.3.

1. *frq*

It may seem paradoxical to claim that *frq* mutations affect the period of FRQ-less rhythms. However, under some conditions, it appears that it is primarily the FLO that is controlling rhythmicity when there is a functional *frq* gene, and mutations in *frq* can affect the FLO, supporting the idea that the FLO functions downstream of the FRQ/WCC feedback loop. In the *chol-1* strain, described above, long-period rhythms are seen when choline is limiting. Long-period rhythms are also seen in a *chol-1 frq* null double mutant, and these rhythms have characteristics similar to the rhythms in the *frq*[+] background: the period is sensitive to the choline concentration and shows a similar pattern of tempera-ture compensation (Lakin-Thomas and Brody, 2000). It is therefore reasonable to assume that the same oscillator is driving conidiation rhythms in both *frq*[+] and *frq* null on limiting choline. As described in Section II.B and in Table 3.1, *frq* mutations can have significant effects on the long period in *chol-1*, indicating that the FLO can be affected by *frq*.

Additional evidence that mutations in the FRQ/WCC feedback loop can affect the FLO comes from the results with menadione in *frq* and *wc* null strains (Brody *et al.*, 2010), as described in Section V.B.8. In the presence of menadione, the period of conidiation rhythms in *frq*[10] and *wc-1* is about 14–15 h, while the period in *wc-2* is about 25 h, indicating that the alleles at these loci influence the period of the FRQ-less rhythm under these conditions.

2. *ult*

The periods of the FRQ-less rhythms produced by geraniol in frq^{10} and *wc-2* mutants are shortened by the addition of the short-period *ult* mutation (Lombardi *et al.*, 2007).

3. *prds*

The lipid-deficient *cel* strain produces long-period conidiation rhythms at low temperature and when supplemented with unsaturated fatty acids. These long-period rhythms continue when *frq* and *wc* null mutations are introduced (Lakin-Thomas and Brody, 2000), and therefore the long-period rhythms in *cel frq*⁺ strains may be said to be driven by the FLO. When the *prd-1* mutation is introduced into *cel*, the period-lengthening effect of fatty acids is abolished, and therefore *prd-1* appears to block the normal functioning of the FLO in *cel* (Lakin-Thomas and Brody, 1985).

The frq^{10} null mutant is rhythmic when supplemented with geraniol (described above), and this system was used to assay the effects of a series of *prd* mutations on the FLO (Lombardi *et al.*, 2007). The *prd-1* and *prd-2* mutations, both long period in a *frq*⁺ background, also lengthen the period in the frq^{10} background. The short-period *prd-6* mutation shortens the frq^{10} period (Lombardi *et al.*, 2007). No change in period is seen with *prd-3* or *prd-4* in the frq^{10} background.

In a similar experiment, triple mutants were constructed between *chol-1* frq^{10} and a series of *prd* mutations, and the effects of the *prds* on the long-period conidiation rhythm in *chol-1* were assayed (Li and Lakin-Thomas, 2010). Both *prd-1* and *prd-2* severely disrupt rhythmicity in choline-depleted cultures, but *prd-3* and *prd-4* have only subtle effects on period and robustness of the conidiation rhythm. These strains were also grown on high choline (which repairs the *chol-1* defect) to assay the effects of the *prd* mutations on the heat-entrainable oscillator in frq^{10}. A similar pattern was seen: both *prd-1* and *prd-2* significantly affect the timing of the heat-entrainable peak, while *prd-3* and *prd-4* have little effect (Li and Lakin-Thomas, 2010).

Both *prd-3* and *prd-4* are known to encode subunits of protein kinases that phosphorylate FRQ protein (see Section III.C.1), and their lack of effect on FRQ-less rhythms is consistent with the assumption that they act primarily on the FRQ/WCC oscillator. The effects of *prd-1* and *prd-2* mutations on multiple FRQ-less rhythms may indicate that these genes act primarily on the FLO. The effects of both *prd-1* and *prd-2* on the period of the conidiation rhythm when FRQ/WCC is functional suggest that their effects on FLO can also affect the period of the intact system.

4. UV90

Using the *chol-1 frq*[10] double mutant, a mutagenesis screen was carried out to look for genes that affect the FRQ-less rhythm on low choline (Li *et al.*, 2011). A new mutation was found, named UV90, that abolishes the FRQ-less conidiation rhythm seen in low-choline cultures. UV90 also severely affects the heat-entrainable conidiation rhythm in choline-sufficient cultures. In a *frq*[+] background, UV90 damps the amplitude of the conidiation rhythm and also damps the FRQ protein rhythm. The response of *frq*[+] cultures to phase resetting by pulses of either light or high temperature is increased in the UV90 mutant background, consistent with a decrease in the amplitude of the circadian oscillator (Li *et al.*, 2011). These results suggest that the UV90 gene product is an important component of the FLO and is also required to maintain the amplitude of the FRQ/WCC oscillator, indicating an influence of the FLO on FRQ/WCC.

E. Multiple independent FLOs, or an integrated system?

The existence of FRQ-less rhythms implies the existence of one or more oscillators (FLOs) that drive these rhythms in the absence of a functional FRQ/WCC oscillator. If we wish to fully understand the circadian system of *Neurospora*, one of the most fundamental and vigorously debated questions at the moment is: how many oscillators make up the complete system? There are several possible interpretations of the data. One view (Shi *et al.*, 2007) describes FLOs as "metabolic oscillators" and proposes that there are different FLOs driving the observed rhythms for every reported condition; these multiple metabolic oscillators (of which there would need to be more than a dozen at the latest count) are not considered to be connected to the circadian system (Shi *et al.*, 2007). Another view proposes multiple FLOs that, under normal circumstances, interact with one another and with the FRQ/WCC to produce a network of coupled oscillators, but that individual FLOs may act independently to drive a particular output (de Paula *et al.*, 2007). If there are multiple oscillators, there is a formal possibility that they may function either "upstream" or "downstream" of FRQ/WCC; in either case, they would need to bypass FRQ/WCC to provide output to the conidiation pathway and biochemical rhythms when FRQ/WCC is disabled. A third view proposes a single FLO, which may be the central rhythm generator for the circadian system, mutually coupled to the FRQ/WCC, which supplies stability, period control, and rhythmic input (Li and Lakin-Thomas, 2010; Roenneberg and Merrow, 1998).

The FRQ-less rhythms described above appear to have different properties, lending support to models in which every reported FRQ-less rhythm is driven by a different FLO. However, properties such as temperature compensation, response to alterations in growth media, and ability to entrain to light/

dark or temperature cycles have not been fully characterized for most of the reported FRQ-less rhythms. Claims of failure to entrain to environmental stimuli are particularly suspect unless a wide range of cycle lengths and stimulus strengths have been tested, as oscillators will fail to entrain to stimuli that are too weak or whose entraining periods are too far from the intrinsic period of the oscillator (Johnson *et al.*, 2003). Because of the many different conditions and experimental paradigms, it is difficult to compare the properties of FRQ-less rhythms reported by different laboratories and impossible to say what they have in common.

Obviously, the best way to address the questions of how many FLOs there are and whether they function independently of the FRQ/WCC oscillator is to identify components of the FLOs. The analysis of genetic interactions in multiple mutant strains can suggest participation in common pathways, as described in Section II.B. The complex interactions between the series of *frq* alleles and the long periods in both *cel* and *chol-1* (Lakin-Thomas, 1998; Lakin-Thomas and Brody, 1985) indicate that the FRQ gene product interacts with the FLO(s) that produce long periods in both *cel* and *chol-1*. The epistasis of *prd-1* over *cel* (Lakin-Thomas and Brody, 1985) places *prd-1* in the same pathway as the FLO in *cel*. The series of interactions among *prd* mutations and *frq* alleles (Morgan and Feldman, 2001; Morgan *et al.*, 2001) place *prd-1* and *prd-2* in the same pathway, interacting with *prd-6* and *frq*. All these interactions point to a system in which FRQ/WCC is not independent of the FLO(s) affected by *cel*, *chol-1*, *prd-1*, *prd-2*, and *prd-6*.

Additional genetic evidence comes from assaying the effects of mutations on more than one FRQ-less rhythm. Four different FRQ-less rhythms have been assayed in *prd-1* and all four were found to be affected: in *cel*, in *chol-1*, with geraniol, and with heat pulses. Three FRQ-less rhythms have been assayed in *prd-2* and all were affected: in *chol-1*, with geraniol, and with heat pulses. Two FRQ-less rhythms are affected by UV90: in *chol-1* and with heat pulses. So far, the evidence indicates that these mutations all affect multiple FRQ-less rhythms. It may be that these gene products participate in several different FLOs, but the simplest explanation is that a single FLO is behind multiple observed rhythms.

It has recently been proposed (Brody *et al.*, 2010) that the many conditions and mutations that allow expression of FRQ-less rhythms may all converge on a pathway that includes ROS and the activation of a RAS–cAMP–protein kinase signaling pathway leading to conidiation. Several of the mutations and treatments that affect the expression of FRQ-less rhythms can be linked to the production of or response to ROS, such as menadione, farnesol, and *sod-1* (Brody *et al.*, 2010). Menadione, by phenocopying *ras*[bd], links ROS to the RAS signaling pathway (Belden *et al.*, 2007a). The clock effects of genes involved in conidiation, such as *rco-1* and *csp-1*, and the effects of caffeine may implicate a cAMP pathway leading to conidiation as a candidate for a component of the FLO (Brody *et al.*, 2010). It is interesting to note that connections are now being reported between the circadian clockworks and the cAMP pathway, and metabolic regulation in general, in other organisms (Bass and Takahashi, 2010; Harrisingh and Nitabach, 2008; Hastings *et al.*, 2008).

A model for the *Neurospora* circadian system is proposed in Fig. 3.2. In this model, the FRQ/WCC feedback loop is coupled to a single FLO. The mechanism of the FLO is unknown and may or may not include a transcription/translation feedback loop. Both oscillators are required in the intact system to maintain the period, temperature compensation, amplitude, and stability of the output. When the FRQ/WCC oscillator is defective in *frq* or *wc* null mutants, the FLO can drive the output at a low amplitude and without the full range of properties such as temperature compensation and period stability. The conditions and genotypes listed above that allow expression of FRQ-less rhythms may act on various parameters of the FLO to increase its amplitude. The differences in properties between different FRQ-less rhythms can be explained by effects of the various conditions and background genotypes on different parameters of the FLO. Versions of this model have been discussed previously (Lakin-Thomas, 2006b; Lakin-Thomas and Brody, 2004; Li and Lakin-Thomas, 2010).

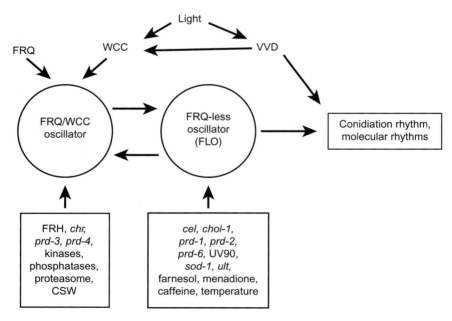

Figure 3.2. Model for the integrated circadian system of *Neurospora*. Circles represent oscillators; arrows represent causal influences. See text for details.

VI. CLOCK-CONTROLLED GENES

A. Output from the circadian oscillator: The hands of the clock

In *Neurospora*, circadian rhythms in several physiological properties have been documented (Bell-Pedersen *et al.*, 1996a; Lakin-Thomas *et al.*, 1990); however, it was the easily observable rhythm in the development of macroconidia that led to further investigation of the *Neurospora* clock mechanisms (Pittendrigh *et al.*, 1959). Mutations that altered the pattern of rhythmic development led to the identification of FRQ/WCC oscillator components (Feldman *et al.*, 1979). Thus, the study of circadian output pathways not only yields information on the cellular functions that are regulated by the clock and the mechanism of this regulation but also is instrumental in discovering oscillator components.

B. Identification and characterization of ccgs

Molecular investigation of circadian output pathways originated in *Neurospora* by using subtractive hybridization to isolate rhythmically expressed genes (Loros *et al.*, 1989). Two genes were identified that expressed mRNA that accumulated with a circadian rhythm with a period that matched the genotype of the strain. In other words, in a wild-type strain, the period of the mRNA rhythm in DD at 25 °C was 22 h, whereas in the long-period frq^7 mutant, the period of the mRNA rhythm was about 29 h. This landmark study provided evidence that the genes were *bone fide* ccgs and provided the criteria to establish a gene as a ccg. Subsequent screens for ccgs (Bell-Pedersen *et al.*, 1996b; Zhu *et al.*, 2001), including microarray studies, led to the identification of over 400 ccgs (Correa *et al.*, 2003; Dong *et al.*, 2008; Nowrousian *et al.*, 2003). These data suggested that as much as 25% of the *Neurospora* transcriptome is under clock control, and that the dominant peak in rhythmic expression for most of the ccgs occurs just before dawn. However, ccgs were found to peak at all possible phases of the day (Correa *et al.*, 2003).

C. The functions of the ccgs provide insights into the role of the clock

The predicted functions of the proteins encoded by the identified *Neurospora* ccgs have yielded key insights into processes that are clock regulated, many of which are conserved in higher eukaryotes, including development, metabolism, cell cycle, transport, and stress responses (Fig. 3.3). For example, genes encoding enzymes involved in carbon and nitrogen metabolism show circadian rhythms in mRNA accumulation, with peaks of expression occurring in the late night to early morning (Correa *et al.*, 2003). In addition, the glycogen phosphorylase, mannitol-1-phosphate dehydrogenase, and a low-affinity glucose transporter

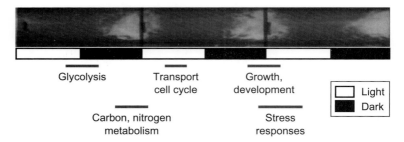

Figure 3.3. Representation of the phase of expression of functional classes of ccgs. (Above) The developmental rhythm in *Neurospora* grown on race tubes in a 12-h light: 12-h dark cycle. Growth is from left to right. The vertical black lines on the tube depict the dark to light transition every 24 h. (Below) Representative functional groups of ccgs with respect to the approximate phases in which they peak in expression.

genes peak in expression in the early night, suggesting that flux into the glycolytic pathway peaks at this time of day to prepare for increased energy requirements related to the consequent development of conidiospores. Several ccgs that are involved in stress responses, including genes encoding antioxidant enzymes that prevent damage due to ROS, are also under clock control, such as glutamate dehydrogenase, glutamine synthetase, oxidoreductase, and catalase. Genes encoding these enzymes peak in expression during the daytime, suggesting that the clock increases antioxidative defense mechanisms to cope with increases in free radicals that can result from light exposure.

In support for a role of the clock in providing a mechanism for anticipating stress responses, the highly conserved *Neurospora* osmosensing p-38 MAPK pathway (the OS pathway), essential for osmotic stress responses, functions as a circadian output pathway that regulates daily rhythms in expression of several downstream target genes of the pathway (de Paula *et al.*, 2007; Vitalini *et al.*, 2007). In this pathway, a circadian signal is relayed to the MAPK OS-2, resulting in rhythmic phosphorylation and activation of the MAPK. Phospho-OS-2, in turn, regulates downstream effector kinases, transcription factors, proteins involved in chromatin remodeling, and translation factors that then regulate rhythmic expression of downstream target genes of the pathway.

Two morning-specific ccgs known to be induced by osmotic stress under control of the OS-MAPK pathway include *ccg-1* and *ccg-9* (Vitalini *et al.*, 2007; Yamashita *et al.*, 2008) (D. Bell-Pedersen, unpublished data). Rhythms in *ccg-1* (a gene encoding a protein of unknown function) are dependent on rhythmic activation of OS-2, and the ATF transcription factor that is activated in response to osmotic stress by phospho-OS-2 binds to the *ccg-1* promoter (Yamashita *et al.*, 2008). The *ccg-9* gene encodes trehalose synthase, which catalyzes the synthesis of the disaccharide trehalose. Trehalose plays an important role in protecting cells from environmental stresses, including osmotic stress (Shinohara *et al.*, 2002). A role in development for *ccg-9* is also suggested by the finding that inactivation of *ccg-9* results in altered conidiophore morphology and abolishes the normal circadian rhythm of conidial development. *ccg-9*-null strains have normal FRQ cycling, phosphorylation, and light induction, indicating that loss of the conidiation rhythm may be a defect in circadian output and is not due to changes in either the FRQ/WCC oscillator or light input into the clock (Shinohara *et al.*, 2002). The phenotype of the *ccg-9* deletion strain differs from *os-2* null mutants (Vitalini *et al.*, 2007), suggesting multiple levels of regulation of *ccg-9*.

Taken together, these data indicate that clock regulation of MAPK signaling pathways provides a mechanism to coordinately control major groups of genes such that they peak at the appropriate times of day to provide a growth and survival advantage to the organism likely by anticipating stresses (de Paula *et al.*, 2007). Consistent with this idea, global gene profiling in *Arabidopsis thaliana* revealed that approximately 68% of the ccgs overlap with genes that are differently regulated in response to osmotic- and cold stresses (Kreps *et al.*, 2002).

D. Signaling time-of-day information from the oscillator to the ccgs

A major goal has been to characterize, at the molecular and biochemical levels, the pathways by which the clock regulates phase-specific gene expression. One mechanism by which the output pathways are rhythmically controlled was believed to be through the rhythmic activity of the WCC transcription factor. Consistent with this idea, many ccgs contain a WCC binding site (LRE) in their promoters (Dong *et al.*, 2008), and WCC was found to bind directly to the promoters of several ccgs, including *vvd, al-3, fl*, and *sub-1* (Bell-Pedersen *et al.*, 2005; Carattoli *et al.*, 1994; Chen *et al.*, 2009; Correa *et al.*, 2003; He and Liu, 2005b; Olmedo *et al.*, 2010b).

To further explore the idea that rhythmic WCC activity is the primary output signal from the FRQ/WCC, global approaches were used to identify the direct targets of the WCC. Using ChIP-sequencing (ChIP-seq) of WC-2 following a light pulse to activate the WCC, more than 400 regions of the chromosome were found to bind the WCC, with about 200 of these binding sites present in predicted promoters of at least one gene (Smith *et al.*, 2010). A consensus WCC binding site (GATCGA) was derived from these studies, with two or more copies typically present in the binding site region. Consistent with previous studies on ccgs, the direct target genes of the WCC were enriched from genes with functions in the cell cycle, transcription, protein binding, and responses to the environment. As expected, most of the direct targets of the WCC were found to be light induced and/or clock regulated, peaking in the early morning. Interestingly, about 20% of the transcription factors known to be present in the *Neurospora* genome were found to be direct targets of the WCC. These 24 transcription factors include known activators and repressors of gene expression, that could, in theory, directly account for the differentially phased ccgs (Fig. 3.4). Further studies on one of the transcription factors, ADV-1, demonstrated that the *adv-1* gene is both light induced and expressed with a circadian rhythm with peak transcript and protein levels in the early morning (Smith *et al.*, 2010). Deletion of *adv-1* had only a minor affect on the development of spores, but similar to the *ccg-9* deletion strain, resulted in the complete loss of the circadian rhythm of spore development. FRQ protein rhythms were unaltered in the *adv-1* deletion strain, suggesting that ADV-1 functions directly downstream of the FRQ/WCC oscillator to regulate the developmental rhythm.

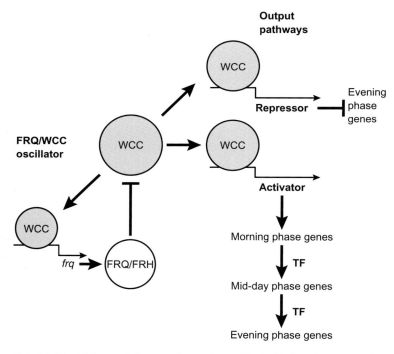

Figure 3.4. Model of differential phase regulation of ccgs. The WCC directly activates the expression of 24 transcription factors in the morning, including activators and repressors of transcription. The repressors would be active during the day, and therefore the targets of the repressors would be expressed in the night. In addition, the activators would turn on the first tier targets during the morning, resulting in morning-specific gene expression. One or more of the first tier targets may include transcription factors that activate or repress the second tier targets to regulate mid-day-specific ccgs, etc.

Taken together, the data suggest a "flat" hierarchical network in which the WCC activates the expression of a suite of second tier transcription factors, which, in turn, activate or repress their individual target genes. Some of these second tier targets may include additional transcription factors that activate or repress the next level. Conservatively, assuming that each of the 24 transcription factor regulates a set of about 20 downstream genes (and that WCC directly binds to regulate rhythms in about 180 additional morning-specific targets), this WCC-directed regulatory network can account for all of the known ccgs and their varied phases. Efforts are now underway to identify the direct targets of these 24 transcription factors using ChIP-seq to comprehensively map the WCC output network. In addition, it is clear that chromatin remodeling, posttranscriptional, translational, and posttranslational mechanisms provide additional combinatorial control of circadian rhythmicity, such as that observed for clock control of the MAPK signaling pathway, and these areas provide fertile ground for future work to decipher the mechanisms for clock control.

E. Output pathways can feed back to regulate oscillator activity

The modulating action of the clock-controlled component VVD on light input pathways provides a clear example of how output pathways can feed back to influence the circadian oscillator (Elvin *et al.*, 2005; Heintzen *et al.*, 2001). In addition to *vvd*, the ccg *prd-4* was shown to influence input to the FRQ/WCC oscillator in *Neurospora* (Gardner and Feldman, 1981; Pregueiro *et al.*, 2006). PRD-4 is an ortholog of mammalian *chk-2*, which is involved in the response to DNA damage (see Section IV). *Prd-4* mRNA accumulates rhythmically, peaking during the day. This regulation would allow cells to respond maximally to DNA damage caused by ionizing radiation during the day, resulting in the arrest of the cell cycle following PRD-4 phosphorylation in response to this damage. PRD-4 feeds back to the clock through an input pathway that signals DNA damage information to the FRQ/WCC oscillator via FRQ phosphorylation. This output to input loop involving *prd-4* is predicted to provide a time-of-day-specific "gate" during which time DNA-damaging agents can affect the phase of the FRQ/WCC oscillator, which would, in turn, affect *prd-4* expression.

In summary, while it is clear that different organisms use their clocks to regulate different biological processes, many of the output signaling pathways, such as the MAPK pathway and certain transcription factors, are conserved. Thus, a detailed understanding of the circadian output network in *Neurospora* will have a major impact on our understanding of circadian regulation in higher eukaryotes, including humans. In addition, these studies may provide key insights into the design of novel therapies for circadian disorders that result in changes in overt circadian rhythms and desynchrony.

VII. OUTLOOK

Great strides have been made in our understanding of the *Neurospora* "clock works" over the past 40 years, moving from biological phenomenology to the molecular and biochemical mechanism of the circadian clock system. We now have considerable detail regarding the nature of the core molecular feedback loops, their light input pathway, and their output. Further, the list of "clock-affecting" mutants continues to grow at a fast pace. More "effectors" are likely to continue to be discovered using the unusual resources that are available for *Neurospora*, such as the gene knockout collections. These genetic tools are likely to be supplemented by cytological and imaging techniques. Below is a partial list of some areas/techniques that are currently being studied, or are on the horizon.

(A) Much remains to be discovered about the mechanism of the oscillator. More details will likely be uncovered regarding clock protein interactions. Perhaps these will be able to be described in biochemical terms, that is, k_m, k_i, flux rates, etc. 3-D crystal structures of the components would be useful in this regard. The nature of the FLO components needs to be established, and how the FLO(s) relate to the FRQ/WCC oscillator. Studying the connection of circadian oscillators to the regulation of the biochemistry of the cell should prove worthwhile, particularly if the FLO(s) are not transcription/translation feedback loops. The role of redox sensing and low molecular weight molecules has been relatively neglected in the clock field. However, a start has been made to establish the light and clock metabolome (http://fungicyc.broadinstitute.org:1555/overviewsWeb/celOv.shtml).

(B) More studies are likely on subcellular localization of clock components. Currently, little is known about the role of subcellular organelles, such as mitochondria, peroxisomes, Golgi, etc., in the clock, although mutations that affect the function of the organelles can alter circadian period.

(C) The mechanism of temperature compensation will continue to be a fruitful area for studies in *Neurospora*. This will be aided by the already available mutants that affect temperature compensation and the identification of the mutated genes. A full description of temperature compensation will likely be aided by mathematical modeling.

(D) Significant progress has been made in understanding the circadian output pathways using new sequencing technologies. Clearly, the complexities of these processes and their links to cellular biochemistry and metabolism will occupy many labs for quite a while.

(E) Chromatin remodeling is an area that will continue to be of interest in understanding rhythmic gene expression. *Neurospora* is a good model system for these analyses as well, as it provides the opportunity to combine genetic, genomic (through ChIP-seq of histone and DNA modifying proteins), and biochemical techniques with imaging of histones using GFP, RFP, etc.

(F) A systems biology approach to monitor rhythmicity in RNA, proteins, and the metabolome will be useful. This has been initiated for the transcriptome (Dong *et al.*, 2008), but additional studies are needed to investigate the proteome and metabolome under different growth/lighting conditions and in specific genetic backgrounds.

(G) Finally, there is a special feature of the *Neurospora* clock system that is fascinating to study and one that may be applicable to understanding synchrony in tissue-specific multioscillator model systems discussed in other chapters. This property is that *Neurospora* grows as a syncytium, with cytoplasm and nuclei streaming throughout the colony. It is intriguing how the growing front of the colony stays synchronized, and how stable the clock signals (phase information) are within a given area of the colony. Experiments involving fluorescent labeling of clock proteins and following their movement in a colony would provide one approach to study this unique feature, and genetics could be used to identify key molecules required for this synchrony.

The progress in understanding the details of the mechanism of the circadian clock, particularly with regard to the oscillator feedback loop, the roles of accessory proteins, temperature compensation, light input, and output, would not have been possible without the use of genetics. *Neurospora* is the undisputed leader in this arena. Importantly, many of these conceptual advances made in *Neurospora* have provided the foundation for our understanding of the clock mechanism in higher organisms. Undoubtedly, genetic tools, coupled with new technologies, will continue to provide the tools needed to further advance the field into uncharted areas of circadian biology.

References

Aronson, B. D., Johnson, K. A., and Dunlap, J. C. (1994a). Circadian clock locus *Frequency*: Protein encoded by a single open reading frame defines period length and temperature compensation. *Proc. Natl. Acad. Sci. USA* **91**, 7683–7687.

Aronson, B. D., Johnson, K. A., Loros, J. J., and Dunlap, J. C. (1994b). Negative feedback defining a circadian clock: Autoregulation of the clock gene-frequency. *Science* **263**, 1578–1584.

Baker, C. L., Kettenbach, A. N., Loros, J. J., Gerber, S. A., and Dunlap, J. C. (2009). Quantitative proteomics reveals a dynamic interactome and phase-specific phosphorylation in the *Neurospora* circadian clock. *Mol. Cell* **34**, 354–363.

Ballario, P., Vittorioso, P., Magrelli, A., Talora, C., Cabibbo, A., and Macino, G. (1996). *White collar-1*, a central regulator of blue light responses in *Neurospora*, is a zinc finger protein. *EMBO J.* **15**, 1650–1657.

Bass, J., and Takahashi, J. S. (2010). Circadian integration of metabolism and energetics. *Science* **330**, 1349–1354.

Belden, W. J., Larrondo, L. F., Froehlich, A. C., Shi, M., Chen, C. H., Loros, J. J., and Dunlap, J. C. (2007a). The band mutation in *Neurospora crassa* is a dominant allele of *ras-1* implicating RAS signaling in circadian output. *Genes Dev.* **21**, 1494–1505.

Belden, W. J., Loros, J. J., and Dunlap, J. C. (2007b). Execution of the circadian negative feedback loop in Neurospora requires the ATP-dependent chromatin-remodeling enzyme CLOCK-SWITCH. *Mol. Cell* **25**, 587–600.

Bell-Pedersen, D., Garceau, N., and Loros, J. J. (1996a). Circadian rhythms in fungi. *J. Genet.* **75**, 387–401.

Bell-Pedersen, D., Shinohara, M. L., Loros, J. J., and Dunlap, J. C. (1996b). Circadian clock-controlled genes isolated from *Neurospora crassa* are late night- to early morning-specific. *Proc. Natl. Acad. Sci. USA* **93**, 13096–13101.

Bell-Pedersen, D., Cassone, V. M., Earnest, D. J., Golden, S. S., Hardin, P. E., Thomas, T. L., and Zoran, M. J. (2005). Circadian rhythms from multiple oscillators: Lessons from diverse organisms. *Nat. Rev. Genet.* **6**, 544–556.

Briggs, W. R. (2007). The LOV domain: A chromophore module servicing multiple photoreceptors. *J. Biomed. Sci.* **14**, 499–504.

Brody, S. (1992). Circadian rhythms in *Neurospora crassa*: The role of mitochondria. *Chronobiol. Int.* **9**, 222–230.

Brody, S., Oelhafen, K., Schneider, K., Perrino, S., Goetz, A., Wang, C., and English, C. (2010). Circadian rhythms in *Neurospora crassa*: Downstream effectors. *Fungal Genet. Biol.* **47**, 159–168.

Brunner, M., and Kaldi, K. (2008). Interlocked feedback loops of the circadian clock of Neurospora crassa. *Mol. Microbiol.* **68**, 255–262.

Campbell, J. W., Enderlin, C. S., and Selitrennikoff, C. (1994). Vectors for expression and modification of cDNA sequences in *Neurospora crassa*. *Fungal Genet. News* **41**, 20–21.

Carattoli, A., Cogoni, C., Morelli, G., and Macino, G. (1994). Molecular characterization of upstream regulatory sequences controlling the photoinduced expression of the albino-3 gene of Neurospora crassa. *Mol. Microbiol.* **13**, 787–795.

Castro-Longoria, E., Ferry, M., Bartnicki-Garcia, S., Hasty, J., and Brody, S. (2010). Circadian rhythms in Neurospora crassa: Dynamics of the clock component frequency visualized using a fluorescent reporter. *Fungal Genet. Biol.* **47**, 332–341.

Cha, J., Chang, S. S., Huang, G., Cheng, P., and Liu, Y. (2008). Control of WHITE COLLAR localization by phosphorylation is a critical step in the circadian negative feedback process. *EMBO J.* **27**, 3246–3255.

Cha, J., Yuan, H., and Liu, Y. (2011). Regulation of the activity and cellular localization of the circadian clock protein FRQ. *J. Biol. Chem.* **286**, 11469–11478.

Chen, C. H., Ringelberg, C. S., Gross, R. H., Dunlap, J. C., and Loros, J. J. (2009). Genome-wide analysis of light-inducible responses reveals hierarchical light signalling in *Neurospora*. *EMBO J.* **28**, 1029–1042.

Chen, C. H., Dunlap, J. C., and Loros, J. J. (2010). *Neurospora* illuminates fungal photoreception. *Fungal Genet. Biol.* **47**, 922–929.

Cheng, P., Yang, Y. H., Heintzen, C., and Liu, Y. (2001a). Coiled-coil domain-mediated FRQ-FRQ interaction is essential for its circadian clock function in *Neurospora*. *EMBO J.* **20**, 101–108.

Cheng, P., Yang, Y. H., and Liu, Y. (2001b). Interlocked feedback loops contribute to the robustness of the *Neurospora* circadian clock. *Proc. Natl. Acad. Sci. USA* **98**, 7408–7413.

Cheng, P., Yang, Y. H., Gardner, K. H., and Liu, Y. (2002). PAS domain-mediated WC-1/WC-2 interaction is essential for maintaining the steady-state level of WC-1 and the function of both proteins in circadian clock and light responses of *Neurospora*. *Mol. Cell. Biol.* **22**, 517–524.

Cheng, P., Yang, Y. H., Wang, L. X., He, Q. Y., and Liu, Y. (2003). WHITE COLLAR-1, a multifunctional *Neurospora* protein involved in the circadian feedback loops, light sensing, and transcription repression of Wc-2. *J. Biol. Chem.* **278**, 3801–3808.

Cheng, P., He, Q., He, Q., Wang, L., and Liu, Y. (2005). Regulation of the *Neurospora* circadian clock by an RNA helicase. *Genes Dev.* **19,** 234–241.

Christensen, M. K., Falkeid, G., Loros, J. J., Dunlap, J. C., Lillo, C., and Ruoff, P. (2004). A nitrate induced *frq*-less oscillator in *Neurospora crassa*. *J. Biol. Rhythms* **19,** 280–286.

Christie, J. M., Salomon, M., Nozue, K., Wada, M., and Briggs, W. R. (1999). LOV (light, oxygen, or voltage) domains of the blue-light photoreceptor phototropin (nph1): Binding sites for the chromophore flavin mononucleotide. *Proc. Natl. Acad. Sci. USA* **96,** 8779–8783.

Colot, H. V., Loros, J. J., and Dunlap, J. C. (2005). Temperature-modulated alternative splicing and promoter use in the circadian clock gene *frequency*. *Mol. Biol. Cell* **16,** 5563–5571.

Colot, H. V., Park, G., Turner, G. E., Ringelberg, C., Crew, C. M., Litvinkova, L., Weiss, R. L., Borkovich, K. A., and Dunlap, J. C. (2006). A high-throughput gene knockout procedure for *Neurospora* reveals functions for multiple transcription factors. *Proc. Natl. Acad. Sci. USA* **103,** 10352–10357.

Correa, A., and Bell-Pedersen, D. (2002). Distinct signaling pathways from the circadian clock participate in regulation of rhythmic conidiospore development in *Neurospora crassa*. *Eukaryot. Cell* **1,** 273–280.

Correa, A., Lewis, A. Z., Greene, A. V., March, I. J., Gomer, R. H., and Bell-Pedersen, D. (2003). Multiple oscillators regulate circadian gene expression in *Neurospora*. *Proc. Natl. Acad. Sci. USA* **100,** 13597–13602.

Crosthwaite, S. K. (2004). Circadian clocks and natural antisense RNA. *FEBS Lett.* **567,** 49–54.

Crosthwaite, S. K., Dunlap, J. C., and Loros, J. J. (1997). Neurospora wc-1 and wc-2: Transcription, photoresponses, and the origins of circadian rhythmicity. *Science* **276,** 763–769.

Davis, R. H. (2000). *Neurospora*: Contributions of a Model Organism. Oxford University Press, Oxford.

de Paula, R. M., Lewis, Z. A., Greene, A. V., Seo, K. S., Morgan, L. W., Vitalini, M. W., Bennett, L., Gomer, R. H., and Bell-Pedersen, D. (2006). Two circadian timing circuits in *Neurospora crassa* cells share components and regulate distinct rhythmic processes. *J. Biol. Rhythms* **21,** 159–168.

de Paula, R. M., Vitalini, M. W., Gomer, R. H., and Bell-Pedersen, D. (2007). Complexity of the *Neurospora crassa* circadian clock system: Multiple loops and oscillators. *Cold Spring Harb. Symp. Quant. Biol.* **72,** 345–351.

Diernfellner, A. C. R., Schafmeier, T., Merrow, M., and Brunner, M. (2005). Molecular mechanism of temperature sensing by the circadian clock of *Neurospora crassa*. *Genes Dev.* **19,** 1968–1973.

Diernfellner, A., Colot, H. V., Dintsis, O., Loros, J. J., Dunlap, J. C., and Brunner, M. (2007). Long and short isoforms of *Neurospora* clock protein FRQ support temperature-compensated circadian rhythms. *FEBS Lett.* **581,** 5759–5764.

Diernfellner, A. C. R., Querfurth, C., Salazar, C., Höfer, T., and Brunner, M. (2009). Phosphorylation modulates rapid nucleocytoplasmic shuttling and cytoplasmic accumulation of *Neurospora* clock protein FRQ on a circadian time scale. *Genes Dev.* **23,** 2192–2200.

Dong, W., Tang, X., Yu, Y., Nilsen, R., Kim, R., Griffith, J., Arnold, J., and Schüttler, H. B. (2008). Systems biology of the clock in *Neurospora crassa*. *PLoS One* **3,** e3105.

Dunlap, J. C., and Loros, J. J. (2006). How fungi keep time: Circadian system in *Neurospora* and other fungi. *Curr. Opin. Microbiol.* **9,** 579–587.

Elvin, M., Loros, J. J., Dunlap, J. C., and Heintzen, C. (2005). The PAS/LOV protein VIVID supports a rapidly dampened daytime oscillator that facilitates entrainment of the *Neurospora* circadian clock. *Genes Dev.* **19,** 2593–2605.

Feldman, J. F., and Hoyle, M. N. (1973). Isolation of circadian clock mutants of *Neurospora crassa*. *Genetics* **75,** 606–613.

Feldman, J. F., Gardner, G. F., and Denison, R. (1979). Genetic analysis of the circadian clock of *Neurospora*. *In* "Biological Rhythms and Their Central Mechanism" (M. Suda, O. Hayaishi, and H. Nakagawa, eds.), pp. 57–66. Elsevier, Amsterdam.

Francis, C. D., and Sargent, M. L. (1979). Effects of temperature perturbations on circadian conidiation in *Neurospora*. *Plant Physiol.* **64,** 1000–1004.

Freitag, M., Hickey, P. C., Raju, N. B., Selker, E. U., and Read, N. D. (2004). GFP as a tool to analyze the organization, dynamics and function of nuclei and microtubules in *Neurospora crassa. Fungal Genet. Biol.* **41,** 897–910.

Froehlich, A. C., Liu, Y., Loros, J. J., and Dunlap, J. C. (2002). White collar-1, a circadian blue light photoreceptor, binding to the *frequency* promoter. *Science* **297,** 815–819.

Froehlich, A. C., Loros, J. J., and Dunlap, J. C. (2003). Rhythmic binding of a WHITE COLLAR-containing complex to the *frequency* promoter is inhibited by FREQUENCY. *Proc. Natl. Acad. Sci. USA* **100,** 5914–5919.

Galagan, J. E., Calvo, S. E., Borkovich, K. A., Selker, E. U., Read, N. D., Jaffe, D., FitzHugh, W., Ma, L. J., Smirnov, S., Purcell, S., *et al.* (2003). The genome sequence of the filamentous fungus *Neurospora crassa. Nature* **422,** 859–868.

Garceau, N. Y., Liu, Y., Loros, J. J., and Dunlap, J. C. (1997). Alternative initiation of translation and time-specific phosphorylation yield multiple forms of the essential clock protein FREQUENCY. *Cell* **89,** 469–476.

Gardner, G. F., and Feldman, J. F. (1981). Temperature compensation of circadian period length in clock mutants of *Neurospora crassa. Plant Physiol.* **68,** 1244–1248.

Geever, R. F., Huiet, L., Baum, J. A., Tyler, B. M., Patel, V. B., Rutledge, B. J., Case, M. E., and Giles, N. H. (1989). DNA sequence, organization and regulation of the qa gene cluster of *Neurospora crassa. J. Mol. Biol.* **207,** 15–34.

Gooch, V. D., Mehra, A., Larrondo, L. F., Fox, J., Touroutoutoudis, M., Loros, J. J., and Dunlap, J. C. (2008). Fully codon-optimized luciferase uncovers novel temperature characteristics of the *Neurospora* clock. *Eukaryot. Cell* **7,** 28–37.

Görl, M., Merrow, M., Huttner, B., Johnson, J., Roenneberg, T., and Brunner, M. (2001). A PEST-like element in FREQUENCY determines the length of the circadian period in *Neurospora crassa. EMBO J.* **20,** 7074–7084.

Granshaw, T., Tsukamoto, M., and Brody, S. (2003). Circadian rhythms in *Neurospora crassa:* Farnesol or geraniol allow expression of rhythmicity in the otherwise arrhythmic strains frq^{10}, *wc-1,* and *wc-2. J. Biol. Rhythms* **18,** 287–296.

Guo, J., Cheng, P., Yuan, H., and Liu, Y. (2009). The exosome regulates circadian gene expression in a posttranscriptional negative feedback loop. *Cell* **138,** 1236–1246.

Guo, J., Cheng, P., and Liu, Y. (2010). Functional significance of FRH in regulating the phosphorylation and stability of *Neurospora* circadian clock protein FRQ. *J. Biol. Chem.* **285,** 11508–11515.

Hardin, P. E., Hall, J. C., and Rosbash, M. (1990). Feedback of the *Drosophila period* gene product on circadian cycling of its messenger RNA levels. *Nature* **343,** 536–540.

Harrisingh, M. C., and Nitabach, M. N. (2008). Circadian rhythms: Integrating circadian timekeeping with cellular physiology. *Science* **320,** 879–880.

Hastings, M. H., Maywood, E. S., and O'Neill, J. S. (2008). Cellular circadian pacemaking and the role of cytosolic rhythms. *Curr. Biol.* **18,** R805–R815.

He, Q., and Liu, Y. (2005a). Degradation of the *Neurospora* circadian clock protein FREQUENCY through the ubiquitin-proteasome pathway. *Biochem. Soc. Trans.* **33,** 953–956.

He, Q., and Liu, Y. (2005b). Molecular mechanism of light responses in *Neurospora*: From light-induced transcription to photoadaptation. *Genes Dev.* **19,** 2888–2899.

He, Q. Y., Cheng, P., Yang, Y. H., Wang, L. X., Gardner, K. H., and Liu, Y. (2002). White collar-1, a DNA binding transcription factor and a light sensor. *Science* **297,** 840–843.

He, Q., Cheng, P., Yang, Y. H., He, Q. Y., Yu, H. T., and Liu, Y. (2003). FWD1-mediated degradation of FREQUENCY in *Neurospora* establishes a conserved mechanism for circadian clock regulation. *EMBO J.* **22,** 4421–4430.

He, Q., Cheng, P., He, Q., and Liu, Y. (2005a). The COP9 signalosome regulated the *Neurospora* circadian clock by controlling the stability of the SCF$^{\text{FWD-1}}$ complex. *Genes Dev.* **19**, 1518–1531.

He, Q., Shu, H., Cheng, P., Chen, S., Wang, L., and Liu, Y. (2005b). Light-independent phosphorylation of WHITE COLLAR-1 regulates its function in the *Neurospora* circadian negative feedback loop. *J. Biol. Chem.* **280**, 17526–17532.

He, Q., Cha, J., Lee, H. C., Yang, Y., and Liu, Y. (2006). CKI and CKII mediate the FREQUENCY-dependent phosphorylation of the WHITE COLLAR complex to close the *Neurospora* circadian negative feedback loop. *Genes Dev.* **20**, 2552–2565.

Heintzen, C., and Liu, Y. (2007). The *Neurospora crassa* circadian clock. *Adv. Genet.* **58**, 25–66.

Heintzen, C., Loros, J. J., and Dunlap, J. C. (2001). The PAS protein VIVID defines a clock-associated feedback loop that represses light input, modulates gating, and regulates clock resetting. *Cell* **104**, 453–464.

Hong, C. I., Ruoff, P., Loros, J. J., and Dunlap, J. C. (2008). Closing the circadian negative feedback loop: FRQ-dependent clearance of WC-1 from the nucleus. *Genes Dev.* **22**, 3196–3204.

Huang, G., Chen, S., Li, S., Cha, J., Long, C., Li, L., He, Q., and Liu, Y. (2007). Protein kinase A and casein kinases mediate sequential phosphorylation events in the circadian negative feedback loop. *Genes Dev.* **21**, 3283–3295.

Iwasaki, H., and Dunlap, J. C. (2000). Microbial circadian oscillatory systems in *Neurospora* and *Synechococcus*: Models for cellular clocks. *Curr. Opin. Microbiol.* **3**, 189–196.

Johnson, C. H., Elliot, J. A., and Foster, R. G. (2003). Entrainment of circadian programs. *Chronobiol. Int.* **20**, 741–774.

Kim, T. S., Logsdon, B. A., Park, S., Mezey, J. G., and Lee, K. (2007). Quantitative trait loci for the circadian clock in *Neurospora crassa*. *Genetics* **177**, 2335–2347.

Konopka, R. J., and Benzer, S. (1971). Clock mutants of *Drosophila melanogaster*. *Proc. Natl. Acad. Sci. USA* **68**, 2112–2116.

Kramer, C., Loros, J. J., Dunlap, J. C., and Crosthwaite, S. K. (2003). Role for antisense RNA in regulating circadian clock function in *Neurospora crassa*. *Nature* **421**, 948–952.

Kreps, J. A., Wu, Y., Chang, H. S., Zhu, T., Wang, X., and Harper, J. F. (2002). Transcriptome changes for *Arabidopsis* in response to salt, osmotic, and cold stress. *Plant Physiol.* **130**, 2129–2141.

Lakin-Thomas, P. L. (1996). Effects of choline depletion on the circadian rhythm in *Neurospora crassa*. *Biol. Rhythm Res.* **27**, 12–30.

Lakin-Thomas, P. L. (1998). Choline depletion, *frq* mutations, and temperature compensation of the circadian rhythm in *Neurospora crassa*. *J. Biol. Rhythms* **13**, 268–277.

Lakin-Thomas, P. L. (2006a). Circadian clock genes *frequency* and *white collar-1* are not essential for entrainment to temperature cycles in *Neurospora crassa*. *Proc. Natl. Acad. Sci. USA* **103**, 4469–4474.

Lakin-Thomas, P. L. (2006b). New models for circadian systems in microorganisms. *FEMS Microbiol. Lett.* **259**, 1–6.

Lakin-Thomas, P. L., and Brody, S. (1985). Circadian rhythms in *Neurospora crassa*: Interactions between clock mutations. *Genetics* **109**, 49–66.

Lakin-Thomas, P. L., and Brody, S. (2000). Circadian rhythms in *Neurospora crassa*: Lipid deficiencies restore robust rhythmicity to null *frequency* and *white-collar* mutants. *Proc. Natl. Acad. Sci. USA* **97**, 256–261.

Lakin-Thomas, P. L., and Brody, S. (2004). Circadian rhythms in microorganisms: New complexities. *Annu. Rev. Microbiol.* **58**, 489–519.

Lakin-Thomas, P. L., Cote, G. G., and Brody, S. (1990). Circadian rhythms in *Neurospora crassa*: Biochemistry and genetics. *Crit. Rev. Microbiol.* **17**, 365–416.

Lakin-Thomas, P. L., Brody, S., and Cote, G. G. (1991). Amplitude model for the effects of mutations and temperature on period and phase resetting of the *Neurospora* circadian oscillator. *J. Biol. Rhythms* **6**, 281–297.

Lakin-Thomas, P. L., Brody, S., and Coté, G. G. (1997). Temperature compensation and membrane composition in *Neurospora crassa*. *Chronobiol. Int.* **14**, 445–454.

Lambreghts, R., Shi, M., Belden, W. J., DeCaprio, D., Park, D., Henn, M. R., Galagan, J. E., Baştürkmen, M., Birren, B. W., Sachs, M. S., *et al.* (2009). A high-density single nucleotide polymorphism map for *Neurospora crassa*. *Genetics* **181**, 767–781.

Lee, K., Loros, J. J., and Dunlap, J. C. (2000). Interconnected feedback loops in the *Neurospora* circadian system. *Science* **289**, 107–110.

Li, S., and Lakin-Thomas, P. L. (2010). Effects of *prd* circadian clock mutations on FRQ-less rhythms in *Neurospora*. *J. Biol. Rhythms* **25**, 71–80.

Li, S., Motavaze, K., Kafes, E., Suntharalingam, S., and Lakin-Thomas, P. (2011). A new mutation affecting FRQ-less rhythms in the circadian system of Neurospora crassa. *PLoS Genetics* **7**, e1002151.

Linden, H., and Macino, G. (1997). *White collar 2*, a partner in blue-light signal transduction, controlling expression of light-regulated genes in *Neurospora crassa*. *EMBO J.* **16**, 98–109.

Liu, Y. (2003). Molecular mechanisms of entrainment in the *Neurospora* circadian clock. *J. Biol. Rhythms* **18**, 195–205.

Liu, Y., and Bell-Pedersen, D. (2006). Circadian rhythms in *Neurospora crassa* and other filamentous fungi. *Eukaryot. Cell* **5**, 1184–1193.

Liu, Y., Garceau, N. Y., Loros, J. J., and Dunlap, J. C. (1997). Thermally regulated translational control of FRQ mediates aspects of temperature responses in the *Neurospora* circadian clock. *Cell* **89**, 477–486.

Liu, Y., Loros, J., and Dunlap, J. C. (2000). Phosphorylation of the *Neurospora* clock protein FREQUENCY determines its degradation rate and strongly influences the period length of the circadian clock. *Proc. Natl. Acad. Sci. USA* **97**, 234–239.

Liu, Y., He, Q., and Cheng, P. (2003). Photoreception in Neurospora: A tale of two White Collar proteins. *Cell. Mol. Life Sci.* **60**, 2131–2138.

Lombardi, L., Schneider, K., Tsukamoto, M., and Brody, S. (2007). Circadian rhythms in *Neurospora crassa*: Clock mutant effects in the absence of a *frq*-based oscillator. *Genetics* **175**, 1175–1183.

Loros, J. J., and Dunlap, J. C. (2001). Genetic and molecular analysis of circadian rhythms in *Neurospora*. *Annu. Rev. Physiol.* **63**, 757–794.

Loros, J. J., and Feldman, J. F. (1986). Loss of temperature compensation of circadian period length in the *frq-9* mutant of *Neurospora crassa*. *J. Biol. Rhythms* **1**, 187–198.

Loros, J. J., Denome, S. A., and Dunlap, J. C. (1989). Molecular cloning of genes under control of the circadian clock in *Neurospora*. *Science* **243**, 385–388.

Luo, C. H., Loros, J. J., and Dunlap, J. C. (1998). Nuclear localization is required for function of the essential clock protein FRQ. *EMBO J.* **17**, 1228–1235.

Mattern, D. L., Forman, L. R., and Brody, S. (1982). Circadian rhythms in *Neurospora crassa*: A mutation affecting temperature compensation. *Proc. Natl. Acad. Sci. USA* **79**, 825–829.

McClung, C. R., Fox, B. A., and Dunlap, J. C. (1989). The *Neurospora* clock gene *frequency* shares a sequence element with the *Drosophila* clock gene *period*. *Nature* **339**, 558–562.

McCluskey, K. (2003). The Fungal Genetics Stock Center: From molds to molecules. *Adv. Appl. Microbiol.* **52**, 245–262.

Mehra, A., Shi, M., Baker, C. L., Colot, H. V., Loros, J. J., and Dunlap, J. C. (2009). A role for casein kinase 2 in the mechanism underlying circadian temperature compensation. *Cell* **137**, 749–760.

Merrow, M., Brunner, M., and Roenneberg, T. (1999). Assignment of circadian function for the *Neurospora* clock gene *frequency*. *Nature* **399**, 584–586.

Merrow, M., Franchi, L., Dragovic, Z., Gorl, M., Johnson, J., Brunner, M., Macino, G., and Roenneberg, T. (2001). Circadian regulation of the light input pathway in *Neurospora crassa*. *EMBO J.* **20**, 307–315.

Michael, T. P., Park, S., Kim, T.-S., Booth, J., Byer, A., Sun, Q., Chory, J., and Lee, K. (2007). Simple sequence repeats provide a substrate for phenotypic variation in the *Neurospora crassa* circadian clock. *PLoS One* **8**, e795.

Morgan, L. W., and Feldman, J. F. (2001). Epistatic and synergistic interactions between circadian clock mutations in *Neurospora crassa*. *Genetics* **159**, 537–543.

Morgan, L. W., Feldman, J. F., and Bell-Pedersen, D. (2001). Genetic interactions between clock mutations in *Neurospora crassa*: Can they help us to understand complexity? *Philos. Trans. R. Soc. Lond. B Biol. Sci.* **356**, 1717–1724.

Nakashima, H. (1981). A liquid culture method for the biochemical analysis of the circadian clock of *Neurospora crassa*. *Plant Cell Physiol.* **22**, 231–238.

Nowrousian, M., Duffield, G. E., Loros, J., and Dunlap, J. (2003). The *frequency* gene is required for temperature-dependent regulation of many clock-controlled genes in *Neurospora crassa*. *Genetics* **164**, 922–933.

Olmedo, M., Navarro-Sampedro, L., Ruger-Herreros, C., Kim, S. R., Jeong, B. K., Lee, B. U., and Corrochano, L. M. (2010a). A role in the regulation of transcription by light for RCO-1 and RCM-1, the *Neurospora* homologs of the yeast Tup1-Ssn6 repressor. *Fungal Genet. Biol.* **47**, 939–952.

Olmedo, M., Ruger-Herreros, C., and Corrochano, L. M. (2010b). Regulation by blue light of the *fluffy* gene encoding a major regulator of conidiation in *Neurospora crassa*. *Genetics* **184**, 651–658.

Perkins, D. D., and Davis, R. H. (2000). *Neurospora* at the millennium. *Fungal Genet. Biol.* **31**, 153–167.

Perkins, D. D., Radford, A., and Sachs, M. S. (2001). The *Neurospora* Compendium. Chromosomal Loci. Academic Press, San Diego.

Pittendrigh, C. S., Bruce, V. G., Rosensweig, N. S., and Rubin, M. L. (1959). A biological clock in *Neurospora*. *Nature* **184**, 169–170.

Pregueiro, A. M., Price-Lloyd, N., Bell-Pedersen, D., Heintzen, C., Loros, J., and Dunlap, J. (2005). Assignment of an essential role for the *Neurospora frequency* gene in circadian entrainment to temperature cycles. *Proc. Natl. Acad. Sci. USA* **102**, 2210–2215.

Pregueiro, A. M., Liu, Q. Y., Baker, C. L., Dunlap, J. C., and Loros, J. J. (2006). The *Neurospora* checkpoint kinase 2: A regulatory link between the circadian and cell cycles. *Science* **313**, 644–649.

Price-Lloyd, N., Elvin, M., and Heintzen, C. (2005). Synchronizing the *Neurospora crassa* circadian clock with the rhythmic environment. *Biochem. Soc. Trans.* **33**, 949–952.

Ramsdale, M., and Lakin-Thomas, P. L. (2000). sn-1,2-diacylglycerol levels in the fungus *Neurospora crassa* display circadian rhythmicity. *J. Biol. Chem.* **275**, 27541–27550.

Roenneberg, T., and Merrow, M. (1998). Molecular circadian oscillators: An alternative hypothesis. *J. Biol. Rhythms* **13**, 167–179.

Roenneberg, T., Dragovic, Z., and Merrow, M. (2005). Demasking biological oscillators: Properties and principles of entrainment exemplified by the *Neurospora* circadian clock. *Proc. Natl. Acad. Sci. USA* **102**, 7742–7747.

Ruoff, P., Loros, J. J., and Dunlap, J. C. (2005). The relationship between FRQ-protein stability and temperature compensation in the *Neurospora* circadian clock. *Proc. Natl. Acad. Sci. USA* **102**, 17681–17686.

Sancar, G., Sancar, C., Brunner, M., and Schafmeier, T. (2009). Activity of the circadian transcription factor White Collar Complex is modulated by phosphorylation of SP-motifs. *FEBS Lett.* **583**, 1833–1840.

Sargent, M. L., and Briggs, W. R. (1967). The effects of light on a circadian rhythm of conidiation in *Neurospora*. *Plant Physiol.* **42**, 1504–1510.

Sargent, M. L., and Kaltenborn, S. H. (1972). Effects of medium composition and carbon dioxide on circadian conidiation in *Neurospora*. *Plant Physiol.* **50**, 171–175.

Sargent, M. L., Briggs, W. R., and Woodward, D. O. (1966). Circadian nature of a rhythm expressed by an invertaseless strain of *Neurospora crassa*. *Plant Physiol.* **41,** 1343–1349.

Schafmeier, T., Haase, A., Káldi, K., Scholz, J., Fuchs, M., and Brunner, M. (2005). Transcriptional feedback of *Neurospora* circadian clock gene by phosphorylation-dependent inactivation of its transcription factor. *Cell* **122,** 1–12.

Schafmeier, T., Kaldi, K., Diernfellner, A., Mohr, C., and Brunner, M. (2006). Phosphorylation-dependent maturation of *Neurospora* circadian clock protein from a nuclear repressor toward a cytoplasmic activator. *Genes Dev.* **20,** 297–306.

Schafmeier, T., Diernfellner, A., Schäfer, A., Dintsis, O., Neiss, A., and Brunner, M. (2008). Circadian activity and abundance rhythms of the Neurospora clock transcription factor WCC associated with rapid nucleo-cytoplasmic shuttling. *Genes Dev.* **22,** 3397–3402.

Schneider, K., Perrino, S., Oelhafen, K., Li, S., Zatsepin, A., Lakin-Thomas, P. L., and Brody, S. (2009). Rhythmic conidiation in constant light in *Vivid* mutants of *Neurospora crassa*. *Genetics* **181,** 917–931.

Schwerdtfeger, C., and Linden, H. (2000). Localization and light-dependent phosphorylation of white collar 1 and 2, the two central components of blue light signaling in *Neurospora crassa*. *Eur. J. Biochem.* **267,** 414–421.

Schwerdtfeger, C., and Linden, H. (2003). VIVID is a flavoprotein and serves as a fungal blue light photoreceptor for photoadaptation. *EMBO J.* **22,** 4846–4855.

Shaw, J., and Brody, S. (2000). Circadian rhythms in Neurospora: A new measurement, the reset zone. *J. Biol. Rhythms* **15,** 225–240.

Shi, M., Larrondo, L. F., Loros, J. J., and Dunlap, J. C. (2007). A developmental cycle masks output from the circadian oscillator under conditions of choline deficiency in *Neurospora*. *Proc. Natl. Acad. Sci. USA* **104,** 20102–20107.

Shi, M., Collett, M., Loros, J. J., and Dunlap, J. C. (2010). FRQ-interacting RNA helicase mediates negative and positive feedback in the *Neurospora* circadian clock. *Genetics* **184,** 351–361.

Shinohara, M. L., Correa, A., Bell-Pedersen, D., Dunlap, J. C., and Loros, J. J. (2002). *Neurospora* clock-controlled gene 9 (*ccg-9*) encodes trehalose synthase: Circadian regulation of stress responses and development. *Eukaryot. Cell* **1,** 33–43.

Shrode, L. B., Lewis, Z. A., White, L. D., Bell-Pedersen, D., and Ebbole, D. J. (2001). *vvd* is required for light adaptation of conidiation-specific genes of *Neurospora crassa*, but not circadian conidiation. *Fungal Genet. Biol.* **32,** 169–181.

Smith, K. M., Sancar, G., Dekhang, R., Sullivan, C. M., Li, S., Tag, A. G., Sancar, C., Bredeweg, E. L., Priest, H. D., McCormick, R. F., *et al.* (2010). Transcription factors in light and circadian clock signaling networks revealed by genomewide mapping of direct targets for *Neurospora* white collar complex. *Eukaryot. Cell* **9,** 1549–1556.

Springer, M. L. (1993). Genetic control of fungal differentiation: The three sporulation pathways of *Neurospora crassa*. *Bioessays* **15,** 365–374.

Talora, C., Franchi, L., Linden, H., Ballario, P., and Macino, G. (1999). Role of a white collar-1-white collar-2 complex in blue-light signal transduction. *EMBO J.* **18,** 4961–4968.

Tang, C. T., Li, S., Long, C., Cha, J., Huang, G., Li, L., Chen, S., and Liu, Y. (2009). Setting the pace of the *Neurospora* circadian clock by multiple independent FRQ phosphorylation events. *Proc. Natl. Acad. Sci. USA* **106,** 10722–10727.

Vitalini, M. W., De Paula, R. M., Goldsmith, C. S., Jones, C. A., Borkovich, K. A., and Bell-Pedersen, D. (2007). Circadian rhythmicity mediated by temporal regulation of the activity of p38 MAPK. *Proc. Natl. Acad. Sci. USA* **104,** 18223–18228.

Vitalini, M. W., Dunlap, J., Heintzen, C., Liu, Y., Loros, J., and Bell-Pedersen, D. (2010). Circadian rhythms. *In* "Cellular and Molecular Biology of Filamentous Fungi" (K. A. Borkovich and D. J. Ebbole, eds.), pp. 442–466. ASM Press, Washington, DC.

Wang, J., Hu, Q., Chen, H., Zhou, Z., Li, W., Wang, Y., Li, S., and He, Q. (2010). Role of individual subunits of the *Neurospora crassa* CSN complex in regulation of deneddylation and stability of cullin proteins. *PLoS Genet.* **6,** 1–15.

Yamashiro, C. T., Ebbole, D. J., Lee, B. U. K., Brown, R. E., Bourland, C., Madi, L., and Yanofsky, C. (1996). Characterization of *rco-1* of *Neurospora crassa*, a pleiotropic gene affecting growth and development that encodes a homolog of *Tup1* of *Saccharomyces cerevisiae*. *Mol. Cell. Biol.* **16,** 6218–6228.

Yamashita, K., Shiozawa, A., Watanabe, S., Fukumori, F., Kimura, M., and Fujimura, M. (2008). ATF-1 transcription factor regulates the expression of *ccg-1* and *cat-1* genes in response to fludioxonil under OS-2 MAP kinase in *Neurospora crassa*. *Fungal Genet. Biol.* **45,** 1562–1569.

Yang, Y. H., Cheng, P., Zhi, G., and Liu, Y. (2001). Identification of a calcium/calmodulin-dependent protein kinase that phosphorylates the *Neurospora* circadian clock protein FREQUENCY. *J. Biol. Chem.* **276,** 41064–41072.

Yang, Y. H., Cheng, P., and Liu, Y. (2002). Regulation of the *Neurospora* circadian clock by casein kinase II. *Genes Dev.* **16,** 994–1006.

Yang, Y., Cheng, P., He, Q., Wang, L., and Liu, Y. (2003). Phosphorylation of FREQUENCY protein by casein kinase II is necessary for the function of the *Neurospora* circadian clock. *Mol. Cell. Biol.* **23,** 6221–6228.

Yang, Y., He, Q., Cheng, P., Wrage, P., Yarden, O., and Liu, Y. (2004). Distinct roles for PP1 and PP2A in the *Neurospora* circadian clock. *Genes Dev.* **18,** 255–260.

Yoshida, Y., Maeda, T., Lee, B., and Hasunuma, K. (2008). Conidiation rhythm and light entrainment in superoxide dismutase mutant in *Neurospora crassa*. *Mol. Genet. Genomics* **279,** 193–202.

Zhu, H., Nowrousian, M., Kupfer, D., Colot, H. V., Berrocal-Tito, G., Lai, H. S., Bell-Pedersen, D., Roe, B. A., Loros, J. J., and Dunlap, J. C. (2001). Analysis of expressed sequence tags from two starvation, time- of-day-specific libraries of *Neurospora crassa* reveals novel clock-controlled genes. *Genetics* **157,** 1057–1065.

Ziv, C., and Yarden, O. (2010). Gene silencing for functional analysis: Assessing RNAi as a tool for manipulation of gene expression. *Methods Mol. Biol.* **638,** 77–100.

4

The Genetics of Plant Clocks

C. Robertson McClung
Department of Biological Sciences, Dartmouth College, Hanover, New Hampshire, USA

Advances in Genetics, Vol. 74
0065-2660/11 $35.00
DOI: 10.1016/B978-0-12-387690-4.00004-0

ABSTRACT

The rotation of the earth on its axis confers the property of dramatic, recurrent, rhythmic environmental change. The rhythmicity of this change from day to night and again to day imparts predictability. As a consequence, most organisms have acquired the capacity to measure time to use this time information to temporally regulate their biology to coordinate with their environment in anticipation of coming change. Circadian rhythms, endogenous rhythms with periods of ~24 h, are driven by an internal circadian clock. This clock integrates temporal information and coordinates of many aspects of biology, including basic metabolism, hormone signaling and responses, and responses to biotic and abiotic stress, making clocks central to "systems biology." This review will first address the extent to which the clock regulates many biological processes. The architecture and mechanisms of the plant circadian oscillator, emphasizing what has been learned from intensive study of the circadian clock in the model plant, *Arabidopsis thaliana*, will be considered. The conservation of clock components in other species will address the extent to which the Arabidopsis model will inform our consideration of plants in general. Finally, studies addressing the role of clocks in fitness will be discussed. Accumulating evidence indicates that the consonance of the endogenous circadian clock with environmental cycles enhances fitness, including both biomass accumulation and reproductive performance. Thus, increased understanding of plant responses to environmental input and to endogenous temporal cues has ecological and agricultural importance.   © 2011, Elsevier Inc.

I. INTRODUCTION

For the first two millennia of study, the science of circadian rhythms was the science of plant circadian rhythms. Daily leaf movements of the tamarind tree, *Tamarindus indicus*, were observed on the island of Tylos (now Bahrain) in the Persian Gulf by Androsthenes during the marches of Alexander the Great in the fourth century BC (Bretzl, 1903). There ensued a lengthy gap in the literature until 1729 when de Mairan (1729) revealed the endogenous origin of the daily leaf movements of the sensitive heliotrope plant (probably *Mimosa pudica*) by demonstrating their persistence in constant darkness.

     More than a century passed before it was realized that these rhythms were only ~24 h; de Candolle (1832) determined that the free-running period of *M. pudica* was 22–23 h. Until the observation of rhythmic conidiation of *Neurospora crassa* in space (Sulzman *et al.*, 1984), this deviation of the free-running period from 24 h was "the strongest single piece of evidence that the overt rhythm is under the control of an endogenous timing mechanism known as a circadian clock" (Feldman, 1982). de Candolle (1832) also showed that the leaf movement rhythm could be entrained to photocycles by inverting the rhythm in response to a reversal in the alternation of light and dark. Temperature compensation, too, was first noted in the leaf movement rhythm of *Phaseolus coccineus* (Bünning, 1931). Thus, each of the defining attributes of circadian rhythms, endogenous origin, persistence in constant conditions, entrainment, and temperature compensation, was first appreciated in the leaf movement rhythms of plants.

Darwin and Darwin (1880) suggested the heritability of circadian rhythms, but this was not experimentally validated until the inheritance of period length among progeny from crosses of parents with distinct period lengths was first reported in *Phaseolus* (Bünning, 1932, 1935).

Circadian rhythms were not scientifically described in animals until nearly the twentieth century, with a description of pigment rhythms in arthropods (Kiesel, 1894). Daily activity rhythms in rats were not described until 1922 (Richter, 1922). However, it is only fair to acknowledge that it was in animals that the molecular era of analysis of the circadian clock began with the identification of mutations conferring altered period length in *Drosophila melanogaster* (Konopka and Benzer, 1971) and with the cloning and characterization of the *per* gene (Bargiello *et al.*, 1984; Reddy *et al.*, 1984). The first mutant higher plant with a defective clock was *timing of cab expression 1* (*toc1*), first described in 1995 (Millar *et al.*, 1995), and *TOC1* was not cloned until 2000 (Strayer *et al.*, 2000).

In this chapter, I briefly summarize our current knowledge of the pervasiveness of rhythmicity in plants, emphasizing transcriptomic studies of rhythmic gene expression. I address the current understanding of the plant oscillator mechanism, as defined largely in studies of the model plant, *Arabidopsis thaliana*. I then describe initial efforts to test the applicability of the Arabidopsis model to other plant species. Finally, I consider the role of plant circadian clocks in enhancing evolutionary fitness.

II. GENOME-WIDE CHARACTERIZATION OF CIRCADIAN-REGULATED GENES AND PROCESSES

Plants are richly rhythmic, and the leaf movement rhythm is only one among many rhythms that include germination, growth, enzyme activity, stomatal movement and gas exchange, photosynthetic activity, flower opening, and fragrance emission (Barak *et al.*, 2000; Cumming and Wagner, 1968; Harmer, 2009; McClung and Kay, 1994; Somers, 1999).

A. Circadian regulation of gene expression

Widespread circadian regulation of gene expression underlies many of these rhythms. The circadian clock controls the transcription and transcript accumulation of many genes, including some, but not all, clock genes, which are required for wild-type oscillator function, as well as many output genes, which are not required to generate the oscillation itself. Microarray analyses indicate that roughly one-third of the Arabidopsis transcriptome cycles in abundance during free run in constant conditions following entrainment to photocycles (Covington et al., 2008), which is consistent with the estimate derived from enhancer trapping experiments (Michael and McClung, 2003). However, a comprehensive exploration of plants grown under a variety of thermocycles, photocycles, and free-run conditions indicates that as much as 90% of the Arabidopsis transcriptome cycles in at least one condition (Michael et al., 2008b). Thus, it is clear that clock control of gene expression is widespread. This offers a cautionary note that time of day needs to be considered in any study of gene expression.

With such an extraordinary penetration into the regulation of gene expression, it should come as no surprise that circadian regulation contributes to many if not most plant processes, beginning with seed dormancy and germination (Penfield and Hall, 2009) and culminating with photoperiodic regulation of flowering time (de Montaigu et al., 2010; Imaizumi, 2010; Song et al., 2010). There is widespread regulation of global signaling mechanisms, including of hormone responses (Covington et al., 2008; Robertson et al., 2009), notably auxin (Covington and Harmer, 2007) and abscisic acid (ABA) (Legnaioli et al., 2009), but including all hormones (Covington et al., 2008; Hanano et al., 2006; Michael et al., 2008a; Mizuno and Yamashino, 2008). In addition, there are rhythms in intracellular calcium signaling (Dodd et al., 2005a; Xu et al., 2007). Such regulation likely contributes to rhythmic growth described in hypocotyl and inflorescence stem elongation (Jouve et al., 1999; Michael et al., 2008a; Nozue et al., 2007). There is considerable interplay between the clock and uptake and homeostasis of nutrients, including iron, magnesium, and nitrogen (Duc et al., 2009; Gutiérrez et al., 2008; Hermans et al., 2010). The clock is emerging as a key player in the coordination of metabolic and signaling pathways (Dodd et al., 2005b, 2007; Fukushima et al., 2009; Graf et al., 2010; Gutiérrez et al., 2008). For example, there is a broad diurnal regulation of the mitochondrial proteome (Lee et al., 2010a). A number of clock-regulated TCP transcription factors interact both with known clock proteins and with cis-regulatory elements associated with nuclear genes encoding mitochondrial proteins and may offer a mechanistic link in the pathway of time-of-day-specific expression of these mitochondrial components (Giraud et al., 2010).

B. Circadian modulation of abiotic and biotic stress responses

Evidence has also accrued supporting a role for the clock in modulating the response to environmental signals. Light is one such example and will be discussed in Section III.D. However, the clock also modulates response to other abiotic stimuli, such as cold (Bieniawska et al., 2008; Fowler et al., 2005;

Nakamichi et al., 2009; Rikin et al., 1993), drought (Legnaioli et al., 2009; Wilkins et al., 2009, 2010), and salt and osmotic stress (Kant et al., 2008; Nakamichi et al., 2009). Suggestive evidence has implicated clock function in the response to biotic stimuli, specifically pathogen resistance (Roden and Ingle, 2009; Walley et al., 2007). Recently, it has been reported that Arabidopsis shows enhanced resistance to oomycete infection at dawn relative to dusk (Wang et al., 2011). Numerous components of R gene-mediated immunity are controlled by CCA1, and resistance to infection is compromised by loss of CCA1 function. It seems possible that the primary defect emerges directly from the loss of clock-gated CCA1 expression rather than consequent perturbation of oscillator function because lhy loss-of-function mutants, which produce similar period shortening to cca1 loss-of-function mutants, do not similarly compromise resistance (Wang et al., 2011). It will be important to determine whether other clock constituents contribute to disease resistance and whether CCA1 (and possibly other clock components) plays a role in the response to pathogens in general.

C. Circadian-regulated transcription: Promoter elements and epigenetic modification

In addition to insight into the diversity of pathways modulated by the clock, a second outcome of the analysis of circadian regulation of the transcriptome has been the identification of promoter elements correlated with phase-specific expression (Covington et al., 2008; Harmer et al., 2000; Michael et al., 2008b). It is particularly interesting that these time-of-day cis-acting modules discovered in Arabidopsis are conserved in papaya, poplar, and rice (Michael et al., 2008b; Zdepski et al., 2008), suggesting that the transcriptional regulatory networks elucidated in model species may be generally applicable among plants. Of obvious interest are the transcription factors that bind to these elements. Here, much less is known and this remains one of the major challenges for the community.

One emerging lesson of the past decade has been that the epigenetic context of promoter sequences is intimately associated with gene expression (Gardner et al., 2011). Thus, coordinate circadian rhythms in transcription of suites of genes are likely to be associated with rhythmic changes in chromatin structure, such as those associated with changes in histone acetylation and deacetylation, or with other modifications, such as histone methylation or sumoylation. For example, acetylation of lysine residues in histones facilitates transcriptional activation by loosening chromatin compaction via the reduction of net negative charge of the histone tails in the nucleosomes or by forming binding sites for bromodomain-containing transcription factors (Lee et al., 2010b). In mammals, histone acetyltransferase (HAT) activity is provided by the oscillator component CLOCK, which itself has HAT activity (Doi et al., 2006), and is also associated with additional HATs such as p300/CBP (Curtis *et al.*, 2004; Etchegaray *et al.*, 2003). Rhythmic histone deacetylase (HDAC) activity is provided by SIRT1 (Nakahata *et al.*, 2008). SIRT1 also deacetylates nonhistone targets, including the clock component PER2 (Asher *et al.*, 2008). The study of chromatin changes associated with circadian gene expression in plants is in its infancy. To date, chromatin changes associated with rhythmic expression of a few clock genes have been studied (see below), but little is known at the mechanistic level. This is certainly an important topic that merits considerably more attention.

III. MOLECULAR ORGANIZATION OF PLANT CIRCADIAN CLOCKS INTO INTERLOCKED TRANSCRIPTIONAL FEEDBACK LOOPS

As is clear from this volume, circadian clocks studied to date largely share a common logic. Oscillations arise from negative feedback loops that include a time delay. Transcription–translation feedback loops (TTFLs) underpin the molecular mechanisms of the circadian clock in eukaryotes studied to date, including fungi, flies, mammals, and plants (Dunlap, 1999; Zhang and Kay, 2010). Initially, these were thought to be relatively simple loops, but it has become clear that most eukaryotic circadian oscillators are based on multiple interlocked TTFLs (Zhang and Kay, 2010). Mathematical analysis has suggested that the increased complexity associated with multiple interlocked loops increases flexibility, which enhances robust entrainment and temperature compensation (Rand *et al.*, 2006). However, as will be discussed below, the clock of the unicellular green alga, *Ostreococcus tauri*, consists of a single feedback loop yet supports robust compensated circadian function that is efficiently entrained to the environment (Corellou *et al.*, 2009; Morant *et al.*, 2010; Thommen *et al.*, 2010). Although common components and architecture are evident in angiosperm and algal clocks, it seems that the two are likely to be distinct in a number of aspects.

A. Transcriptional regulation

The Arabidopsis circadian clock has been described as consisting of three main TTFLs, a central loop and two interlocked loops termed "morning" and "evening" based on the times of day in which loop components are maximally expressed (Harmer, 2009; McClung, 2008; McClung and Gutiérrez, 2010; Pruneda-Paz and Kay, 2010). However, each of these loops has multiple components, and many more individual loops are becoming evident. Figure 4.1 presents a simplified view of the Arabidopsis clock that emphasizes transcriptional relationships. As will be discussed below, posttranscriptional and posttranslational mechanisms also play critical roles.

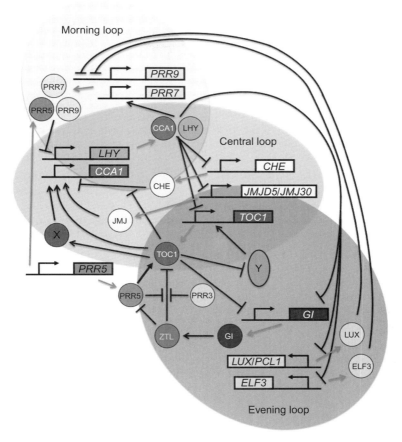

Figure 4.1. A working model of the Arabidopsis circadian clock. The Arabidopsis circadian clock consists of a series of interconnected feedback loops. This model, almost certainly oversimplified, emphasizes transcriptional relationships among clock genes and proteins. Genes are coded as rectangles and proteins are coded as circles. Gray lines from genes to protein represent transcription, transcript processing, and translation. Because the identity of Y remains unknown, it is simply represented as an oval. Regulatory interactions are in black, with arrowheads indicating positive regulation and perpendicular lines indicating negative regulation. Inspired by Harmer (2009) and modified from McClung and Gutiérrez (2010). (For color version of this figure, the reader is referred to the Web version of this book.)

The central loop, first proposed a decade ago (Alabadí *et al.*, 2001), consists of a Pseudo-Response Regulator (PRR), TIMING OF CAB EXPRESSION 1 (TOC1), and two single Myb domain transcription factors, CIRCADIAN

CLOCK-ASSOCIATED 1 (CCA1) and LATE ELONGATED HYPOCOTYL (LHY). *TOC1* was identified in a forward genetic screen on the basis of short circadian period as monitored with a firefly luciferase reporter gene driven from promoter of the *CHLOROPHYLL a/b-BINDING PROTEIN* (also known as *LIGHT-HARVESTING COMPLEX B, LHCB*) gene (*proCAB:LUC*) (Millar et al., 1995). *CCA1* and *LHY* were identified as important for clock function because of arrhythmia of plants in which CCA1 or LHY was constitutively over-expressed (Schaffer et al., 1998; Wang and Tobin, 1998). A key test of an oscillator component is the prediction that its abundance determines phase. Thus, a pulse of an oscillator component should shift phase. This has been accomplished for a number of clock genes, such as Neurospora *FRQ* (Aronson et al., 1994). Recently, it has been shown that pulses of *CCA1* and *LHY* expression driven by an inducible promoter shift the phase of the Arabidopsis clock, confirming them as clock components (Knowles et al., 2008).

Loss-of-function mutants and overexpression studies indicate that TOC1 is a positive regulator of *CCA1* and *LHY*, while CCA1 and LHY are negative regulators of *TOC1* and of themselves (Alabadí et al., 2001, 2002; Mizoguchi et al., 2002; Mizuno and Nakamichi, 2005). Although our understanding is still far from complete, many molecular details are now known. CCA1 and LHY bind as homo- and heterodimers (Lu et al., 2009; Yakir et al., 2009) to the *TOC1* promoter to inhibit its expression. Rhythmic binding of CCA1 and presumably LHY to the evening element (EE; Harmer et al., 2000) in the *TOC1* promoter antagonizes histone H3 acetylation, associated with open chromatin structure (Perales and Más, 2007). Pharmacological inhibition of histone deacetylation alters the waveform of *TOC1* mRNA abundance (Perales and Más, 2007). Histone H3 deacetylation at the *TOC1* promoter is opposed by REVEILLE 8 (RVE8), a CCA1 paralog, which binds to the *TOC1* promoter and promotes histone H3 acetylation (Farinas and Mas, 2011). The identities of the associated HATs and HDACs are not yet known.

TOC1 is a positive regulator of *CCA1* and *LHY*, although until recently was not thought to possess intrinsic DNA-binding activity. However, it has recently been established that the C-terminal CCT (for CONSTANS, CONSTANS-LIKE, and TOC1) motif of CONSTANS has DNA-binding activity *in vitro* (Tiwari et al., 2010). Thus, the binding of TOC1 to the *CCA1* promoter (Pruneda-Paz et al., 2009) may be direct. In addition, CCA1 HIKING EXPEDITION (CHE), a TCP transcription factor, also binds to the *CCA1* promoter and negatively regulates *CCA1* (Pruneda-Paz et al., 2009). TOC1 and CHE interact, but the details of this interaction are incompletely known (Pruneda-Paz et al., 2009). As with *TOC1*, *CCA1* expression changes are associated with changes in chromatin (Ni et al., 2009).

In the "morning" loop, CCA1 and LHY are positive regulators of two *TOC1* relatives, *PRR7* and *PRR9* (Farré et al., 2005; Harmer and Kay, 2005; Mizuno and Nakamichi, 2005). In turn, PRR5, PRR7, and PRR9 bind to the

CCA1 and LHY promoters to repress expression (Nakamichi et al., 2010). Peak binding of the three proteins occurs sequentially from early morning (PRR9) through mid-day (PRR7) to late afternoon/early night (PRR5). The sequential expression patterns of the three PRR proteins extend the temporal window over which CCA1 and LHY expression is repressed, offering a mechanistic explanation for their partially redundant function shown through genetic analysis of loss-of-function mutations (Eriksson et al., 2003; Farré et al., 2005; Michael et al., 2003; Nakamichi et al., 2005a,b, 2007; Salomé and McClung, 2005).

The "evening" loop is complicated and incompletely defined. CCA1 and LHY negatively regulate GI, a negative regulator of TOC1. The relationship between TOC1 and GI is complex. Overexpression of TOC1 strongly represses GI mRNA (Makino et al., 2002). toc1–2 loss of function has little effect on the amplitude of GI mRNA accumulation (Ito et al., 2008; Martin-Tryon et al., 2007). Double mutant analyses showed additivity in the short period and other phenotypes of toc1 and gi loss-of-function mutants, suggesting that negative regulation of GI by TOC1 is likely indirect (Martin-Tryon et al., 2007). The initial three-loop model suggested that TOC1 is activated by an evening component, Y, that included GI (Locke et al., 2006). However, as discussed in more detail below, a major function of GI is stabilization of ZTL, which targets TOC1 for proteasomal degradation, making GI a repressor of TOC1 protein (Kim et al., 2007b). Nonetheless, GI remains an indirect activator of TOC1 mRNA because the GI-regulated degradation of TOC1 protein relieves inhibition of Y, enhancing the activation of TOC1 mRNA accumulation (Pokhilko et al., 2010). Thus, a recent model distinguishes GI from unknown component, Y, as the direct activator of TOC1 mRNA accumulation (Pokhilko et al., 2010). GI has been shown to function as a scaffold protein in transcription complexes at the CO promoter (Sawa et al., 2007), so it remains possible that GI may play additional roles in the transcriptional activation of TOC1 and other genes.

Although the basic outlines of three loops are defined, it is unlikely that any is fully understood in terms of either essential components or regulatory interactions. For example, although a number of players have been identified, how CCA1 and LHY are regulated is not fully described. It is becoming clear that CCA1 and LHY are regulated differently. Although PRR5, PRR7, and PRR9 each bind to both CCA1 and LHY promoters (Nakamichi et al., 2010), another negative regulator, CHE, binds to the CCA1, but not to the LHY promoter (Pruneda-Paz et al., 2009). Similarly, TIC encodes a nuclear protein of unknown function necessary for proper LHY, but not for CCA1, expression (Ding et al., 2007).

Proper regulation of CCA1 and LHY requires additional evening-expressed clock genes, including LUX ARRHYTHMO/PHYTOCLOCK 1 (LUX/PCL1) (Hazen et al., 2005; Onai and Ishiura, 2005), EARLY FLOWERING 3 (ELF3) (Kikis et al., 2005), and ELF4 (Kikis et al., 2005; McWatters et al.,

2007). Recent work has identified *PRR9* as the regulatory target of both LUX/ PCL1 and ELF3, thereby defining new regulatory interactions linking evening-expressed genes to the morning loop (Dixon *et al.*, 2011; Helfer *et al.*, 2011).

LUX/PCL1 was independently identified in two forward genetic screens as required for circadian rhythmicity; loss of function results in arrhythmicity (Hazen *et al.*, 2005; Onai and Ishiura, 2005). *LUX/PCL1* encodes a Myb domain transcription factor. The *in vitro* DNA-binding specificity of LUX was recently characterized via universal protein-binding microarrays (Helfer *et al.*, 2011). Chromatin immunoprecipitation shows that LUX binds to the *PRR9* promoter and acts as a negative regulator. PRR9, together with PRR7 and PRR5, comprises a suite of PRRs that negatively regulate *CCA1* and *LHY* during the day (Nakamichi *et al.*, 2010). Thus, LUX/PCL1 is an indirect positive regulator of *CCA1* via repression of *PRR9*, a *CCA1* repressor. There must be additional targets of LUX, however, because PRR9 overexpression, which would be one consequence of loss of LUX function, results in short period and is insufficient to confer arrhythmicity (Matsushika *et al.*, 2002). LUX binds to and inhibits its own promoter *in vivo*, defining a negative autoregulatory loop (Helfer *et al.*, 2011). There are likely to be additional LUX targets, which may be identified by genomic scale analysis of LUX binding.

ELF3 was identified on the basis of altered flowering time (Zagotta *et al.*, 1996) and established as a negative regulator of light signaling to the clock (Covington *et al.*, 2001; Hicks *et al.*, 2001; Liu *et al.*, 2001; McWatters *et al.*, 2000). More recently, ELF3 has been shown to be a positive regulator of *CCA1* and *LHY* (Kikis *et al.*, 2005) and established as an essential component of the clock (Thines and Harmon, 2010). ELF3 binds to the *PRR9* promoter as a negative regulator (Dixon *et al.*, 2011). Thus, ELF3, like LUX, is an indirect activator of *CCA1* and *LHY* via its repression of *PRR9*, a repressor of *CCA1* and *LHY*. Again as with LUX/PCL1, a fuller investigation of ELF3 binding targets will be necessary to fully understand its role in the clock.

Although the model of the clock is rapidly acquiring new regulatory relationships (Fig. 4.1), such as the recruitment of PRR5, PRR7, and PRR9 to *CCA1* and *LHY*, as well as of LUX/PCL1 and ELF3 to *PRR9*, the lack of transcriptional activators and of direct positive regulatory relationships is conspicuous. It seems improbable that the clock relies exclusively on transcriptional repression, so perhaps this indicates a great deal of redundancy among transcriptional activators. If this is the case, loss-of-function genetic screens will need to be supplemented with gain-of-function approaches, and genetics approaches will need to be augmented with biochemical and molecular biological approaches.

Recent work has further illuminated chromatin changes associated with *CCA1* and *LHY* expression. Jumonji-C (Jmj-C) domain-containing proteins act as histone demethylases in chromatin remodeling (Tsukada *et al.*, 2006), and a Jmj-C domain-containing protein has recently been identified as a likely regulator of *CCA1* and *LHY* expression (Jones et al., 2010; Lu et al., 2011). One study considered genes that were highly co-expressed with *TOC1* in a broad series of microarray experiments and identified *JMJD5* (At3G20810) as both highly co-expressed with *TOC1* and clock regulated (Jones et al., 2010). The second study addressed the same gene, which they call *JMJ30*, on the basis of its robust rhythm in transcript abundance (Lu et al., 2011). Both studies show that loss of JMJD5/JMJ30 function confers short circadian period in the white light. In addition, Jones et al. (2010) showed conditional arrhythmicity of *jmjd5* mutants in bright red light, although period was wild type in the dark. Over-expression of JMJD5/JMJ30 delays flowering and, like loss of function, shortens circadian period (Lu et al., 2011). Jones et al. (2010) found that the expression of *CCA1* and *LHY* is altered in *jmjd5* mutants. *CCA1* and *LHY* cycling show reduced amplitude in *jmjd5* mutants, with peak expression considerably reduced, suggesting that JMJD5/JMJ30 is a positive regulator of *CCA1* and *LHY*. Human JMJD5 demethylates histone H3 dimethylated at lysine 36, which is a repressive modification found associated with both *CCA1* and *LHY* (Oh et al., 2008), so it is possible that JMJD5 acts directly on *CCA1* and *LHY* chromatin to activate transcription. *GI* and *TOC1* expression cycle with reduced amplitude in *jmjd5* mutants. This is possibly a secondary consequence of the altered expression of *CCA1* and *LHY*, which are negative regulators of *GI* and *TOC1*. However, the other study did not see these waveform changes in *CCA1* and *LHY* in loss-of-function *jmj30* mutants, so it seems that further study will be necessary to reconcile these disparate observations and resolve the mechanism of action of *JMJD5/JMJ30*. Overexpression of CCA1 and LHY downregulates *JMJ30/JMJD5* mRNA accumulation, indicating they are negative regulators of *JMJ30/JMJD5* expression. Chromatin immunoprecipitation detected CCA1 and LHY bound to an EE-containing region in the *JMJ30/JMJD5* promoter (Lu et al., 2011), showing that CCA1 and LHY are direct negative regulators of *JMJ30/JMJD5*. Collectively, these results suggest a new negative regulatory loop between *CCA1/LHY* and *JMJ30/JMJD5*.

Other histone modifications, such as H3K4 and H3K9 methylation, are also correlated with clock-regulated transcription in mammals (Brown et al., 2005). As described above, histone methylation has been recently implicated in the regulation of *CCA1* and *LHY*, but little is known about the role of histone methylation in clock-regulated transcriptional control in plants. However, a role for methylation in the downregulation of the flowering repressor, *FLOWERING LOCUS C* (*FLC*), is well documented. Protein arginine methyltransferase 5 (PRMT5) catalyzes the symmetric dimethylation of histone H4 (H4R3sme2) at the *FLC* promoter, which decreases expression to promote flowering (Pei et al., 2007; Schmitz et al., 2008; Wang et al., 2007). Because period lengthens in response to increased *FLC* expression (Edwards et al., 2006; Salathia et al., 2006; Swarup et al., 1999), one might expect that loss of PRMT5 function would also lengthen period via increased *FLC* expression. Indeed, *prmt5* mutants exhibit long period (Hong *et al.*, 2010; Sanchez *et al.*, 2010), although the period lengthening is greater than might be expected based on increased *FLC* expression. Moreover, the *prmt5 flc* double mutant retains long period (Sanchez *et al.*, 2010), indicating that the period lengthening associated with loss of *PRMT5* function is not mediated entirely through histone modification and concomitant changes in *FLC* gene expression.

B. Posttranscriptional regulation

Transcriptional regulation is important in clock function, but it is clear that posttranscriptional regulation is also an essential constituent of the clock mechanism (Staiger and Koster, 2011). First, clock-regulated transcripts are over-represented among unstable transcripts, suggesting that transcript instability might contribute to circadian waveform (Gutiérrez *et al.*, 2002). The clock regulates the degradation of a subset of transcripts via the downstream (DST) instability determinant pathway (Lidder *et al.*, 2005). Light regulates the stability of the *CCA1* transcript, offering a new route for light input to set clock phase (Yakir *et al.*, 2007). Light has also been shown to affect the translation of *LHY* mRNA into protein (Kim *et al.*, 2003).

Alternative splicing is rapidly emerging as an important mechanism in expanding the proteomes of eukaryotes (Nilsen and Graveley, 2010) and has been encountered as a mechanism to regulate expression of several clock genes, including *CCA1*, *ELF3*, and *PRR9* (Filichkin *et al.*, 2010; Hazen *et al.*, 2009; Sanchez *et al.*, 2010), as well as *GLYCINE-RICH PROTEIN 7* (*GRP7*) and *GRP8* (Schöning *et al.*, 2008; Staiger *et al.*, 2003), components of a circadian slave oscillator implicated in the promotion of flowering (Streitner *et al.*, 2008). The latter is an example of a negative autoregulatory loop involving alternative splicing. *GRP7* and *GRP8* (also designated *COLD CIRCADIAN REGULATED 2* and *CCR1*, respectively) transcriptions are clock regulated (Carpenter *et al.*, 1994; Heintzen *et al.*, 1994). When the GRP7 and GRP8 proteins attain a threshold abundance, they bind to their own mRNAs to promote alternative splicing yielding mRNAs with premature termination codons, which target the alternatively spliced mRNAs for non-sense-mediated decay (Heintzen *et al.*, 1997; Schöning *et al.*, 2007, 2008). GRP7 and GRP8 form an output pathway regulating responses to biotic and abiotic stresses (Carpenter *et al.*, 1994; Fu *et al.*, 2007; Kim *et al.*, 2008; Schmidt *et al.*, 2011).

The detailed molecular mechanisms by which alternative splicing occurs within circadian networks are only beginning to be appreciated, but important new insights have recently emerged. Among the genes with altered expression in *prmt5* mutants, the morning loop components *PRR9* and *PRR7*, and the evening loop component, *GI*, show increased amplitude as well as long period and are potential direct or indirect targets of PRMT5 activity (Sanchez *et al.*, 2010). In contrast, central loop components *CCA1*, *LHY*, and *TOC1* are relatively unaffected in terms of amplitude and only exhibit long period (Hong *et al.*, 2010). How might the loss of PRMT5 function affect PRR7 and PRR9 expression? In addition to methylating histones, PRMT5 methylates Sm proteins, constituents of small nuclear ribonucleoprotein components of the spliceosome, and inhibition of methylation disrupts pre-mRNA splicing (Pahlich *et al.*, 2006). Loss of *prmt5* function disrupts alternative splicing of *PRR9* (Sanchez *et al.*, 2010), suggesting that methylation of splicing factors by PRMT5 is critical to clock function. Methylation of splicing factors enhances recruitment of the spliceosome to non-consensus 5′ splice donor sites, and the introns improperly spliced in *prmt5*, such as intron 2 of *PRR9*, diverge from consensus (Sanchez *et al.*, 2010). In the example of *PRR9*, in the absence of PRMT5 function, an alternative 5′ splice donor sequence for intron 2 that lies 8 nt downstream is selected, resulting in a frameshift in the resultant transcript and a premature truncation of translation (Fig. 4.2). There is a second alternative splice in *PRR9* in which intron 3 is sometimes retained, again resulting in premature termination of translation, but this does not seem to be related to PRMT5 function. This offers a plausible mechanism for a role of PRMT5 in modifying 5′ splice site selection via Sm protein methylation (Sanchez *et al.*, 2010). However, one cannot exclude a role for histone methylation being relevant as well because it has recently been shown that histone modifications, including methylation of K residues in H3, regulate alternative splicing of a number of human genes (Luco *et al.*, 2011).

C. Protein localization, modification, and stability

Posttranslational processes, notably phosphorylation and proteolysis, play critical roles in all clock systems (Gallego and Virshup, 2007; Kojima *et al.*, 2011). It has been known for some time that casein kinase 2 phosphorylates CCA1 and LHY (Sugano *et al.*, 1998, 1999), and that this phosphorylation is necessary for CCA1 function (Daniel *et al.*, 2004). More recently, it has been established that all five PRRs are phosphorylated with functionally significant consequences, although the responsible kinases have not been identified (Fujiwara *et al.*, 2008). Recently, it has been shown that proteasomal degradation is necessary at all circadian phases for clock function in *O. tauri* (van Ooijen *et al.*, 2011). In Arabidopsis, PRR5 and TOC1 are targeted for proteasome-mediated degradation through interaction with the F-box protein ZEITLUPE (ZTL) (Fujiwara *et al.*, 2008; Kiba *et al.*, 2007; Más *et al.*, 2003), which assembles *in vivo* with known core components to form an SCF complex (Han *et al.*, 2004; Harmon *et al.*, 2008; Risseeuw *et al.*, 2003). Phosphorylation of TOC1 and PRR3 promotes their

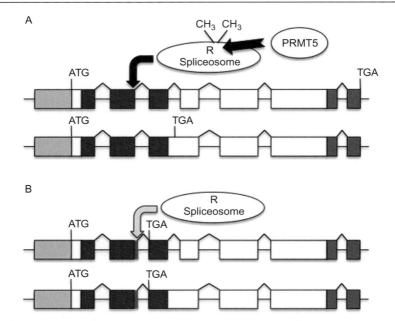

Figure 4.2. PRMT5 modulates splicing of the *PRR9* transcript. (A) PRMT5 methylates an arginine (R) residue(s) in a spliceosomal Sm protein, altering the interaction with 5' splice donor sequences of intron 2 of the primary *PRR9* transcript, indicated by the thin horizontal line. Splicing events are indicated above the exons. (B) In the absence of PRMT5 function, the spliceosome chooses an alternate 5' splice donor sequence that is 8 nt downstream (8 nt is indicated in red), which confers a frameshift and introduces a premature nonsense codon. (A and B, lower cartoons) In a second alternative splicing event, intron 3 is sometimes retained in the transcript, again resulting in a premature translational stop, although PRMT5 does not influence this alternative splicing event. Thus, only the splicing events in the uppermost cartoon yield a transcript capable of being translated to produce full-length, functional PRR9 protein. Boxes indicate exons. The gray shading indicates the 5'UTR. The pseudo receiver (PR) domain is colored blue and the CCT domain is colored green. Start and stop codons are indicated. (For interpretation of the references to color in this figure legend, the reader is referred to the Web version of this chapter.)

interaction, which blocks the interaction of TOC1 with ZTL (Fujiwara *et al.*, 2008). Thus, PRR3 may stabilize TOC1 by sequestering it from ZTL (Para *et al.*, 2007). A second PRR, PRR5, also binds to and stabilizes TOC1, although this PRR5–TOC1 interaction itself is not altered by the phosphorylation status of both partners (Wang *et al.*, 2010). This interaction affects more than the interaction of PRR5 with ZTL because PRR5–TOC1 interaction promotes nuclear accumulation of TOC1 (Wang *et al.*, 2010). This resembles the interactions of TIMELESS-PERIOD, CYCLE-CLOCK, and PERIOD 2-

CRYPTOCHROME in Drosophila and mammals; in each case, interaction with the former promotes nuclear accumulation of the latter (Maurer et al., 2009; Meyer et al., 2006; Miyazaki et al., 2001). Interaction with PRR5 promotes TOC1 phosphorylation, which enhances the interaction of TOC1 with ZTL, but nuclear localization of the PRR5–TOC1 complex sequesters both proteins from cytoplasmic ZTL (Wang et al., 2010). Thus, PRR5 could modulate TOC1 degradation by sequestering it in the nucleus, away from cytoplasmic ZTL (Fujiwara et al., 2008). However, it has not been shown that the PRR5-dependent phosphorylation sites on TOC1 are those that promote the TOC1/ZTL interaction. Phosphorylation of PRR5 promotes its interaction with ZTL, which leads to PRR5 degradation. Unlike with CCA1, there is an almost total lack of detailed knowledge of which sites on the PRRs are phosphorylated, in which order, and by which kinases, and this represents a major gap in our knowledge.

The evening loop is necessary for ZTL-mediated degradation of TOC1. Rhythmic GI transcription drives rhythmic GI protein accumulation in the evening. GI interacts with the LOV (light, oxygen, voltage) domain of ZTL, which stabilizes ZTL (Kim et al., 2007b). Thus, the stabilization of ZTL by GI provides a molecular explanation for the rhythm in ZTL protein abundance despite the lack of cycling in ZTL transcript abundance (Somers et al., 2000, 2004). The ZTL–GI interaction is dramatically enhanced by blue light, and this enhancement is abolished by mutations of ZTL that disrupt LOV domain photochemistry. This demonstrates ZTL to be a blue light photoreceptor that mediates direct light input into the clock (Kim et al., 2007b). After dark, the ZTL-LOV domain undergoes a conformational shift, reducing the affinity of ZTL for GI. ZTL and GI dissociate and ZTL is now able to interact with TOC1 and PRR5, targeting them for proteasomal degradation. Genetic analysis has made it clear that ZTL is the primary F-box protein responsible for degradation of TOC1 and PRR5. However, ZTL is a member of a small family, including LOV, KELCH PROTEIN 2 (LKP2) and FLAVIN, KELCH, F-BOX 1 (FKF1). FKF1 plays a critical role in the photoperiodic flowering time pathway (Imaizumi et al., 2003, 2005; Nelson et al., 2000). Loss of function of either LKP2 or FKF1 has only minor effects on circadian function. However, clock defects are more pronounced in a *ztl fkf1* double mutant and even more pronounced in a *ztl fkf1 lkp2* triple mutant than in the *ztl* single mutant. This shows that there is partial redundancy of function among the three proteins and establishes that both LKP2 and FKF1 are capable of targeting TOC1 and PRR5 for proteasomal degradation (Baudry et al., 2010).

The two remaining PRRs, PRR7 and PRR9, function in the "morning loop" as negative regulators of CCA1 and LHY (Nakamichi et al., 2010). Both proteins show progressive phosphorylation as the day progresses although, as noted above, details of phosphorylation sites and the responsible kinases are not known (Fujiwara et al., 2008). By parallel with PRR5 and TOC1, it seems likely that phosphorylated PRR7 and PRR9 are targeted for proteasomal degradation. However, mechanistic details such as the identity of a putative F-box protein or other ubiquitin ligase complex are lacking.

D. Entrainment

Light and temperature are the two predominant input signals entraining the circadian clock to environmental cycles. Both phytochrome and cryptochrome photoreceptors mediate entrainment (Devlin and Kay, 2001; Nagy et al., 1993; Somers et al., 1998a) via the induction of a number of clock genes, including CCA1 (Wang and Tobin, 1998), LHY (Martínez-García et al., 2000), PRR9 (Farré et al., 2005), and GI (Locke et al., 2005). As discussed above, light also affects the posttranscriptional stability of the CCA1 transcript (Yakir et al., 2007) and the translation of LHY (Kim et al., 2003). The establishment of ZTL as a blue light photoreceptor, as discussed above, provides a mechanism for light entrainment via regulated degradation of the ZTL targets, TOC1 and PRR5 (Fujiwara et al., 2008; Kim et al., 2007b).

As is implicit in the phase response curve, the clock modulates its own sensitivity to light, for example, by gating phytochrome signaling (Anderson et al., 1997). In part, this is because of the clock-gated expression of photoreceptors (Bognár et al., 1999; Hall et al., 2001; Kircher et al., 2002). In addition, numerous genes, many themselves clock regulated, modify (typically to attenuate) light signaling in time-of-day-specific fashion. These genes include ELF3 (Covington et al., 2001; McWatters et al., 2000), ELF4 (Doyle et al., 2002; Kikis et al., 2005; McWatters et al., 2007), TIC (Hall et al., 2003), LIGHT-INSENSITIVE PERIOD 1 (LIP1) (Kevei et al., 2007), FAR-RED-ELONGATED HYPOCOTYL 3 (FHY3) (Allen et al., 2006), and XAP5 CIRCADIAN TIMEKEEPER (XCT) (Martin-Tryon and Harmer, 2008).

The Arabidopsis clock is efficiently entrained to temperature cycles of as little as 4° in amplitude (Heintzen et al., 1994; Michael and McClung, 2002; Somers et al., 1998b), although the identity of the primary temperature-sensing mechanism for this and, indeed, for any plant response is not established (McClung and Davis, 2010). Some components necessary for temperature entrainment have been identified. The prr7 prr9 double mutant cannot be entrained to 22°/12° thermocycles, yet retains the capacity to respond to photocycles when grown at a constant temperature (Salomé and McClung, 2005). However, this defect is conditional as prr7 prr9 seedlings entrain to 28°/22° thermocycles (Salomé et al., 2010). Thus, there must be multiple routes of temperature input to the clock.

Clock phase is modulated by, in addition to light and temperature, organic nitrogen intermediates (Gutiérrez et al., 2008) and levels of hormones, including cytokinin, brassinosteroid, and ABA (Hanano et al., 2006; Legnaioli et al., 2009; Salomé et al., 2006). These intermediates and hormones are themselves under circadian regulation (Harmer, 2009; McClung, 2006; Pruneda-Paz and Kay, 2010). Similarly, the clock not only modulates intracellular calcium levels but also receives input from intracellular calcium via a cADPR-based feedback loop (Dodd et al., 2007). Thus, the clock is monitoring at least some of its output pathways and using the status of these outputs as inputs to modulate the status of central oscillator components.

Although circadian rhythms in respiration have been observed in dry onion seeds (Bryant, 1972), the clock in Arabidopsis seeds is arrested in an evening phase (Penfield and Hall, 2009). Typically, the germinating seedling is exposed to and rapidly entrained by light or temperature cycles following germination. However, no cycles are necessary for a population of seedlings to manifest synchronized rhythms because imbibition (hydration of the seed) itself acts as a strong entraining cue (Salomé et al., 2008; Zhong et al., 1998), although the mechanism of this entrainment is unclear.

E. Temperature compensation

Temperature compensation is one of the defining characters of circadian rhythms and refers to the relative insensitivity of period length to ambient temperature over a range of physiologically relevant temperatures. The rate of a biochemical reaction typically doubles with an increase in temperature of $10\,^{\circ}C$ (expressed as $Q_{10} = 2$), whereas the rate of the circadian clock, the inverse of its period, typically is buffered against changes in temperature ($Q_{10} \sim 1$). Hypotheses to explain temperature compensation have invoked several opposing reactions that are individually not compensated, with temperature compensation as an emergent property (Ruoff et al., 2005). It is thought that temperature compensation is not necessarily an intrinsic property of the circadian oscillator, but rather derives from the recruitment of other genes that regulate the oscillator components (Edwards et al., 2005; Mehra et al., 2009).

The mechanistic basis of temperature compensation is incompletely defined, although it is becoming clear that multiple pathways contribute. FLOWERING LOCUS C (FLC) is important for temperature compensation at high temperatures (Edwards et al., 2006). However, many Arabidopsis accessions carry nonfunctional flc alleles due to the selective advantage under certain conditions of the resulting early flowering phenotype (Lempe et al., 2005; Shindo et al., 2005), so genes other than FLC must be important for temperature compensation.

The evening loop buffers free-running period against changes in ambient temperature, with GI and other evening-expressed genes modulating the expression of CCA1 and LHY (Gould et al., 2006). Another evening-expressed gene, ZTL, has also been proposed to contribute to temperature compensation (Edwards et al., 2005). Similarly, the morning loop plays a role in temperature compensation. PRR7 and PRR9, negative regulators of CCA1 and LHY transcription (Nakamichi *et al.*, 2010), are important components of the temperature compensation mechanism via modulation of the expression levels of *CCA1* and *LHY* (Salomé *et al.*, 2010). Simultaneous inactivation of both *PRR7* and *PRR9* has no effect on period at low temperature but results in extreme period lengthening at high temperature (overcompensation). Inactivation of *CCA1* and *LHY* fully suppresses the overcompensation defects of *prr7 prr9* double mutants, indicating that the temperature-dependent period lengthening of *prr7 prr9* is due to hyperactivation of *CCA1* and *LHY* (Salomé *et al.*, 2010).

CCA1 and LHY are at least partially redundant, but the degree of redundancy varies as a function of temperature, with CCA1 more important for period maintenance at low temperatures and LHY more important at high temperatures (Gould *et al.*, 2006). The emphasized role of CCA1 at low temperatures may be explained by posttranslational modulation of CCA1 activity by CK2-mediated phosphorylation (Portolés and Más, 2010). The binding of CCA1 to its target promoters, including those of a number of clock genes (*PRR7*, *PRR9*, *TOC1*, and *LUX*), increases with temperature, but CK2-mediated phosphorylation of CCA1, which also increases with temperature, decreases CCA1 promoter binding activity (Portolés and Más, 2010). Thus, there is a temperature-dependent balance between CCA1-binding activity and opposing CK2-mediated phosphorylation, and at low temperatures, CK2-mediated phosphorylation is reduced, so CCA1-binding activity is enhanced. Thus, CK2 contributes to temperature compensation. CK2 has also been recruited to play an important, although mechanistically distinct, role in temperature compensation in *N. crassa* (Mehra *et al.*, 2009).

IV. CONSERVATION OF CLOCK GENES

A key justification of the study of model organisms is that information gleaned in the model will prove relevant to other less tractable organisms that are of more practical significance. In Arabidopsis, the circadian clock regulates much of the transcriptome, which turns out to be the case in other plants including the close relative, papaya, as well as more distantly related plants, such as poplar, rice, and maize (Hayes *et al.*, 2010; Khan *et al.*, 2010; Michael *et al.*, 2008b; Zdepski *et al.*, 2008). The identification of conserved time-of-day-specific *cis*-regulatory sequences in two of these diverse taxa argues in favor of a conserved mechanistic basis for clock regulation of gene expression (Michael *et al.*, 2008b; Zdepski *et al.*, 2008). This supports the case for continued investigation of the Arabidopsis clock at ever increasing resolution, as well as for studies translating our knowledge of the Arabidopsis clock into other species.

As described above, the Arabidopsis clock comprises multiple interlocked feedback loops. A number of studies have identified Arabidopsis clock gene orthologs in order to probe clock function in diverse species including rice

(Murakami et al., 2003, 2005, 2007), duckweed (Serikawa et al., 2008), the common ice plant (Boxall et al., 2005), and chestnut (Ramos et al., 2005). Phylogenetic analyses indicate conservation of the CCA1/LHY and PRR gene families among the angiosperms, including both monocots and dicots (Murakami et al., 2003; Takata et al., 2009, 2010). From this, one may infer that the common ancestor of the monocots and eudicots had the basic components necessary for the construction of a circadian clock with multiple interlocked feedback loops prior to the separation of these groups 200 million years ago, supporting the hypothesis that the mechanistic model of the Arabidopsis oscillator will prove a useful model for most agricultural species. A number of gene duplications have occurred in plant lineages, with consequent opportunities for sub- and neofunctionalization among clock genes (McClung, 2010). Differential expression of CCA1 copies in Arabidopsis hybrids and allotetraploids has been suggested to contribute to hybrid vigor, providing impetus to these investigations (Ni et al., 2009). It will be particularly interesting to explore functional changes in clock gene expression and function associated with recent polyploidization events in agricultural species, such as the Brassicas, which have triploidized since the divergence from Arabidopsis (Parkin et al., 2005). Initial studies indicate not only an expansion of clock gene families but also general conservation of clock function (Kim et al., 2007a; Lou et al., 2011; Xu et al., 2010).

CCA1-, LHY-, TOC1-, and LUX-like genes are conserved in the green algae, O. tauri and Chlamydomonas reinhardtii, and functional studies of these genes suggest that the central loop of the Arabidopsis clock is conserved in these distant relatives (Corellou et al., 2009; Matsuo et al., 2008). CCA1 and LHY are conserved and function in the clock of the moss, Physcomitrella patens (Okada et al., 2009). However, additional studies show that the algal clock differs significantly from that of Arabidopsis. The O. tauri and C. reinhardtii genomes lack additional PRR-like genes and lack orthologs of ELF3, ELF4, GI, and LUX (Corellou et al., 2009; Matsuo et al., 2008). In addition, forward genetic studies have identified many genes required for clock function that lack apparent orthologs in higher plants (Iliev et al., 2006; Matsuo et al., 2008). The O. tauri genome encodes a novel LOV sensing containing histidine kinase (LOV-HK) that functions as a new class of eukaryotic blue light receptor essential for clock function (Djouani-Tahri et al., 2011). Curiously, at least some of its clock function is independent of blue light photoreception.

The O. tauri clock has been investigated in some detail and differs in several significant ways from that of Arabidopsis. First, it seems that a single feedback loop, corresponding to the core CCA1/TOC1 loop of the Arabidopsis clock, encodes the oscillator responsible for clock activity; the evening and morning loops are absent (Corellou et al., 2009). This presents the opportunity to model a simpler oscillator system (Morant et al., 2010; Thommen et al., 2010). Moreover, the robustness and entrainment properties of the O. tauri clock question the necessity of the multiloop structure to achieve these properties, as has been postulated based on modeling of the Arabidopsis clock (Troein *et al.*, 2009). It seems that a combination of circadian gating of light-dependent mechanisms together with multiple light-dependent reactions can confer both robustness and flexibility to a circadian clock with only one feedback loop (Troein *et al.*, 2011).

However, the most startling property of the circadian clock in O. *tauri* is the persistence of a circadian rhythm in the absence of transcription and translation (O'Neill *et al.*, 2011). O. *tauri* is an obligate photoautotroph, and rhythms as assessed by luciferase gene fusions do not persist in continuous dark. Indeed, transcription, as measured by radiolabeled UTP incorporation, ceases in extended dark. Nonetheless, when dark-adapted cultures are reilluminated, luminescence rhythms resume at a circadian phase not solely specified by the time of illumination, suggesting that O. *tauri* keeps time in the dark in the absence of transcription. This observation prompted an investigation of the rhythmicity of the oxidation state of peroxiredoxin (PRX), a member of a ubiquitous family of antioxidant proteins. A circadian rhythm in PRX oxidation state persists in human red blood cells, and the enucleate state of the red blood cell dictates that this rhythm must be independent of transcription (O'Neill and Reddy, 2011). Similarly, O. *tauri* manifests a circadian rhythm in PRX oxidation state that, like the rhythm in red blood cells, is entrainable and temperature compensated (O'Neill and Reddy, 2011; O'Neill *et al.*, 2011). Inhibition of transcription and translation fails to disrupt the circadian rhythm in PRX oxidation state. Thus, the rhythm in PRX oxidation state must persist based on posttranslational mechanisms. This recalls the circadian rhythms in photosynthesis and in stalk optical density detected in enucleated *Acetabularia crenulata*, *Acetabularia major*, and *Acetabularia mediterranea*, other algae (Sweeney and Haxo, 1961; Woolum, 1991). Moreover, the PRX oxidation state rhythm in O. *tauri* is reminiscent of the rhythm in phosphorylation of cyanobacterial KaiC that can be reconstituted *in vitro* in a simple mixture of three proteins, KaiA, KaiB, and KaiC, with ATP (Nakajima *et al.*, 2005). PRX is, like the cyanobacterial KaiB, a member of the thioredoxin-like superfamily (Johnson *et al.*, 2008). PRX and Kai rhythms might represent a vestige of an early metabolic clock present in the common ancestor of bacteria and eukaryotes (O'Neill *et al.*, 2011).

V. CIRCADIAN CLOCKS ENHANCE FITNESS

Evolution has occurred on a rotating world characterized by drastic but predictable changes in the environment associated with the alternation of day and night. One premise to the study of circadian rhythms is that the circadian clock enables the anticipation of dawn and dusk, coordinates organismal biology with the temporal environment, and, thus, enhances fitness (Resco *et al.*, 2009;

Yerushalmi and Green, 2009). Because the clock regulation of plant biology is so widespread, any growth and survival advantage can potentially accrue from many sources.

Consistent with the hypothesis that a clock consonant with the environmental cycle enhances fitness, it has been shown that a functioning circadian clock enhances survival and biomass accumulation (Dodd *et al.*, 2005b; Green *et al.*, 2002; Yerushalmi *et al.*, 2011). This has been suggested to result from increased chlorophyll content and enhanced photosynthetic capacity (Dodd *et al.*, 2005b). It has also been shown that the circadian clock allows the optimization of nightly starch utilization (Graf *et al.*, 2010). A recent study grew a genotypically heterogeneous F2 population segregating null and wild-type alleles of *PRR5* and *PRR9* under short and long day/night lengths (*T*-cycles) of 20 and 28 h and found that the circadian period of F3 seedlings was positively correlated with the *T*-cycle imposed on their F2 parents (Yerushalmi *et al.*, 2011). Average period of the F3 plants was shorter in the progeny of segregating F2 population plants grown under 20-h cycles and longer in the progeny of plants grown under 28-h cycles. The frequency of the long period *prr9* allele was maintained in the progeny of the F2 plants grown under 28-h cycles but was significantly decreased in the progeny of the F2 population grown under 20-h cycles, demonstrating selection against the long period *prr9* allele in 20-h cycles. This demonstrates strong selective pressure for circadian rhythms that resonate with the environmentally imposed period (Yerushalmi *et al.*, 2011). However, this study, as well as an earlier study of Dodd *et al.* (2005b), used day/night lengths that deviate considerably from the 24-h cycle encountered in nature. The short period *toc1–2* and long period *ztl-3* mutants both grow better in 24-h day/night cycles than in days (20 or 28 h, respectively) matching their periods (Graf *et al.*, 2010). It will be important to investigate the clock effects on fitness in natural day lengths.

The investigation of the contribution of the circadian clock to fitness will also require increased consideration of real-world issues, such as season and weather. An intriguing study allowed clocks to evolve *in silico* under conditions of multiple photoperiods, to mimic seasonality, and of noisy timing of the light signal, to mimic weather. The combination of seasonal variation and weather (noise) favored the evolution of network complexity (Troein *et al.*, 2009). Similarly, network complexity was found to be necessary for efficient tracking of both dawn and dusk (Edwards *et al.*, 2010). Expanding these studies to include temperature and other abiotic and biotic perturbations seems likely to further emphasize the need for complexity in clock networks. It will be important to move these experiments into the field, with common garden experiments at multiple geographic sites (e.g., Wilczek *et al.*, 2009). It will also be important to reconcile these arguments that favor the emergence of complexity in clock networks with the retention of robustness and flexibility despite the apparent simplicity of the *O. tauri* clock (Troein *et al.*, 2011).

VI. CONCLUDING REMARKS

Progress in understanding the molecular basis of plant circadian rhythmicity is accelerating. The logic and architecture of the higher plant circadian clock are consistent with metazoan clocks, but largely distinct components have been recruited to fulfill critical clock functions. As in metazoans, the higher plant clock consists of a set of interlocked feedback loops. The regulatory interactions among these loops are intricate and new ones continue to be revealed, adding to the complexity of our clock model. Although the model of the clock is rapidly acquiring new regulatory relationships, the lack of direct positive transcriptional regulators is conspicuous and needs to be addressed. Transcriptome scale experiments have revealed the breadth of clock regulation of plant processes. There now needs to be detailed attention paid to selected output pathways to dissect the regulatory logic by which the clock orchestrates metabolism and physiology. Data now support the hypothesis that a functioning clock contributes to fitness, but there are few detailed explanations of the mechanistic basis for this relationship. There is a need to increase agricultural productivity in the face of an eroding arable land base and environmental degradation, including that due to global climate change (Brown and Funk, 2008). It seems reasonable to propose that modulation of clock function has the potential to contribute to enhancement of plant productivity. This optimism is supported by the implication of clock function in the heterosis (hybrid vigor) encountered in hybrids and allopolyploids (Ni et al., 2009). All in all, it is a good time to study the plant circadian clock.

Acknowledgment

My work on circadian rhythms is funded by grants from the National Science Foundation (IOS 0605736, IOS 0923752, IOS 0960803, and IOS-1025965).

References

Alabadí, D., Oyama, T., Yanovsky, M. J., Harmon, F. G., Más, P., and Kay, S. A. (2001). Reciprocal regulation between TOC1 and LHY/CCA1 within the Arabidopsis circadian clock. Science 293, 880–883.

Alabadí, D., Yanovsky, M. J., Más, P., Harmer, S. L., and Kay, S. A. (2002). Critical role for CCA1 and LHY in maintaining circadian rhythmicity in Arabidopsis. Curr. Biol. 12, 757–761.

Allen, T., Koustenis, A., Theodorou, G., Somers, D. E., Kay, S. A., Whitelam, G. C., and Devlin, P. F. (2006). Arabidopsis FHY3 specifically gates phytochrome signaling to the circadian clock. Plant Cell 18, 2506–2516.

Anderson, S. L., Somers, D. E., Millar, A. J., Hanson, K., Chory, J., and Kay, S. A. (1997). Attenuation of phytochrome A and B signaling pathways by the Arabidopsis circadian clock. Plant Cell 9, 1727–1743.

Aronson, B. D., Johnson, K. A., Loros, J. J., and Dunlap, J. C. (1994). Negative feedback defining a circadian clock: Autoregulation of the clock gene *frequency*. *Science* **263**, 1578–1584.

Asher, G., Gatfield, D., Stratmann, M., Reinke, H., Dibner, C., Kreppel, F., Mostoslavsky, R., Alt, F. W., and Schibler, U. (2008). SIRT1 regulates circadian clock gene expression through PER2 deacetylation. *Cell* **134**, 317–328.

Barak, S., Tobin, E. M., Andronis, C., Sugano, S., and Green, R. M. (2000). All in good time: The Arabidopsis circadian clock. *Trends Plant Sci.* **5**, 517–522.

Bargiello, T., Jackson, F., and Young, M. (1984). Restoration of circadian behavioural rhythms by gene transfer in Drosophila. *Nature* **312**, 752–754.

Baudry, A., Ito, S., Song, Y. H., Strait, A. A., Kiba, T., Lu, S., Henriques, R., Pruneda-Paz, J. L., Chua, N.-H., Tobin, E. M., Kay, S. A., and Imaizumi, T. (2010). F-box proteins FKF1 and LKP2 act in concert with ZEITLUPE to control Arabidopsis clock progression. *Plant Cell* **22**, 606–622.

Bieniawska, Z., Espinoza, C., Schlereth, A., Sulpice, R., Hincha, D. K., and Hannah, M. A. (2008). Disruption of the Arabidopsis circadian clock is responsible for extensive variation in the cold-responsive transcriptome. *Plant Physiol.* **147**, 263–279.

Bognár, L. K., Hall, A., Ádám, É., Thain, S. C., Nagy, F., and Millar, A. J. (1999). The circadian clock controls the expression pattern of the circadian input photoreceptor, phytochrome B. *Proc. Natl. Acad. Sci. USA* **96**, 14652–14657.

Boxall, S. F., Foster, J. M., Bohnert, H. J., Cushman, J. C., Nimmo, H. G., and Hartwell, J. (2005). Conservation and divergence of circadian clock operation in a stress-inducible crassulacean acid metabolism species reveals clock compensation against stress. *Plant Physiol.* **137**, 969–982.

Bretzl, H. (1903). Botanische Forschungen des Alexanderzuges. B.G. Teubner, Leipzig.

Brown, M. E., and Funk, C. C. (2008). Food security under climate change. *Science* **319**, 580–581.

Brown, S. A., Ripperger, J., Kadener, S., Fleury-Olela, F., Vilbois, F., Rosbash, M., and Schibler, U. (2005). PERIOD1-associated proteins modulate the negative limb of the mammalian circadian oscillator. *Science* **308**, 693–696.

Bryant, T. R. (1972). Gas exchange in dry seeds: Circadian rhythmicity in the absence of DNA replication, transcription, and translation. *Science* **178**, 634–636.

Bünning, E. (1931). Untersuchungen über die autonomen tagesperiodischen Bewungen der Primär-blätter von *Phaseolus multiflorus*. *Jahrb. Wiss. Bot.* **75**, 439–480.

Bünning, E. (1932). Über die Erblichket der Tagesperiodizitat bei den Phaseolus-Blättern. *Jahrb. Wiss. Bot.* **77**, 283–320.

Bünning, E. (1935). Zur Kenntnis der erblichen Tagesperioizität bei den Primärblätter von Phaseolus multiflorus. *Jahrb. Wiss. Bot.* **81**, 411–418.

Carpenter, C. D., Kreps, J. A., and Simon, A. E. (1994). Genes encoding glycine-rich *Arabidopsis thaliana* proteins with RNA-binding motifs are influenced by cold treatment and an endogenous circadian rhythm. *Plant Physiol.* **104**, 1015–1025.

Corellou, F., Schwartz, C., Motta, J.-P., Djouani-Tahri, E. B., Sanchez, F., and Bouget, F.-Y. (2009). Clocks in the green lineage: Comparative functional analysis of the circadian architecture of the picoeukaryote *Ostreococcus*. *Plant Cell* **21**, 3436–3449.

Covington, M. F., and Harmer, S. L. (2007). The circadian clock regulates auxin signaling and responses in *Arabidopsis*. *PLoS Biol.* **5**, e222.

Covington, M. F., Panda, S., Liu, X. L., Strayer, C. A., Wagner, D. R., and Kay, S. A. (2001). ELF3 modulates resetting of the circadian clock in Arabidopsis. *Plant Cell* **13**, 1305–1316.

Covington, M. F., Maloof, J. N., Straume, M., Kay, S. A., and Harmer, S. L. (2008). Global transcriptome analysis reveals circadian regulation of key pathways in plant growth and development. *Genome Biol.* **9**, R130.

Cumming, B. G., and Wagner, E. (1968). Rhythmic processes in plants. *Annu. Rev. Plant Physiol.* **19**, 381–416.

Curtis, A. M., Seo, S.-b., Westgate, E. J., Rudic, R. D., Smyth, E. M., Chakravarti, D., FitzGerald, G. A., and McNamara, P. (2004). Histone acetyltransferase-dependent chromatin remodeling and the vascular clock. *J. Biol. Chem.* **279,** 7091–7097.

Daniel, X., Sugano, S., and Tobin, E. M. (2004). CK2 phosphorylation of CCA1 is necessary for its circadian oscillator function in Arabidopsis. *Proc. Natl. Acad. Sci. USA* **101,** 3292–3297.

Darwin, C., and Darwin, F. (1880). The Power of Movement in Plants. Murray, London.

de Candolle, A. P. (1832). Physiologie Végétale. Bechet Jeune, Paris.

de Mairan, J. (1729). Observation botanique. *Hist. Acad. R. Sci.* 35–36.

de Montaigu, A., Tóth, R., and Coupland, G. (2010). Plant development goes like clockwork. *Trends Genet.* **26,** 296–306.

Devlin, P. F., and Kay, S. A. (2001). Circadian photoperception. *Annu. Rev. Physiol.* **63,** 677–694.

Ding, Z., Millar, A. J., Davis, A. M., and Davis, S. J. (2007). TIME FOR COFFEE encodes a nuclear regulator in the *Arabidopsis thaliana* circadian clock. *Plant Cell* **19,** 1522–1536.

Dixon, L. E., Knox, K., Kozma-Bognar, L., Southern, M. M., Pokhilko, A., and Millar, A. J. (2011). Temporal repression of core circadian genes is mediated through EARLY FLOWERING 3 in Arabidopsis. *Curr. Biol.* **21,** 120–125.

Djouani-Tahri, E.-B., Christie, J. M., Sanchez-Ferandin, S., Sanchez, F., Bouget, F.-Y., and Corellou, F. (2011). A eukaryotic LOV-histidine kinase with circadian clock function in the picoalga Ostreococcus. *Plant J.* **65,** 578–588.

Dodd, A. N., Love, J., and Webb, A. A. R. (2005a). The plant clock shows its metal: Circadian regulation of cytosolic free Ca^{2+}. *Trends Plant Sci.* **10,** 15–21.

Dodd, A. N., Salathia, N., Hall, A., Kevei, E., Toth, R., Nagy, F., Hibberd, J. M., Millar, A. J., and Webb, A. A. R. (2005b). Plant circadian clocks increase photosynthesis, growth, survival, and competitive advantage. *Science* **309,** 630–633.

Dodd, A. N., Gardner, M. J., Hotta, C. T., Hubbard, K. E., Dalchau, N., Love, J., Assie, J.-M., Robertson, F. C., Jakobsen, M. K., Gonçalves, J., Sanders, D., and Webb, A. A. R. (2007). The Arabidopsis circadian clock incorporates a cADPR-based feedback loop. *Science* **318,** 1789–1792.

Doi, M., Hirayama, J., and Sassone-Corsi, P. (2006). Circadian regulator CLOCK is a histone acetyltransferase. *Cell* **125,** 497–508.

Doyle, M. R., Davis, S. J., Bastow, R. M., McWatters, H. G., Kozma-Bognar, L., Nagy, F., Millar, A. J., and Amasino, R. M. (2002). The ELF4 gene controls circadian rhythms and flowering time in *Arabidopsis thaliana*. *Nature* **419,** 74–77.

Duc, C., Cellier, F., Lobréaux, S., Briat, J.-F., and Gaymard, F. (2009). Regulation of iron homeostasis in *Arabidopsis thaliana* by the clock regulator TIME FOR COFFEE. *J. Biol. Chem.* **284,** 36271–36281.

Dunlap, J. C. (1999). Molecular bases for circadian clocks. *Cell* **96,** 271–290.

Edwards, K. D., Lynn, J. R., Gyula, P., Nagy, F., and Millar, A. J. (2005). Natural allelic variation in the temperature compensation mechanisms of the *Arabidopsis thaliana* circadian clock. *Genetics* **170,** 387–400.

Edwards, K. D., Anderson, P. E., Hall, A., Salathia, N. S., Locke, J. C. W., Lynn, J. R., Straume, M., Smith, J. Q., and Millar, A. J. (2006). FLOWERING LOCUS C mediates natural variation in the high-temperature response of the Arabidopsis circadian clock. *Plant Cell* **18,** 639–650.

Edwards, K. D., Akman, O. E., Knox, K., Lumsden, P. J., Thomson, A. W., Brown, P. E., Pokhilko, A., Kozma-Bognar, L., Nagy, F., Rand, D. A., and Millar, A. J. (2010). Quantitative analysis of regulatory flexibility under changing environmental conditions. *Mol. Syst. Biol.* **6,** 424.

Eriksson, M. E., Hanano, S., Southern, M. M., Hall, A., and Millar, A. J. (2003). Response regulator homologues have complementary, light-dependent functions in the *Arabidopsis* circadian clock. *Planta* **218,** 159–162.

Etchegaray, J.-P., Lee, C., Wade, P. A., and Reppert, S. M. (2003). Rhythmic histone acetylation underlies transcription in the mammalian circadian clock. *Nature* **421,** 177–182.

Farinas, B., and Mas, P. (2011). Functional implication of the MYB transcription factor RVE8/LCL5 in the circadian control of histone acetylation. *Plant J.* **66,** 318–329.

Farré, E. M., Harmer, S. L., Harmon, F. G., Yanovsky, M. J., and Kay, S. A. (2005). Overlapping and distinct roles of PRR7 and PRR9 in the Arabidopsis circadian clock. *Curr. Biol.* **15,** 47–54.

Feldman, J. F. (1982). Genetic approaches to circadian clocks. *Annu. Rev. Plant Physiol.* **33,** 583–608.

Filichkin, S. A., Priest, H. D., Givan, S. A., Shen, R., Bryant, D. W., Fox, S. E., Wong, W.-K., and Mockler, T. C. (2010). Genome-wide mapping of alternative splicing in *Arabidopsis thaliana*. *Genome Res.* **20,** 45–58.

Fowler, S. G., Cook, D., and Thomashow, M. F. (2005). Low temperature induction of Arabidopsis CBF1, 2, and 3 is gated by the circadian clock. *Plant Physiol.* **137,** 961–968.

Fu, Z. Q., Guo, M., Jeong, B.r., Tian, F., Elthon, T. E., Cerny, R. L., Staiger, D., and Alfano, J. R. (2007). A type III effector ADP-ribosylates RNA-binding proteins and quells plant immunity. *Nature* **447,** 284–288.

Fujiwara, S., Wang, L., Han, L., Suh, S. S., Salomé, P. A., McClung, C. R., and Somers, D. E. (2008). Post-translational regulation of the circadian clock through selective proteolysis and phosphory-lation of pseudo-response regulator proteins. *J. Biol. Chem.* **283,** 23073–23083.

Fukushima, A., Kusano, M., Nakamichi, N., Kobayashi, M., Hayashi, N., Sakakibara, H., Mizuno, T., and Saito, K. (2009). Impact of clock-associated *Arabidopsis* pseudo-response regulators in meta-bolic coordination. *Proc. Natl. Acad. Sci. USA* **106,** 7251–7256.

Gallego, M., and Virshup, D. M. (2007). Post-translational modifications regulate the ticking of the circadian clock. *Nat. Rev. Mol. Cell Biol.* **8,** 139–148.

Gardner, K. E., Allis, C. D., and Strahl, B. D. (2011). Operating on chromatin, a colorful language where context matters. *J. Mol. Biol.* **409,** 36–46.

Giraud, E., Ng, S., Carrie, C., Duncan, O., Low, J., Lee, C. P., Van Aken, O., Millar, A. H., Murcha, M., and Whelan, J. (2010). TCP transcription factors link the regulation of genes encoding mitochondrial proteins with the circadian clock in *Arabidopsis thaliana*. *Plant Cell* **22,** 3921–3934.

Gould, P. D., Locke, J. C. W., Larue, C., Southern, M. M., Davis, S. J., Hanano, S., Moyle, R., Milich, R., Putterill, J., Millar, A. J., and Hall, A. (2006). The molecular basis of temperature compensation in the *Arabidopsis* circadian clock. *Plant Cell* **18,** 1177–1187.

Graf, A., Schlereth, A., Stitt, M., and Smith, A. M. (2010). Circadian control of carbohydrate availability for growth in *Arabidopsis* plants at night. *Proc. Natl. Acad. Sci. USA* **107,** 9458–9463.

Green, R. M., Tingay, S., Wang, Z.-Y., and Tobin, E. M. (2002). Circadian rhythms confer a higher level of fitness to Arabidopsis plants. *Plant Physiol.* **129,** 576–584.

Gutiérrez, R. A., Ewing, R. M., Cherry, J. M., and Green, P. J. (2002). Identification of unstable transcripts in Arabidopsis by cDNA microarray analysis: Rapid decay is associated with a group of touch- and specific clock-controlled genes. *Proc. Natl. Acad. Sci. USA* **99,** 11513–11518.

Gutiérrez, R. A., Stokes, T. L., Thum, K., Xu, X., Obertello, M., Katari, M. S., Tanurdzic, M., Dean, A., Nero, D. C., McClung, C. R., and Coruzzi, G. M. (2008). Systems approach identifies an organic nitrogen-responsive gene network that is regulated by the master clock control gene *CCA1*. *Proc. Natl. Acad. Sci. USA* **105,** 4939–4944.

Hall, A., Bastow, R. M., Davis, S. J., Hanano, S., McWatters, H. G., Hibberd, V., Doyle, M. R., Sung, S., Halliday, K. J., Amasino, R. M., and Millar, A. J. (2003). The TIME FOR COFFEE gene maintains the amplitude and timing of Arabidopsis circadian clocks. *Plant Cell* **15,** 2719–2729.

Hall, A., Bognár, L. K., Tóth, R., Nagy, F., and Millar, A. J. (2001). Conditional circadian regulation of PHYTOCHROME A gene expression. *Plant Physiol.* **127,** 1808–1818.

Han, L., Mason, M., Risseeuw, E. P., Crosby, W. L., and Somers, D. E. (2004). Formation of an SCF^{ZTL} complex is required for proper regulation of circadian timing. *Plant J.* **40**, 291–301.

Hanano, S., Domagalska, M. A., Nagy, F., and Davis, S. J. (2006). Multiple phytohormones influence distinct parameters of the plant circadian clock. *Genes Cells* **11**, 1381–1392.

Harmer, S. L. (2009). The circadian system in higher plants. *Annu. Rev. Plant Biol.* **60**, 357–377.

Harmer, S. L., and Kay, S. A. (2005). Positive and negative factors confer phase-specific circadian regulation of transcription in Arabidopsis. *Plant Cell* **17**, 1926–1940.

Harmer, S. L., Hogenesch, J. B., Straume, M., Chang, H. S., Han, B., Zhu, T., Wang, X., Kreps, J. A., and Kay, S. A. (2000). Orchestrated transcription of key pathways in Arabidopsis by the circadian clock. *Science* **290**, 2110–2113.

Harmon, F., Imaizumi, T., and Gray, W. M. (2008). CUL1 regulates TOC1 protein stability in the Arabidopsis circadian clock. *Plant J.* **55**, 568–579.

Hayes, K. R., Beatty, M., Meng, X., Simmons, C. R., Habben, J. E., and Danilevskaya, O. N. (2010). Maize global transcriptomics reveals pervasive leaf diurnal rhythms but rhythms in developing ears are largely limited to the core oscillator. *PLoS One* **5**, e12887.

Hazen, S. P., Schultz, T. F., Pruneda-Paz, J. L., Borevitz, J. O., Ecker, J. R., and Kay, S. A. (2005). *LUX ARRHYTHMO* encodes a Myb domain protein essential for circadian rhythms. *Proc. Natl. Acad. Sci. USA* **102**, 10387–10392.

Hazen, S. P., Naef, F., Quisel, T., Gendron, J. M., Chen, H., Ecker, J. R., Borevitz, J. O., and Kay, S. A. (2009). Exploring the transcriptional landscape of plant circadian rhythms using genome tiling arrays. *Genome Biol.* **10**, R17.1–R17.12.

Heintzen, C., Melzer, S., Fischer, R., Kappeler, S., Apel, K., and Staiger, D. (1994). A light- and temperature-entrained circadian clock controls expression of transcripts encoding nuclear proteins with homology to RNA-binding proteins in meristematic tissue. *Plant J.* **5**, 799–813.

Heintzen, C., Nater, M., Apel, K., and Staiger, D. (1997). AtGRP7, a nuclear RNA-binding protein as a component of a circadian-regulated negative feedback loop in *Arabidopsis thaliana. Proc. Natl. Acad. Sci. USA* **94**, 8515–8520.

Helfer, A., Nusinow, D. A., Chow, B. Y., Gehrke, A. R., Bulyk, M. L., and Kay, S. A. (2011). *LUX ARRHYTHMO* encodes a nighttime repressor of circadian gene expression in the *Arabidopsis* core clock. *Curr. Biol.* **21**, 126–133.

Hermans, C., Vuylsteke, M., Coppens, F., Craciun, A., Inzé, D., and Verbruggen, N. (2010). Early transcriptomic changes induced by magnesium deficiency in *Arabidopsis thaliana* reveal the alteration of circadian clock gene expression in roots and the triggering of abscisic acid-responsive genes. *New Phytol.* **187**, 119–131.

Hicks, K. A., Albertson, T. M., and Wagner, D. R. (2001). *EARLY FLOWERING 3* encodes a novel protein that regulates circadian clock function and flowering in Arabidopsis. *Plant Cell* **13**, 1281–1292.

Hong, S., Song, H.-R., Lutz, K., Kerstetter, R. A., Michael, T. P., and McClung, C. R. (2010). Type II protein arginine methyltransferase 5 (PRMT5) is required for circadian period determination in *Arabidopsis thaliana. Proc. Natl. Acad. Sci. USA* **107**, 21211–21216.

Iliev, D., Voytsekh, O., Schmidt, E. M., Fiedler, M., Nykytenko, A., and Mittag, M. (2006). A heteromeric RNA-binding protein is involved in maintaining acrophase and period of the circadian clock. *Plant Physiol.* **142**, 797–806.

Imaizumi, T. (2010). Arabidopsis circadian clock and photoperiodism: Time to think about location. *Curr. Opin. Plant Biol.* **13**, 83–89.

Imaizumi, T., Schultz, T. F., Harmon, F. G., Ho, L. A., and Kay, S. A. (2005). FKF1 F-box protein mediates cyclic degradation of a repressor of CONSTANS in Arabidopsis. *Science* **309**, 293–297.

Imaizumi, T., Tran, H. G., Swartz, T. E., Briggs, W. R., and Kay, S. A. (2003). FKF1 is essential for photoperiodic-specific light signalling in Arabidopsis. *Nature* **426**, 302–306.

Ito, S., Niwa, Y., Nakamichi, N., Kawamura, H., Yamashino, T., and Mizuno, T. (2008). Insight into missing genetic links between two evening-expressed pseudo-response regulator genes *TOC1* and *PRR5* in the circadian clock-controlled circuitry in *Arabidopsis thaliana*. *Plant Cell Physiol.* **49**, 201–213.

Johnson, C. H., Mori, T., and Xu, Y. (2008). A cyanobacterial circadian clockwork. *Curr. Biol.* **18**, R816–R825.

Jones, M. A., Covington, M. F., DiTacchio, L., Vollmers, C., Panda, S., and Harmer, S. L. (2010). Jumonji domain protein JMJD5 functions in both the plant and human circadian systems. *Proc. Natl. Acad. Sci. USA* **107**, 21623–21628.

Jouve, L., Gaspar, T., Kevers, C., Greppin, H., and Agosti, R. D. (1999). Involvement of indole-3-acetic acid in the circadian growth of the first internode of *Arabidopsis*. *Planta* **209**, 136–142.

Kant, P., Gordon, M., Kant, S., Zolla, G., Davydov, O., Heimer, Y. M., Chalifa-Caspi, V., Shaked, R., and Barak, S. (2008). Functional-genomics-based identification of genes that regulate Arabidopsis responses to multiple abiotic stresses. *Plant Cell Environ.* **31**, 697–714.

Kevei, É., Gyula, P., Fehér, B., Tóth, R., Viczián, A., Kircher, S., Rea, D., Dorjgotov, D., Schäfer, E., Millar, A. J., Kozma-Bognár, L., and Nagy, F. (2007). *Arabidopsis thaliana* circadian clock is regulated by the small GTPase LIP1. *Curr. Biol.* **17**, 1456–1464.

Khan, S., Rowe, S. C., and Harmon, F. G. (2010). Coordination of the maize transcriptome by a conserved circadian clock. *BMC Plant Biol.* **10**, 126.

Kiba, T., Henriques, R., Sakakibara, H., and Chua, N.-H. (2007). Targeted degradation of PSEUDO-RESPONSE REGULATOR5 by a SCFZTL complex regulates clock function and photomorphogenesis in *Arabidopsis thaliana*. *Plant Cell* **19**, 2516–2530.

Kiesel, A. (1894). Untersuchungen zur Physiologie des facettierten Auges. *Sitzungsber. Akad. Wiss. Wien* **103**, 97–139.

Kikis, E. A., Khanna, R., and Quail, P. H. (2005). ELF4 is a phytochrome-regulated component of a negative-feedback loop involving the central oscillator components CCA1 and LHY. *Plant J.* **44**, 300–313.

Kim, J.-Y., Song, H.-R., Taylor, B. L., and Carré, I. A. (2003). Light-regulated translation mediates gated induction of the *Arabidopsis* clock protein LHY. *EMBO J.* **22**, 935–944.

Kim, J. A., Yang, T.-J., Kim, J. S., Park, J. Y., Kwon, S.-J., Lim, M.-H., Jin, M., Lee, S. C., Lee, S. I., Choi, B.-S., Um, S.-H., Kim, H.-I., *et al.* (2007a). Isolation of circadian-associated genes in *Brassica rapa* by comparative genomics with *Arabidopsis thaliana*. *Mol. Cells* **23**, 145–153.

Kim, W.-Y., Fujiwara, S., Suh, S.-S., Kim, J., Kim, Y., Han, L., David, K., Putterill, J., Nam, H. G., and Somers, D. E. (2007b). ZEITLUPE is a circadian photoreceptor stabilized by GIGANTEA in blue light. *Nature* **449**, 356–360.

Kim, J. S., Jung, H. J., Lee, H. J., Kim, K. A., Goh, C.-H., Woo, Y., Oh, S. H., Han, Y. S., and Kang, H. (2008). Glycine-rich RNA-binding protein7 affects abiotic stress responses by regulating stomata opening and closing in Arabidopsis thaliana. *Plant J.* **55**, 455–466.

Kircher, S., Gil, P., Kozma-Bognár, L., Fejes, E., Speth, V., Husselstein-Muller, T., Bauer, D., Ádám, É., Schäfer, E., and Nagy, F. (2002). Nucleocytoplasmic partitioning of the plant photoreceptors Phytochrome A, B, C, D, and E is regulated differentially by light and exhibits a diurnal rhythm. *Plant Cell* **14**, 1541–1555.

Knowles, S. M., Lu, S. X., and Tobin, E. M. (2008). Testing time: Can ethanol-induced pulses of proposed oscillator components phase shift rhythms in Arabidopsis? *J. Biol. Rhythms* **23**, 463–471.

Kojima, S., Shingle, D. L., and Green, C. B. (2011). Post-transcriptional control of circadian rhythms. *J. Cell Sci.* **124**, 311–320.

Konopka, R., and Benzer, S. (1971). Clock mutants of *Drosophila melanogaster*. *Proc. Natl. Acad. Sci. USA* **68**, 2112–2116.

Lee, C. P., Eubel, H., and Millar, A. H. (2010a). Diurnal changes in mitochondrial function reveal daily optimization of light and dark respiratory metabolism in Arabidopsis. *Mol. Cell. Proteomics* **9,** 2125–2139.

Lee, J.-S., Smith, E., and Shilatifard, A. (2010b). The language of histone crosstalk. *Cell* **142,** 682–685.

Legnaioli, T., Cuevas, J., and Mas, P. (2009). TOC1 functions as a molecular switch connecting the circadian clock with plant responses to drought. *EMBO J.* **28,** 3745–3757.

Lempe, J., Balasubramanian, S., Sureshkumar, S., Singh, A., Schmid, M., and Weigel, D. (2005). Diversity of flowering responses in wild *Arabidopsis thaliana* strains. *PLoS Genet.* **1,** 109–118.

Lidder, P., Gutiérrez, R. A., Salomé, P. A., McClung, C. R., and Green, P. J. (2005). Circadian control of messenger RNA stability. Association with a sequence-specific messenger RNA decay pathway. *Plant Physiol.* **138,** 2374–2385.

Liu, X. L., Covington, M. F., Fankhauser, C., Chory, J., and Wagner, D. R. (2001). ELF3 encodes a circadian clock–regulated nuclear protein that functions in an Arabidopsis PHYB signal transduction pathway. *Plant Cell* **13,** 1293–1304.

Locke, J. C. W., Southern, M. M., Kozma-Bognar, L., Hibberd, V., Brown, P. E., Turner, M. S., and Millar, A. J. (2005). Extension of a genetic network model by iterative experimentation and mathematical analysis. *Mol. Syst. Biol.* **1,** 2005.0013.

Locke, J. C. W., Kozma-Bognár, L., Gould, P. D., Fehér, B., Kevei, É., Nagy, F., Turner, M. S., Hall, A., and Millar, A. J. (2006). Experimental validation of a predicted feedback loop in the multi-oscillator clock of *Arabidopsis thaliana*. *Mol. Syst. Biol.* **2,** 59.

Lou, P., Xie, Q., Xu, X., Edwards, C. E., Brock, M. T., Weinig, C., and McClung, C. R. (2011). Genetic architecture of the circadian clock and flowering time in *Brassica rapa*. *Theor. Appl. Genet.* **123,** 397–409.

Lu, S. X., Knowles, S. M., Andronis, C., Ong, M. S., and Tobin, E. M. (2009). CIRCADIAN CLOCK ASSOCIATED1 and LATE ELONGATED HYPOCOTYL function synergistically in the circadian clock of Arabidopsis. *Plant Physiol.* **150,** 834–843.

Lu, S. X., Knowles, S. M., Webb, C. J., Celaya, R. B., Cha, C., Siu, J. P., and Tobin, E. M. (2011). The Jumonji C domain-containing protein JMJ30 regulates period length in the Arabidopsis circadian clock. *Plant Physiol.* **155,** 906–915.

Luco, R. F., Allo, M., Schor, I. E., Kornblihtt, A. R., and Misteli, T. (2011). Epigenetics in alternative pre-mRNA splicing. *Cell* **144,** 16–26.

Makino, S., Matsushika, A., Kojima, M., Yamashino, T., and Mizuno, T. (2002). The APRR1/TOC1 quintet implicated in circadian rhythms of *Arabidopsis thaliana*: I. Characterization with APRR1-overexpressing plants. *Plant Cell Physiol.* **43,** 58–69.

Martínez-García, J. F., Huq, E., and Quail, P. H. (2000). Direct targeting of light signals to a promoter element-bound transcription factor. *Science* **288,** 859–863.

Martin-Tryon, E. L., and Harmer, S. L. (2008). XAP5 CIRCADIAN TIMEKEEPER coordinates light signals for proper timing of photomorphogenesis and the circadian clock in Arabidopsis. *Plant Cell* **20,** 1244–1259.

Martin-Tryon, E. L., Kreps, J. A., and Harmer, S. L. (2007). GIGANTEA acts in blue light signaling and has biochemically separable roles in circadian clock and flowering time regulation. *Plant Physiol.* **143,** 473–486.

Más, P., Kim, W.-Y., Somers, D. E., and Kay, S. A. (2003). Targeted degradation of TOC1 by ZTL modulates circadian function in *Arabidopsis thaliana*. *Nature* **426,** 567–570.

Matsuo, T., Okamoto, K., Onai, K., Niwa, Y., Shimogawara, K., and Ishiura, M. (2008). A systematic forward genetic analysis identified components of the *Chlamydomonas* circadian system. *Genes Dev.* **22,** 918–930.

Matsushika, A., Imamura, A., Yamashino, T., and Mizuno, T. (2002). Aberrant expression of the light-inducible and circadian-regulated *APRR9* gene belonging to the circadian-associated APRR1/TOC1 quintet results in the phenotype of early flowering in *Arabidopsis thaliana*. *Plant Cell Physiol.* **43**, 833–843.

Maurer, C., Huang, H. C., and Weber, F. (2009). Cytoplasmic interaction with CYCLE promotes the post-translational processing of the circadian CLOCK protein. *FEBS Lett.* **583**, 1561–1566.

McClung, C. R. (2006). Plant circadian rhythms. *Plant Cell* **18**, 792–803.

McClung, C. R. (2008). Comes a time. *Curr. Opin. Plant Biol.* **11**, 514–520.

McClung, C. R. (2010). A modern circadian clock in the common angiosperm ancestor of monocots and eudicots. *BMC Biol.* **8**, 55.

McClung, C. R., and Davis, S. J. (2010). Ambient thermometers in plants: From physiological outputs towards mechanisms of thermal sensing. *Curr. Biol.* **20**, R1086–R1092.

McClung, C. R., and Gutiérrez, R. A. (2010). Network news: Prime time for systems biology of the plant circadian clock. *Curr. Opin. Genet. Dev.* **20**, 588–598.

McClung, C. R., and Kay, S. A. (1994). Circadian rhythms in *Arabidopsis thaliana*. In "Arabidopsis" (E. M. Meyerowitz and C. R. Somerville, eds.), pp. 615–637. Cold Spring Harbor Laboratory Press, Plainview, NY.

McWatters, H. G., Bastow, R. M., Hall, A., and Millar, A. J. (2000). The *ELF3 zeitnehmer* regulates light signalling to the circadian clock. *Nature* **408**, 716–720.

McWatters, H. G., Kolmos, E., Hall, A., Doyle, M. R., Amasino, R. M., Gyula, P., Nagy, F., Millar, A. J., and Davis, S. J. (2007). ELF4 is required for oscillatory properties of the circadian clock. *Plant Physiol.* **144**, 391–401.

Mehra, A., Shi, M., Baker, C. L., Colot, H. V., Loros, J. J., and Dunlap, J. C. (2009). A role for casein kinase 2 in the mechanism underlying circadian temperature compensation. *Cell* **137**, 749–760.

Meyer, P., Saez, L., and Young, M. W. (2006). PER-TIM interactions in living *Drosophila* cells: An interval timer for the circadian clock. *Genes Dev.* **311**, 226–229.

Michael, T. P., and McClung, C. R. (2002). Phase-specific circadian clock regulatory elements in *Arabidopsis thaliana*. *Plant Physiol.* **130**, 627–638.

Michael, T. P., and McClung, C. R. (2003). Enhancer trapping reveals widespread circadian clock transcriptional control in *Arabidopsis thaliana*. *Plant Physiol.* **132**, 629–639.

Michael, T. P., Breton, G., Hazen, S. P., Priest, H., Mockler, T. C., Kay, S. A., and Chory, J. (2008a). A morning-specific phytohormone gene expression program underlying rhythmic plant growth. *PLoS Biol.* **6**, 1887–1898.

Michael, T. P., Mockler, T. C., Breton, G., McEntee, C., Byer, A., Trout, J. D., Hazen, S. P., Shen, R., Priest, H. D., Sullivan, C. M., Givan, S. A., Yanovsky, M., et al. (2008b). Network discovery pipeline elucidates conserved time-of-day-specific cis-regulatory modules. *PLoS Genet.* **4**, e14.

Michael, T. P., Salomé, P. A., Yu, H. J., Spencer, T. R., Sharp, E. L., Alonso, J. M., Ecker, J. R., and McClung, C. R. (2003). Enhanced fitness conferred by naturally occurring variation in the circadian clock. *Science* **302**, 1049–1053.

Millar, A. J., Carré, I. A., Strayer, C. A., Chua, N.-H., and Kay, S. A. (1995). Circadian clock mutants in *Arabidopsis* identified by luciferase imaging. *Science* **267**, 1161–1163.

Miyazaki, K., Mesaki, M., and Ishida, N. (2001). Nuclear entry mechanism of rat PER2 (rPER2): Role of rPER2 in nuclear localization of CRY protein. *Mol. Cell. Biol.* **21**, 6651–6659.

Mizoguchi, T., Wheatley, K., Hanzawa, Y., Wright, L., Mizoguchi, M., Song, H.-R., Carré, I. A., and Coupland, G. (2002). LHY and CCA1 are partially redundant genes required to maintain circadian rhythms in Arabidopsis. *Dev. Cell* **2**, 629–641.

Mizuno, T., and Nakamichi, N. (2005). Pseudo-response regulators (PRRs) or true oscillator components (TOCs). *Plant Cell Physiol.* **46**, 677–685.

Mizuno, T., and Yamashino, T. (2008). Comparative transcriptome of diurnally oscillating genes and hormone-responsive genes in *Arabidopsis thaliana*: Insight into circadian clock-controlled daily responses to common ambient stresses in plants. *Plant Cell Physiol.* **49,** 481–487.

Morant, P.-E., Thommen, Q., Pfeuty, B., Vandermoere, C., Corellou, F., Bouget, F. Y., and Lefranc, M. (2010). A robust two-gene oscillator at the core of *Ostreococcus tauri* circadian clock. *Chaos* **20,** 045108-1-12.

Murakami, M., Ashikari, M., Miura, K., Yamashino, T., and Mizuno, T. (2003). The evolutionarily conserved OsPRR quintet: Rice Pseudo-Response Regulators implicated in circadian rhythm. *Plant Cell Physiol.* **44,** 1229–1236.

Murakami, M., Matsushika, A., Ashikari, M., Yamashino, T., and Mizuno, T. (2005). Circadian-associated rice pseudo response regulators (OsPRRs): insight into the control of flowering time. *Biosci. Biotechnol. Biochem.* **69,** 410–414.

Murakami, M., Tago, Y., Yamashino, T., and Mizuno, T. (2007). Comparative overviews of clock-associated genes of *Arabidopsis thaliana* and *Oryza sativa*. *Plant Cell Physiol.* **48,** 110–121.

Nagy, F., Fejes, E., Wehmeyer, B., Dallman, G., and Schafer, E. (1993). The circadian oscillator is regulated by a very low fluence response of phytochrome in wheat. *Proc. Natl. Acad. Sci. USA* **90,** 6290–6294.

Nakahata, Y., Kaluzova, M., Grimaldi, B., Sahar, S., Hirayama, J., Chen, D., Guarente, L. P., and Sassone-Corsi, P. (2008). The NAD^+-dependent deacetylase SIRT1 modulates CLOCK-mediated chromatin remodeling and circadian control. *Cell* **134,** 329–340.

Nakajima, M., Imai, K., Ito, H., Nishiwaki, T., Murayama, Y., Iwasaki, H., Oyama, T., and Kondo, T. (2005). Reconstitution of circadian oscillation of cyanobacterial KaiC phosphorylation in vitro. *Science* **308,** 414–415.

Nakamichi, N., Kita, M., Ito, S., Sato, E., Yamashino, T., and Mizuno, T. (2005a). PSEUDO-RESPONSE REGULATORS, PRR9, PRR7 and PRR5, together play essential roles close to the circadian clock of *Arabidopsis thaliana*. *Plant Cell Physiol.* **46,** 686–698.

Nakamichi, N., Kita, M., Ito, S., Sato, E., Yamashino, T., and Mizuno, T. (2005b). The Arabidopsis Pseudo-Response Regulators, PRR5 and PRR7, coordinately play essential roles for circadian clock function. *Plant Cell Physiol.* **46,** 609–619.

Nakamichi, N., Kita, M., Niinuma, K., Ito, S., Yamashino, T., Mizoguchi, T., and Mizuno, T. (2007). Arabidopsis clock-associated Pseudo-Response Regulators PRR9, PRR7 and PRR5 coordinately and positively regulate flowering time through the canonical CONSTANS-dependent photoperiodic pathway. *Plant Cell Physiol.* **48,** 822–832.

Nakamichi, N., Kusano, M., Fukushima, A., Kita, M., Ito, S., Yamashino, T., Saito, K., Sakakibara, H., and Mizuno, T. (2009). Transcript profiling of an Arabidopsis PSEUDO RE-SPONSE REGULATOR arrhythmic triple mutant reveals a role for the circadian clock in cold stress response. *Plant Cell Physiol.* **50,** 447–462.

Nakamichi, N., Kiba, T., Henriques, R., Mizuno, T., Chua, N.-H., and Sakakibara, H. (2010). PSEUDO-RESPONSE REGULATORS 9, 7 and 5 are transcriptional repressors in the Arabidopsis circadian clock. *Plant Cell* **22,** 594–605.

Nelson, D. C., Lasswell, J., Rogg, L. E., Cohen, M. A., and Bartel, B. (2000). *FKF1*, a clock-controlled gene that regulates the transition to flowering in *Arabidopsis*. *Cell* **101,** 331–340.

Ni, Z., Kim, E.-D., Ha, M., Lackey, E., Liu, J., Zhang, Y., Sun, Q., and Chen, Z. J. (2009). Altered circadian rhythms regulate growth vigour in hybrids and allopolyploids. *Nature* **457,** 327–331.

Nilsen, T. W., and Graveley, B. R. (2010). Expansion of the eukaryotic proteome by alternative splicing. *Nature* **463,** 457–463.

Nozue, K., Covington, M. F., Duek, P. D., Lorrain, S., Fankhauser, C., Harmer, S. L., and Maloof, J. N. (2007). Rhythmic growth explained by coincidence between internal and external cues. *Nature* **448,** 358–361.

Oh, S., Park, S., and van Nocker, S. (2008). Genic and global functions for Paf1C in chromatin modification and gene expression in Arabidopsis. *PLoS Genet.* **4,** e1000077.

Okada, R., Kondo, S., Satbhai, S. B., Yamaguchi, N., Tsukuda, M., and Aoki, S. (2009). Functional characterization of *CCA1/LHY* homolog genes, *PpCCA1a* and *PpCCA1b*, in the moss *Physcomitrella patens*. *Plant J.* **60,** 551–563.

Onai, K., and Ishiura, M. (2005). *PHYTOCLOCK1* encoding a novel GARP protein essential for the *Arabidopsis* circadian clock. *Genes Cells* **10,** 963–972.

O'Neill, J. S., and Reddy, A. B. (2011). Circadian clocks in human red blood cells. *Nature* **469,** 498–503.

O'Neill, J. S., van Ooijen, G., Dixon, L. E., Troein, C., Corellou, F., Bouget, F. Y., Reddy, A. B., and Millar, A. J. (2011). Circadian rhythms persist without transcription in a eukaryote. *Nature* **469,** 554–558.

Pahlich, S., Zakaryan, R. P., and Gehring, H. (2006). Protein arginine methylation: Cellular functions and methods of analysis. *Biochim. Biophys. Acta* **1764,** 1890–1903.

Para, A., Farré, E. M., Imaizumi, T., Pruneda-Paz, J. L., Harmon, F. G., and Kay, S. A. (2007). PRR3 is a vascular regulator of TOC1 stability in the Arabidopsis circadian clock. *Plant Cell* **19,** 3462–3473.

Parkin, I. A. P., Gulden, S. M., Sharpe, A. G., Lukens, L., Trick, M., Osborn, T. C., and Lydiate, D. J. (2005). Segmental structure of the *Brassica napus* genome based on comparative analysis with *Arabidopsis thaliana*. *Genetics* **171,** 765–781.

Pei, Y., Niu, L., Lu, F., Liu, C., Zhai, J., Kong, X., and Cao, X. (2007). Mutations in the Type II protein arginine methyltransferase AtPRMT5 result in pleiotropic developmental defects in Arabidopsis. *Plant Physiol.* **144,** 1913–1923.

Penfield, S., and Hall, A. (2009). A role for multiple circadian clock genes in the response to signals that break seed dormancy in Arabidopsis. *Plant Cell* **21,** 1722–1732.

Perales, M., and Más, P. (2007). A functional link between rhythmic changes in chromatin structure and the Arabidopsis biological clock. *Plant Cell* **19,** 2111–2123.

Pokhilko, A., Hodge, S. K., Stratford, K., Knox, K., Edwards, K. D., Thomson, A. W., Mizuno, T., and Millar, A. J. (2010). Data assimilation constrains new connections and components in a complex, eukaryotic circadian clock model. *Mol. Syst. Biol.* **6,** 416.

Portolés, S., and Más, P. (2010). The functional interplay between protein kinase CK2 and CCA1 transcriptional activity is essential for clock temperature compensation in Arabidopsis. *PLoS Genet.* **6,** e1001201.

Pruneda-Paz, J. L., and Kay, S. A. (2010). An expanding universe of circadian networks in higher plants. *Trends Plant Sci.* **15,** 259–265.

Pruneda-Paz, J. L., Breton, G., Para, A., and Kay, S. A. (2009). A functional genomics approach reveals CHE as a novel component of the Arabidopsis circadian clock. *Science* **323,** 1481–1485.

Ramos, A., Perez-Solis, E., Ibanez, C., Casado, R., Collada, C., Gomez, L., Aragoncillo, C., and Allona, I. (2005). Winter disruption of the circadian clock in chestnut. *Proc. Natl. Acad. Sci. USA* **102,** 7037–7042.

Rand, D. A., Shulgin, B. V., Salazar, J. D., and Millar, A. J. (2006). Uncovering the design principles of circadian clocks: Mathematical analysis of flexibility and evolutionary goals. *J. Theor. Biol.* **238,** 616–635.

Reddy, P., Zehring, W. A., Wheeler, D. A., Pirotta, V., Hadfield, C., Hall, J. C., and Rosbash, M. (1984). Molecular analysis of the *period* locus in *Drosophila melanogaster* and identification of a transcript involved in biological rhythms. *Cell* **38,** 701–710.

Resco, V., Hartwell, J., and Hall, A. (2009). Ecological implications of plants' ability to tell the time. *Ecol. Lett.* **12,** 583–592.

Richter, C. P. (1922). A behavioristic study of the activity of the rat. *Comp. Psychol. Monogr.* **1,** 1–55.

Rikin, A., Dillwith, J. W., and Bergman, D. K. (1993). Correlation between the circadian-rhythm of resistance to extreme temperatures and changes in fatty-acid composition in cotton seedlings. *Plant Physiol.* **101,** 31–36.

Risseeuw, E. P., Daskalchuk, T. E., Banks, T. W., Liu, E., Cotelesage, J., Hellmann, H., Estelle, M., Somers, D. E., and Crosby, W. L. (2003). Protein interaction analysis of SCF ubiquitin E3 ligase subunits from Arabidopsis. *Plant J.* **34,** 753–767.

Robertson, F., Skeffington, A., Gardner, M., and Webb, A. A. R. (2009). Interactions between circadian and hormonal signalling in plants. *Plant Mol. Biol.* **69,** 419–427.

Roden, L. C., and Ingle, R. A. (2009). Lights, rhythms, infection: The role of light and the circadian clock in determining the outcome of plant–pathogen interactions. *Plant Cell* **21,** 2546–2552.

Ruoff, P., Loros, J. J., and Dunlap, J. C. (2005). The relationship between FRQ-protein stability and temperature compensation in the *Neurospora* circadian clock. *Proc. Natl. Acad. Sci. USA* **102,** 17681–17686.

Salathia, N., Davis, S. J., Lynn, J. R., Michaels, S. D., Amasino, R. M., and Millar, A. J. (2006). FLOWERING LOCUS C-dependent and -independent regulation of the circadian clock by the autonomous and vernalization pathways. *BMC Plant Biol.* **6,** 10.

Salomé, P. A., and McClung, C. R. (2005). *PSEUDO-RESPONSE REGULATOR 7 and 9* are partially redundant genes essential for the temperature responsiveness of the Arabidopsis circadian clock. *Plant Cell* **17,** 791–803.

Salomé, P. A., To, J. P. C., Kieber, J. J., and McClung, C. R. (2006). *Arabidopsis* Response Regulators ARR3 and ARR4 play cytokinin-independent roles in the control of circadian period. *Plant Cell* **18,** 55–69.

Salomé, P. A., Xie, Q., and McClung, C. R. (2008). Circadian timekeeping during early Arabidopsis development. *Plant Physiol.* **147,** 1110–1125.

Salomé, P. A., Weigel, D., and McClung, C. R. (2010). The role of the *Arabidopsis* morning loop components CCA1, LHY, PRR7 and PRR9 in temperature compensation. *Plant Cell* **22,** 3650–3661.

Sanchez, S. E., Petrillo, E., Beckwith, E. J., Zhang, X., Rugnone, M. L., Hernando, C. E., Cuevas, J. C., Godoy Herz, M. A., Depetris-Chauvin, A., Simpson, C. G., Brown, J. W. S., Cerdán, P. D., et al. (2010). A methyl transferase links the circadian clock to the regulation of alternative splicing. *Nature* **468,** 112–116.

Sawa, M., Nusinow, D. A., Kay, S. A., and Imaizumi, T. (2007). FKF1 and GIGANTEA complex formation is required for day-length measurement in Arabidopsis. *Science* **318,** 261–265.

Schaffer, R., Ramsay, N., Samach, A., Corden, S., Putterill, J., Carré, I. A., and Coupland, G. (1998). The *late elongated hypocotyl* mutation of *Arabidopsis* disrupts circadian rhythms and the photoperiodic control of flowering. *Cell* **93,** 1219–1229.

Schmidt, F., Marnef, A., Cheung, M.-K., Wilson, I., Hancock, J., Staiger, D., and Ladomery, M. (2011). A proteomic analysis of oligo(dT)-bound mRNP containing oxidative stress-induced *Arabidopsis thaliana* RNA-binding proteins ATGRP7 and ATGRP8. *Mol. Biol. Rep.* **37,** 839–845.

Schmitz, R. J., Sung, S., and Amasino, R. M. (2008). Histone arginine methylation is required for vernalization-induced epigenetic silencing of FLC in winter-annual *Arabidopsis thaliana*. *Proc. Natl. Acad. Sci. USA* **105,** 411–416.

Schöning, J. C., Streitner, C., Meyer, I. M., Gao, Y., and Staiger, D. (2008). Reciprocal regulation of glycine-rich RNA-binding proteins via an interlocked feedback loop coupling alternative splicing to nonsense-mediated decay in Arabidopsis. *Nucleic Acids Res.* **36,** 6977–6987.

Schöning, J. C., Streitner, C., Page, D. R., Hennig, S., Uchida, K., Wolf, E., Furuya, M., and Staiger, D. (2007). Auto-regulation of the circadian slave oscillator component AtGRP7 and regulation of its targets is impaired by a single RNA recognition motif point mutation. *Plant J.* **52,** 1119–1130.

Serikawa, M., Miwa, K., Kondo, T., and Oyama, T. (2008). Functional conservation of clock-related genes in flowering plants: Overexpression and RNAi analyses of the circadian rhythm in the monocotyledon *Lemna gibba*. *Plant Physiol.* **146**, 1952–1963.

Shindo, C., Aranzana, M. J., Lister, C., Baxter, C., Nicholls, C., Nordborg, M., and Dean, C. (2005). Role of *FRIGIDA* and *FLOWERING LOCUS C* in determining variation in flowering time of Arabidopsis. *Plant Physiol.* **138**, 1163–1173.

Somers, D. (1999). The physiology and molecular bases of the plant circadian clock. *Plant Physiol.* **121**, 9–19.

Somers, D. E., Devlin, P., and Kay, S. A. (1998a). Phytochromes and cryptochromes in the entrainment of the *Arabidopsis* circadian clock. *Science* **282**, 1488–1490.

Somers, D. E., Webb, A. A. R., Pearson, M., and Kay, S. A. (1998b). The short-period mutant, *toc1-1*, alters circadian clock regulation of multiple outputs throughout development in *Arabidopsis thaliana*. *Development* **125**, 485–494.

Somers, D. E., Schultz, T. F., Milnamow, M., and Kay, S. A. (2000). ZEITLUPE encodes a novel clock-associated PAS protein from Arabidopsis. *Cell* **101**, 319–329.

Somers, D. E., Kim, W. Y., and Geng, R. (2004). The F-Box protein ZEITLUPE confers dosage-dependent control on the circadian clock, photomorphogenesis, and flowering time. *Plant Cell* **16**, 769–782.

Song, Y. H., Ito, S., and Imaizumi, T. (2010). Similarities in the circadian clock and photoperiodism in plants. *Curr. Opin. Plant Biol.* **13**, 594–603.

Staiger, D., and Koster, T. (2011). Spotlight on post-transcriptional control in the circadian system. *Cell. Mol. Life Sci.* **68**, 71–83.

Staiger, D., Zecca, L., Kirk, D. A. W., Apel, K., and Eckstein, L. (2003). The circadian clock regulated RNA-binding protein AtGRP7 autoregulates its expression by influencing alternative splicing of its own pre-mRNA. *Plant J.* **33**, 361–371.

Strayer, C., Oyama, T., Schultz, T. F., Raman, R., Somers, D. E., Más, P., Panda, S., Kreps, J. A., and Kay, S. A. (2000). Cloning of the *Arabidopsis* clock gene *TOC1*, an autoregulatory response regulator homolog. *Science* **289**, 768–771.

Streitner, C., Danisman, S., Wehrle, F., Schöning, J. C., Alfano, J. R., and Staiger, D. (2008). The small glycine-rich RNA binding protein AtGRP7 promotes floral transition in Arabidopsis thaliana. *Plant J.* **56**, 239–250.

Sugano, S., Andronis, C., Green, R. M., Wang, Z.-Y., and Tobin, E. M. (1998). Protein kinase CK2 interacts with and phosphorylates the *Arabidopsis* circadian clock-associated 1 protein. *Proc. Natl. Acad. Sci. USA* **95**, 11020–11025.

Sugano, S., Andronis, C., Ong, M. S., Green, R. M., and Tobin, E. M. (1999). The protein kinase CK2 is involved in regulation of circadian rhythms in Arabidopsis. *Proc. Natl. Acad. Sci. USA* **96**, 12362–12366.

Sulzman, F. M., Ellman, D., Fuller, C. A., Moore-Ede, M. C., and Wassmer, G. (1984). *Neurospora* circadian rhythms in space: A reexamination of the endogenous-exogenous question. *Science* **225**, 232–234.

Swarup, K., Alonso-Blanco, C., Lynn, J. R., Michaels, S. D., Amasino, R. M., Koornneef, M., and Millar, A. J. (1999). Natural allelic variation identifies new genes in the Arabidopsis circadian system. *Plant J.* **20**, 67–77.

Sweeney, B. M., and Haxo, F. T. (1961). Persistence of a photosynthetic rhythm in enucleated Acetabularia. *Science* **134**, 1361–1363.

Takata, N., Saito, S., Saito, C. T., Nanjo, T., Shinohara, K., and Uemura, M. (2009). Molecular phylogeny and expression of poplar circadian clock genes, *LHY1* and *LHY2*. *New Phytol.* **181**, 808–819.

Takata, N., Saito, S., Saito, C. T., and Uemura, M. (2010). Phylogenetic footprint of the plant clock system in angiosperms: evolutionary processes of Pseudo-Response Regulators. *BMC Evol. Biol.* **10**, 126.

Thines, B., and Harmon, F. G. (2010). Ambient temperature response establishes ELF3 as a required component of the core Arabidopsis circadian clock. *Proc. Natl. Acad. Sci. USA* **107**, 3257–3262.

Thommen, Q., Pfeuty, B., Morant, P.-E., Corellou, F., Bouget, F. o.-Y., and Lefranc, M. (2010). Robustness of circadian clocks to daylight fluctuations: Hints from the picoeucaryote *Ostreococcus tauri*. *PLoS Comput. Biol.* **6**, e1000990-1-15.

Tiwari, S. B., Shen, Y., Chang, H.-C., Hou, Y., Harris, A., Ma, S. F., McPartland, M., Hymus, G. J., Adam, L., Marion, C., Belachew, A., Repetti, P. P., *et al.* (2010). The flowering time regulator CONSTANS is recruited to the *FLOWERING LOCUS T* promoter via a unique cis-element. *New Phytol.* **187**, 57–66.

Troein, C., Locke, J. C. W., Turner, M. S., and Millar, A. J. (2009). Weather and seasons together demand complex biological clocks. *Curr. Biol.* **19**, 1961–1964.

Troein, C., Corellou, F., Dixon, L. E., van Ooijen, G., O'Neill, J. S., Bouget, F.-Y., and Millar, A. J. (2011). Multiple light inputs to a simple clock circuit allow complex biological rhythms. *Plant J.* **66**, 375–385.

Tsukada, Y., Fang, J., Erdjument-Bromage, H., Warren, M. E., Borchers, C. H., Tempst, P., and Zhang, Y. (2006). Histone demethylation by a family of JmjC domain-containing proteins. *Nature* **439**, 811–816.

van Ooijen, G., Dixon, L. E., Troein, C., and Millar, A. J. (2011). Proteasome function is required for biological timing throughout the 24-hour cycle. *Curr. Biol.* **21**, 869–875.

Walley, J. W., Coughlan, S., Hudson, M. E., Covington, M. F., Kaspi, R., Banu, G., Harmer, S. L., and Dehesh, K. (2007). Mechanical stress induces biotic and abiotic stress responses via a novel cis-element. *PLoS Genet.* **3**, 1800–1812.

Wang, Z.-Y., and Tobin, E. M. (1998). Constitutive expression of the *CIRCADIAN CLOCK ASSOCIATED 1* (*CCA1*) gene disrupts circadian rhythms and suppresses its own expression. *Cell* **93**, 1207–1217.

Wang, X., Zhang, Y., Ma, Q., Zhang, Z., Xue, Y., Bao, S., and Chong, K. (2007). SKB1-mediated symmetric dimethylation of histone H4R3 controls flowering time in Arabidopsis. *EMBO J.* **26**, 1934–1941.

Wang, L., Fujiwara, S., and Somers, D. E. (2010). PRR5 regulates phosphorylation, nuclear import and subnuclear localization of TOC1 in the Arabidopsis circadian clock. *EMBO J.* **29**, 1903–1915.

Wang, W., Barnaby, J. Y., Tada, Y., Li, H., Tör, M., Caldelari, D., Lee, D.-u., Fu, X.-D., and Dong, X. (2011). Timing of plant immune responses by a central circadian regulator. *Nature* **470**, 110–114.

Wilczek, A. M., Roe, J. L., Knapp, M. C., Cooper, M. D., Lopez-Gallego, C., Martin, L. J., Muir, C. D., Sim, S., Walker, A., Anderson, J., Egan, J. F., Moyers, B. T., *et al.* (2009). Effects of genetic perturbation on seasonal life history plasticity. *Science* **323**, 930–934.

Wilkins, O., Bräutigam, K., and Campbell, M. M. (2010). Time of day shapes Arabidopsis drought transcriptomes. *Plant J.* **63**, 715–727.

Wilkins, O., Waldron, L., Nahal, H., Provart, N. J., and Campbell, M. M. (2009). Genotype and time of day shape the Populus drought response. *Plant J.* **60**, 703–715.

Woolum, J. C. (1991). A re-examination of the role of the nucleus in generating the circadian rhythm in Acetabularia. *J. Biol. Rhythms* **6**, 129–136.

Xu, X., Hotta, C. T., Dodd, A. N., Love, J., Sharrock, R., Lee, Y. W., Xie, Q., Johnson, C. H., and Webb, A. A. R. (2007). Distinct light and clock modulation of cytosolic free Ca2+ oscillations and rhythmic *CHLOROPHYLL A/B BINDING PROTEIN2* promoter activity in Arabidopsis. *Plant Cell* **19**, 3474–3490.

Xu, X., Xie, Q., and McClung, C. R. (2010). Robust circadian rhythms of gene expression in Brassica rapa tissue culture. *Plant Physiol.* **153**, 841–850.

Yakir, E., Hilman, D., Hassidim, M., and Green, R. M. (2007). *CIRCADIAN CLOCK ASSOCIATED1* transcript stability and the entrainment of the circadian clock in Arabidopsis. *Plant Physiol.* **145**, 925–932.

Yakir, E., Hilman, D., Kron, I., Hassidim, M., Melamed-Book, N., and Green, R. M. (2009). Posttranslational regulation of CIRCADIAN CLOCK ASSOCIATED1 in the circadian oscillator of Arabidopsis. *Plant Physiol.* **150**, 844–857.

Yerushalmi, S., and Green, R. M. (2009). Evidence for the adaptive significance of circadian rhythms. *Ecol. Lett.* **12**, 970–981.

Yerushalmi, S., Yakir, E., and Green, R. M. (2011). Circadian clocks and adaptation in Arabidopsis. *Mol. Ecol.* **20**, 1155–1165.

Zagotta, M. T., Hicks, K. A., Jacobs, C. I., Young, J. C., Hangarter, R. P., and Meeks-Wagner, D. R. (1996). The *Arabidopsis* ELF3 gene regulates vegetative photomorphogenesis and the photoperiodic induction of flowering. *Plant J.* **10**, 691–702.

Zdepski, A., Wang, W., Priest, H. D., Ali, F., Alam, M., Mockler, T. C., and Michael, T. P. (2008). Conserved daily transcriptional programs in *Carica papaya. Trop. Plant Biol.* **1**, 236–245.

Zhang, E. E., and Kay, S. A. (2010). Clocks not winding down: Unravelling circadian networks. *Nat. Rev. Mol. Cell Biol.* **11**, 764–776.

Zhong, H. H., Painter, J. E., Salomé, P. A., Straume, M., and McClung, C. R. (1998). Imbibition, but not release from stratification, sets the circadian clock in Arabidopsis seedlings. *Plant Cell* **10**, 2005–2017.

5

Molecular Genetic Analysis of Circadian Timekeeping in *Drosophila*

Paul E. Hardin

Department of Biology and Center for Biological Clocks Research, Texas A&M University, College Station, Texas, USA

ABSTRACT

A genetic screen for mutants that alter circadian rhythms in *Drosophila* identified the first clock gene—the *period* (*per*) gene. The *per* gene is a central player within a transcriptional feedback loop that represents the core mechanism for keeping circadian time in *Drosophila* and other animals. The *per* feedback loop, or core loop, is interlocked with the *Clock* (*Clk*) feedback loop, but whether the *Clk*

Advances in Genetics, Vol. 74
0065-2660/11 $35.00
DOI: 10.1016/B978-0-12-387690-4.00005-2

feedback loop contributes to circadian timekeeping is not known. A series of distinct molecular events are thought to control transcriptional feedback in the core loop. The time it takes to complete these events should take much less than 24 h, thus delays must be imposed at different steps within the core loop. As new clock genes are identified, the molecular mechanisms responsible for these delays have been revealed in ever-increasing detail and provide an in-depth accounting of how transcriptional feedback loops keep circadian time. The phase of these feedback loops shifts to maintain synchrony with environmental cycles, the most reliable of which is light. Although a great deal is known about cell-autonomous mechanisms of light-induced phase shifting by CRYPTOCHROME (CRY), much less is known about non-cell autonomous mechanisms. CRY mediates phase shifts through an uncharacterized mechanism in certain brain oscillator neurons and carries out a dual role as a photoreceptor and transcription factor in other tissues. Here, I review how transcriptional feedback loops function to keep time in *Drosophila*, how they impose delays to maintain a 24-h cycle, and how they maintain synchrony with environmental light:dark cycles. The transcriptional feedback loops that keep time in *Drosophila* are well conserved in other animals, thus what we learn about these loops in *Drosophila* should continue to provide insight into the operation of analogous transcriptional feedback loops in other animals. © 2011, Elsevier Inc.

I. INTRODUCTION

Research on the fruit fly *Drosophila melanogaster* has had a profound impact on our understanding of the circadian timekeeping mechanism. Groundbreaking studies by Ron Konopka and Seymour Benzer identified the first "clock gene," *period* (*per*), in a screen for mutants with altered free-running (in constant darkness or DD) circadian periods in the rhythm of adult emergence (Konopka and Benzer, 1971). As *per* mutants alter circadian period, *per* was thought to be integrally involved in keeping circadian time, but determining how *per* contributed to circadian timekeeping did not take off until the *per* gene was isolated at Brandeis University in the labs of Michael Rosbash and Jeff Hall and at Rockefeller University in Michael Young's lab (Bargiello and Young, 1984; Bargiello et al., 1984; Reddy et al., 1984; Zehring et al., 1984). The PER protein sequence did not offer many clues about its role in the clock because its only distinguishing features were a stretch of threonine–glycine repeats, later shown to be involved in adapting to different thermal environments (Sawyer et al., 1997), and a region similar to a portion of the *Drosophila* SINGLE-MINDED (SIM) and mammalian aryl hydrocarbon receptor nuclear translocator (Arnt) proteins, termed the Per–Arnt–Sim (PAS) domain (Nambu et al., 1991), that mediates protein–protein interactions (Huang et al., 1993). However, the discovery that *per* mRNA and protein cycle in a circadian manner (Hardin *et al.*, 1990; Siwicki *et al.*, 1988), and that PER protein is required for cycling of *per* mRNA (Hardin *et al.*, 1990), suggested that *per* contributes to circadian timekeeping via a feedback loop in which PER protein controls rhythms in *per* mRNA expression (Hardin *et al.*, 1990). Studies demonstrating transcriptional control of *per* mRNA cycling and PER-dependent inhibition of *per* mRNA expression further refined the role of PER in this feedback loop as a transcriptional repressor (Hardin *et al.*, 1992; Zeng *et al.*, 1994). Subsequent studies not only support the view that this transcriptional feedback loop keeps circadian time in *Drosophila* but also show that similar feedback loops keep circadian time in diverse eukaryotic species including plants, fungi, and animals, the latter of which even shares critical feedback loop components such as *per* (for reviews, see Bell-Pedersen *et al.*, 2005; Dunlap, 1999; Young and Kay, 2001).

Although *per* is an essential feedback loop component, many other genes are required to sustain the *per* feedback loop. In Section II of this chapter, I explain the roles other genes play within the *per* feedback loop and describe how the *per* feedback loop relates to other interlocked feedback loops. As the *per* feedback loop has many components and is inextricably linked to other feedbacks loops, I refer to the *per* feedback loop as the "core loop" and to the combined core and interlocked feedback loops as "circadian feedback loops." The different steps required to construct a transcriptional feedback loop (such as the core loop) can be completed in far less than ~ 24 h, indicating that potent mechanisms have evolved to impart delays in feedback regulation. Section III focuses on how posttranscriptional regulation of key feedback loop components produces delays in transcriptional feedback that set the pace of the circadian oscillator. Entrainment of circadian feedback loops to environmental light–dark cycles is critical for driving physiological, metabolic, and behavioral rhythms at the appropriate time of day. Unlike the case in mammals, light can directly entrain circadian feedback loops in peripheral tissues from *Drosophila*, thus there is no intermediary "master" pacemaker in the *Drosophila* brain that relays light information to the periphery. However, light can entrain the network of *Drosophila* brain pacemaker neurons via multiple mechanisms depending on which cells detect the light. Section IV discusses how light entrains circadian feedback loops in different cells and tissues.

In Section V of this review, I summarize the main conclusions from each section, point out where there are gaps in our understanding, and discuss how filling these gaps may explain how these circadian feedback loops account for basic features of circadian clock such as the 24-h period and entrainment to environmental cycles. The general consensus is that circadian feedback loops sit at the heart of the circadian timekeeping mechanism in eukaryotes. However, experiments demonstrating that circadian rhythms in the phosphorylation/dephosphorylation of KaiC protein in cyanobacteria can be reconstituted in a test tube (Nakajima et al., 2005), and rhythms in the oxidation of peroxiredoxins in a primitive eukaryotic alga Ostreococcus tauri and red blood cells occur in the absence of transcription (O'Neill and Reddy, 2011; O'Neill et al., 2011), reveal that other circadian timekeeping mechanisms exist in eukaryotes that do not require transcriptional feedback. I discuss the role of circadian feedback loops in circadian timekeeping in light of these new results.

II. THE CIRCADIAN FEEDBACK LOOPS OF *DROSOPHILA*

The *per* feedback loop suggested a mechanism for keeping circadian time: once *per* transcription is initiated during mid-day, the levels of *per* mRNA rise until early evening, when accumulating levels of PER protein repress *per* transcription, thereby reducing the levels of *per* mRNA until the early day, when PER protein is eliminated and the next round of *per* transcription begins. Although this feedback loop provided a framework for how *per* contributes to circadian time-keeping, it raised many questions. For instance, PER feeds back to repress transcription of its own gene, but does PER achieve this by binding DNA at specific sites to repress transcription, competing with *per* activators for DNA-binding sites, or binding to *per* activators to inhibit their DNA binding? Does PER feed back to repress transcription alone, or are other factors involved in insuring that feedback occurs at the correct time of day? What genes are responsible for activating *per* transcription? A combination of approaches including genetic screens, molecular interaction assays, *per* promoter analysis, and molecular searches for clock gene orthologs were used to answer these questions.

Genetic screens uncovered many additional clock genes including *timeless* (*tim*) (Sehgal et al., 1994), *Clock* (*Clk*) (Allada et al., 1998), *cycle* (*cyc*) (Rutila et al., 1998), *doubletime* (*dbt*) (Kloss et al., 1998; Price et al., 1998), *shaggy* (*sgg*) (Martinek et al., 2001), casein kinase 2 (CK2) subunits (Akten et al., 2003; Lin et al., 2002a), and *cryptochrome* (*cry*) (Stanewsky et al., 1998). A screen for proteins that interact with PER also identified *tim* (Gekakis et al., 1995), which binds to the PAS domain of PER. Dissecting the *per* promoter for transcriptional regulatory elements identified a canonical E-box element (in this case 5'-CACGTG-3') that is required to activate *per* transcription (Hao et al., 1997) and was later found to be a conserved "circadian" regulatory element for many animal genes (reviewed in Bell-Pedersen et al., 2005; Hardin, 2004; Young and Kay, 2001). The first genetic screen for mouse clock genes identified *Clock* (King et al., 1997a,b; Vitaterna et al., 1994), which is a member of the basic-helix-loop-helix-PAS (bHLH-PAS) transcription factor family that typically binds E-box elements to activate transcription. Protein coding sequences from mouse *Clock* were used to recover the *Drosophila Clock* ortholog from a cDNA library screen (Darlington et al., 1998), thereby complementing the identification of *Clk* via genetic screening (Allada *et al.*, 1998) and expressed sequence tags (Bae *et al.*, 1998). The discovery of additional clock genes and regulatory sequences not only support the *per* feedback loop model but also add to its mechanistic detail. I briefly summarize clock gene function within the core feedback loop and refer you to a number of reviews for more detailed descriptions (Allada and Chung, 2010; Hardin, 2005; Zheng and Sehgal, 2008).

A. The core feedback loop

In the core feedback loop (Fig. 5.1), *per* and *tim* transcription are activated from ~ZT4 to ~ZT18 (Zeitgeber Time, or ZT, refers to time in hours during a light–dark cycle, where ZT0 is lights on and ZT12 is lights off) when CLK and its heterodimeric bHLH-PAS partner CYC bind E-boxes in the *per* and *tim* promoters (Allada *et al.*, 1998; Darlington *et al.*, 1998; Hao *et al.*, 1997; Rutila *et al.*, 1998). PER and TIM proteins start accumulating in the cytoplasm at ~ZT12, about 6–8 h after their respective mRNAs. This lag in PER and TIM protein accumulation is thought to be caused by the combined effects of phosphorylation-dependent degradation of PER by DBT, a homolog of mammalian casein kinase 1ε (Kloss *et al.*, 1998, 2001; Price *et al.*, 1998), and stabilization of PER–DBT complexes by TIM (Price *et al.*, 1995), which enables cytoplasmic accumulation of DBT–PER–TIM complexes (Curtin *et al.*, 1995; Gekakis *et al.*, 1995; Zeng *et al.*, 1996). However, recent data (described below) suggest that translational regulation may mediate the delay in PER cytoplasmic accumulation rather than DBT-dependent destabilization of PER (Chiu *et al.*, 2008; Lim *et al.*, 2011). Phosphorylation of PER by CK2 and TIM by SGG, a homolog of mammalian glycogen synthase kinase 3, promotes nuclear localization of PER–DBT and TIM (Akten *et al.*, 2003; Lin *et al.*, 2002a; Martinek *et al.*, 2001). PER and TIM phosphorylation are counterbalanced by protein phosphatase 2a (PP2a) and protein phosphatase 1 (PP1)-mediated dephosphorylation, respectively, which stabilize PER and TIM and alter their nuclear localization (Fang *et al.*, 2007; Sathyanarayanan *et al.*, 2004). Once in the nucleus, PER–DBT complexes (and/or PER–TIM–DBT complexes) then bind CLK, promote CLK phosphorylation, and release CLK–CYC from E-boxes to inhibit transcription from ~ZT18 to ~ZT4 (Bae *et al.*, 2000; Lee *et al.*, 1998, 1999; Menet *et al.*, 2010; Yu *et al.*, 2006). At ZT0, lights turn on and induce TIM degradation (Hunter-Ensor *et al.*, 1996; Lee *et al.*, 1996; Myers *et al.*, 1996; Zeng *et al.*, 1996), thus "deprotecting" PER. Progressive phosphorylation of PER by DBT in the nucleus ultimately triggers binding of the E3 ubiquitin ligase SLIMB, which targets PER for degradation in the proteasome by ~ZT4 (Grima *et al.*, 2002; Kloss *et al.*, 2001; Ko *et al.*, 2002; Naidoo *et al.*, 1999). Once PER is degraded, hypophosphorylated CLK accumulates and CLK–CYC binds E-boxes to initiate another cycle of *per* and *tim* transcription.

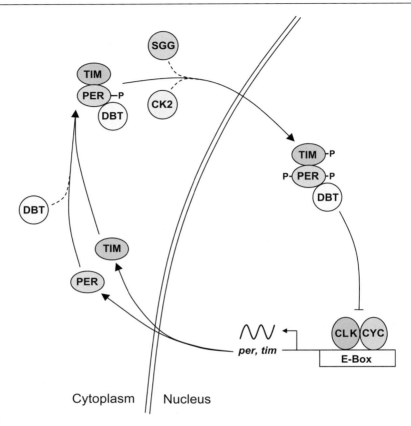

Figure 5.1. The core feedback loop. All gene, regulatory element, and protein names are as defined in the text. Double line, nuclear envelope; sinusoidal line, mRNA rhythm; solid arrows, synthesis, assembly and/or localization steps; blocked line, repression; protein regulatory factor activity, step employing protein phosphorylation; P, phosphorylation site(s). See text for detailed description. (See Color Insert.)

The core feedback loop nicely explains the regulation of cycling mRNAs that peak in abundance during the early evening. However, not all cycling gene expression peaks in the early evening. The first example of this in flies was *Clk* mRNA, which peaks during the early morning (Bae *et al.*, 1998; Darlington *et al.*, 1998). Rhythms in *Clk* mRNA levels are also dependent on the core feedback loop, as these rhythms were abolished in *per*[01] and *tim*[01] null mutants (Bae *et al.*, 1998). How these rhythms were abolished was intriguing; whereas *per* and *tim* mRNAs remain at relatively high levels in *per*[01] and *tim*[01] mutants (Hardin *et al.*, 1990; Sehgal *et al.*, 1995), *Clk* mRNA falls to constant low levels (Bae *et al.*, 1998). Although PER and TIM function to repress *per* and *tim* transcription (Sehgal *et al.*, 1994; Zeng *et al.*, 1994), the low levels of *Clk* mRNA in *per*^{01} and *tim*^{01} flies suggested that PER and TIM somehow activate *Clk* transcription (Bae *et al.*, 1998). As *per* and *tim* expressions are abolished in severe loss of function *Clk*^{Jrk} and *cyc*^{01} mutants (Allada *et al.*, 1998; Rutila *et al.*, 1998), *Clk* mRNA levels were expected to also be low in *Clk*^{Jrk} and *cyc*^{01} flies due to the loss of PER and TIM. However, *Clk* mRNA was expressed at peak levels in *Clk*^{Jrk} and *cyc*^{01} mutants, which suggested that CLK–CYC somehow represses *Clk* expression, in contrast to its traditional role as a transcription activator (Glossop *et al.*, 1999). A model was developed to explain these puzzling results that invoked a second "*Clk* feedback loop" that interlocked with the core feedback loop (Glossop *et al.*, 1999). The subsequent identification of the PAR transcription factors *vrille* (*vri*) and PAR domain protein 1ε and δ (*Pdp1ε/δ*) as components of the *Clk* feedback loop strongly supports the interlocked feedback loop model (Cyran *et al.*, 2003; Glossop *et al.*, 2003; Zheng *et al.*, 2009). I briefly outline the important features of the *Clk* feedback loop and refer you to other reviews for a more detailed description (Allada and Chung, 2010; Hardin, 2005; Zheng and Sehgal, 2008).

B. Interlocked feedback loops

In the *Clk* loop (Fig. 5.2), CLK–CYC binds E-boxes to activate *vri* transcription between ~ZT4 and ZT16 (Blau and Young, 1999; Cyran *et al.*, 2003). VRI protein accumulates in phase with *vri* mRNA, ultimately peaking in abundance at ~ZT14. As VRI levels rise, VRI binds VRI/PDP1-boxes (V/P-boxes) in the *Clk* promoter, thereby repressing *Clk* transcription (Cyran *et al.*, 2003; Glossop *et al.*, 2003). As PER–DBT and PER–TIM–DBT complexes feed back to inhibit CLK–CYC-dependent transcription from ~ZT16 to ZT4, *vri* mRNA and protein decline to low levels, thus permitting activation of *Clk* transcription. Mutants that disrupt CLK–CYC transcriptional activity (e.g., *Clk*^{Jrk}, *cyc*^{01}) exhibit constant high levels of *Clk* mRNA (Glossop *et al.*, 1999), indicating that *Clk* is constitutively activated independent of circadian oscillator function. However, another PAR transcription factor, PDP1ε/δ, also plays a role in *Clk* activation (Benito *et al.*, 2007; Cyran *et al.*, 2003; Zheng *et al.*, 2009). CLK–CYC binds E-boxes to activate *Pdp1ε/δ* between ZT4 and ZT16 (Cyran *et al.*, 2003). PDP1ε/δ rises to peak levels ~ZT18, several hours after VRI levels peak, and binds V/P-boxes to activate *Clk* transcription (Cyran *et al.*, 2003). The extent to which *Pdp1ε/δ* activates *Clk* transcription is debatable, as altering *Pdp1ε/δ* levels via RNA interference and overexpression has mild effects on *Clk* mRNA levels compared to mutants that eliminate *Pdp1ε/δ* specifically or all *Pdp1* isoforms (Benito *et al.*, 2007; Cyran *et al.*, 2003; Zheng *et al.*, 2009). Perhaps *vri* and *Pdp1ε/δ* function to enhance *Clk* mRNA amplitude once developmental

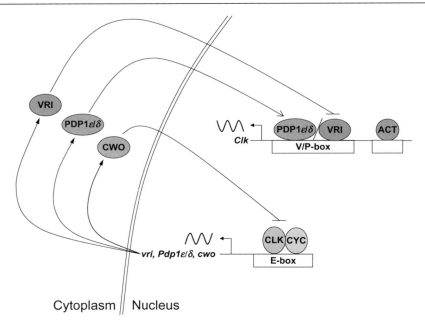

Cytoplasm ‖ Nucleus

Figure 5.2. The *Clk* feedback loop. All gene, regulatory element, and protein names are as defined in the text. All symbols are as defined in Fig. 5.1. ACT, *Clk* activator; open arrow, transcription activation; antiphase sinusoidal line, antiphase mRNA rhythm; backslash, binding by one or the other protein. See text for detailed description. (See Color Insert.)

activators establish the *Clk* expression pattern (Houl *et al.*, 2008). Once PER complexes are degraded during mid-day, the next cycle of *vri* transcription and VRI-dependent repression is initiated.

In addition to the *per* and *Clk* feedback loops, another feedback loop that is controlled by the bHLH-orange transcription inhibitor CLOCKWORK ORANGE (CWO) has been proposed (Fig. 5.2). In this feedback loop, CLK–CYC binds E-boxes to activate *cwo* transcription, and CWO then feeds back to inhibit CLK–CYC transcription (Kadener *et al.*, 2007; Lim *et al.*, 2007; Matsumoto *et al.*, 2007). Inhibition of CLK–CYC occurs through competition for binding E-boxes (Kadener *et al.*, 2007; Lim *et al.*, 2007; Matsumoto *et al.*, 2007) and thus independently reinforces inhibition by PER complexes. This model for CWO function is primarily based on *in vitro* and *Drosophila* Schneider 2 (S2) cell culture data, which contrasts with *cwo* mutant data showing period lengthening and lower levels of *per*, *tim*, *vri*, and *Pdp1ε/δ* levels (Kadener *et al.*, 2007; Lim *et al.*, 2007; Matsumoto *et al.*, 2007; Richier *et al.*, 2008). This *in vivo* data imply that *cwo* is necessary for high-level transcription of CLK–CYC-activated genes. Additional studies that document the phase of CWO protein cycling and DNA binding will help to resolve the role of CWO in the circadian oscillator.

Although the *per* and *Clk* feedback loops control mRNA cycling in opposite phases of the circadian cycle, they are not equally important for circadian oscillator function (Hardin, 2006). The *per* loop is required for *Clk* loop function, as CLK–CYC-dependent activation and PER–TIM–DBT-dependent repression control rhythms in *vri* transcription. However, the *Clk* loop is not necessary for *per* loop function, as reversing *Clk* mRNA cycling has little effect on molecular or behavioral rhythms (Kim et al., 2002), and rhythms in *Clk* mRNA levels do not drive rhythms in CLK levels or transcriptional activity (Yu et al., 2006). If *Clk* mRNA rhythms are not important for core feedback loop function, then what is the importance of the *Clk* feedback loop? Microarray studies show that ∼ 10% of *Drosophila* head mRNAs are rhythmically expressed (Ceriani et al., 2002; Claridge-Chang et al., 2001; Keegan et al., 2007; Lin et al., 2002b; McDonald and Rosbash, 2001; Ueda et al., 2002; Wijnen et al., 2006), and a large fraction of those are expressed with *Clk*-like peak times near dawn. Reducing, eliminating, or increasing *Pdp1ε/δ* expression abolishes behavioral rhythms (Benito et al., 2007; Zheng et al., 2009), which suggests that *Pdp1ε/δ* controls output gene expression. Indeed, *Pdp1ε/δ* controls rhythmic expression of *takeout* (*to*), which regulates courtship behavior (Dauwalder et al., 2002), but such regulation appears to be indirect as no canonical V/P-boxes are present near the *to* promoter (Benito et al., 2010). Although *vri* overexpression abolishes oscillator function due to repression of *Clk* transcription (Cyran et al., 2003; Glossop et al., 2003), it has not been possible to test whether *vri* is required for oscillator function because *vri* null mutants are lethal (Cyran et al., 2003; George and Terracol, 1997). Perhaps *vri*, like *Pdp1ε/δ*, also is required to control rhythmic transcription of output genes that peak near dawn. Given the importance of the core loop, it is imperative that the mechanisms governing rhythmic transcription within this loop are defined.

III. POSTTRANSCRIPTIONAL REGULATION OF RHYTHMIC TRANSCRIPTION

To keep circadian time, the various molecular events within the core feedback loop must be completed in ∼ 24 h. However, the transcriptional activation and elongation, transcript processing and transport to the cytosol, protein synthesis and accumulation, nuclear localization, transcriptional repression, and repressor degradation that mediate feedback loop function collectively take much less than 24 h to complete. Consequently, delays must be imposed at one or more steps in the core loop to achieve a 24-h cycle. As described in Section V, there is good evidence that

PER stability, nuclear localization, and transcriptional repression are controlled at the posttranscriptional level by TIM binding and kinases and phosphatases that control the phosphorylation state of PER and TIM. In this section, I review recent studies that begin to reveal the molecular consequences of PER–TIM interactions and site-specific PER phosphorylation by different kinases. In addition, I discuss the regulation and function of rhythms in CLK phosphorylation and other forms of posttranscriptional regulation that govern timekeeping by the core loop.

A. PER phosphorylation and translational control

PER phosphorylation increases as PER accumulates during the night and peaks as PER is degraded in the proteasome a few hours after dawn (Edery et al., 1994; Naidoo et al., 1999). DBT binds to PER and promotes PER degradation (Kim et al., 2007; Kloss et al., 1998, 2001; Price et al., 1998), whereas TIM binds to PER and prevents PER degradation (Kloss et al., 1998, 2001; Ko et al., 2002; Price et al., 1998). Much progress has been made in understanding the regulation of PER degradation in the nucleus (Fig. 5.3C). At this time of the circadian cycle, PER is complexed with DBT and CLK–CYC to repress transcription (Kloss et al., 2001; Lee et al., 1999; Yu and Hardin, 2006). PER is progressively phosphorylated during this timeframe, but its SLIMB F-box protein (homolog of mammalian B-TrCP)-induced degradation is delayed for several hours even after light or clock regulated removal of TIM (Grima et al., 2002; Ko et al., 2002). The reason for this delay is that DBT phosphorylation at PER serine 47 (S47) is the final step in a series of DBT phosphorylation events that produce an atypical SLIMB-binding site (Chiu et al., 2008). Once SLIMB binds, PER is ubiquitinated and rapidly degraded (Chiu et al., 2008), thereby releasing transcriptional repression.

When DBT is coexpressed with PER in S2 cells, DBT phosphorylates PER at many sites before the ultimate phosphorylation event on S47 (Chiu et al., 2008; Kivimae et al., 2008). However, PER is phosphorylated by other kinases in the presence or absence of DBT in S2 cells (Chiu et al., 2008; Kivimae et al., 2008). For instance, PER is phosphorylated by CK2 at S149, S151, and S153 to promote nuclear localization of PER complexes (Chiu et al., 2008; Lin et al., 2005) (Fig. 5.3B). PER is also phosphorylated at multiple consensus proline-directed kinase target sites (Chiu et al., 2008), in which a proline follows the phosphorylated serine or threonine. One such proline-directed site at S661 is phosphorylated by an as-yet unidentified kinase, but rather than promoting PER degradation, S661 phosphorylation primes phosphorylation of S657 by SGG to promote PER nuclear localization (Ko et al., 2010). SGG was thought to promote nuclear localization of PER complexes by phosphorylating TIM (Martinek et al., 2001), but SGG directly interacts with and stabilizes CRY, which, in turn, stabilizes TIM (Stoleru et al., 2007), and consequently PER, thus providing increased substrate for CK2 and SGG to effect PER nuclear localization.

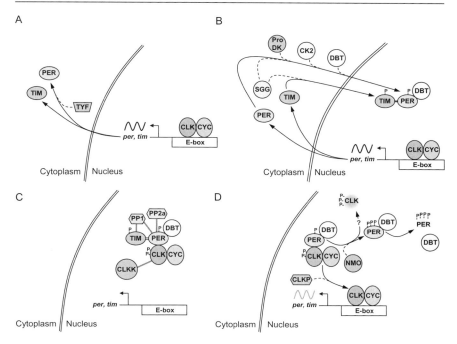

Figure 5.3. Posttranscriptional regulatory steps within the core feedback loop. All gene, regulatory element, and protein names are as defined in the text. All symbols are as defined in Fig. 5.1. (A) Delay in PER synthesis from ~ZT6 to ZT12. (B) Movement of PER and TIM into the nucleus from ~ZT16 to ZT20. (C) Stabilization of nuclear PER as PER complexes repress CLK–CYC transcription from ~ZT16 to ZT3. (D) Release of PER repression and reactivation of CLK–CYC transcription from ~ZT3 to ZT6. Double bar, stabilizing activity; CLKK, CLK kinase; ProDK, proline-directed kinase; gray bars, kinase and phosphatase targets; faded protein symbols, protein degradation; ?, putative direct effect; gray sinusoidal line, initiation of transcription. See text for detailed description. (See Color Insert.)

Moreover, CK2 and SGG are found predominantly in the cytoplasm (Lin *et al.*, 2002a; Yuan *et al.*, 2005), consistent with their role in promoting PER nuclear localization. SGG phosphorylation promotes PER nuclear localization but not PER degradation (Ko *et al.*, 2010), which implies that these nuclear localization and degradation are programmed by independent phosphorylation cascades. The lynchpin for PER nuclear localization (or at least an important player) is the proline-directed kinase that acts in the cytoplasm to prime SGG phosphorylation by phosphorylating S661. Identifying this proline-directed kinase and determining the relationship between SGG- and CK2-dependent phosphorylation will provide a clearer picture of how PER nuclear localization is regulated.

PER is phosphorylated at S661 and several additional consensus proline-directed kinase sites (Chiu et al., 2008; Kivimae et al., 2008). One such site at S596 is situated within the "short-period domain" of PER, a region spanning amino acids S585 to Y601 that contains mutants that shorten circadian period including per^S and per^T (Baylies et al., 1987, 1992; Konopka et al., 1994; Rutila et al., 1992; Yu et al., 1987). Disrupting the only proline-directed phosphorylation site in this region by mutating Y597 results in short-period rhythms (Baylies et al., 1992), suggesting that phosphorylation at this site normally acts to lengthen (or delay) the circadian cycle. Indeed, recent work demonstrates that NEMO (NMO) kinase phosphorylates S596 (Chiu et al., 2011), and nmo mutant and nmo RNAi knockdown flies show short-period rhythms (Chiu et al., 2011; Yu et al., 2011) (Fig. 5.3D). Mutating S595 to A also shortens circadian period and gates the phosphorylation of nearby residues S589 (the original per^S mutant site), S595, and S593 by DBT (Chiu et al., 2011). Phosphorylation of PER by NMO at S596 delays the phosphorylation of S47 by DBT (Chiu et al., 2011), and thus PER degradation, consistent with a delay in PER degradation in the morning when NMO is overexpressed (Yu et al., 2011). A model proposed by Edery and colleagues postulates that phosphorylation of PER by NMO and DBT within the "short-period domain" alters the conformation of PER, thereby inhibiting the phosphorylation of sites required for SLIMB binding and delaying PER degradation (Chiu et al., 2011).

In addition to the delay in PER degradation, PER protein accumulation lags 6–8 h behind per mRNA accumulation. This lag was thought to be produced by DBT-induced PER degradation and protection by TIM. However, PER is not phosphorylated at the key S47 site until after dawn (Chiu et al., 2008), thus arguing against the same SLIMB-dependent mechanism functioning in the cytoplasm. The delayed accumulation of PER in the cytoplasm could be achieved by accelerating PER degradation or impeding PER synthesis (Fig. 5.3A). DBT associates with PER as PER accumulates in the cytoplasm, thus it is possible that cytoplasmically localized F-box proteins could bind phosphorylated PER and promote degradation. Alternatively, there may be a delay in the translation of per mRNA. Recent work on the twenty-four (tyf) gene suggests that PER accumulation is translationally controlled (Lim et al., 2011). TYF associates with the 5'-cap binding complex, polyA binding protein, and per mRNA, and loss of tyf reduces PER accumulation in ventralateral neurons (LNvs) in the fly brain, which are necessary and sufficient for locomotor activity rhythms (Frisch et al., 1994; Grima et al., 2004; Renn et al., 1999; Stoleru et al., 2004), suggesting that TYF promotes PER translation in LNvs (Lim et al., 2011). This loss of tyf function result implies that per mRNA is poorly translated and/or under active translational repression, and that TYF enhances translation efficiency or removes the translation block. The extent to which regulation of per mRNA translation contributes to the delay in PER cytoplasmic accumulation is not known, but such regulation could play a major role in determining the pace of the circadian oscillator. TYF appears to function only in LNvs (Lim *et al.*, 2011), thus translational regulation by different factors or other posttranscriptional regulatory processes act to delay the cytoplasmic accumulation of PER in other oscillator cells. Translational regulation of other oscillator components is mediated by microRNAs. The microRNA *bantam* targets sequences in the 3′-UTR of *Clk* mRNA, and mutating these *bantam* target sites greatly decreases rescue of behavioral rhythms by *Clk* transgenes (Kadener *et al.*, 2009). Translation of *vri* and *cwo* mRNAs are also thought to be regulated by microRNAs (Kadener *et al.*, 2009), and a number of microRNAs are rhythmically expressed (Yang *et al.*, 2008).

As PER accumulates, it is phosphorylated and enters the nucleus, where TIM plays an important role in stabilizing phosphorylated PER (Fig. 5.3B). TIM likely stabilizes PER by inhibiting phosphorylation at sites that promote SLIMB binding and degradation. How TIM inhibits PER phosphorylation at these sites is not well characterized, but recent analysis suggests that TIM may inhibit PER phosphorylation by delivering PP1 to PER (Fang *et al.*, 2007). PER is also dephosphorylated by PP2a (Sathyanarayanan *et al.*, 2004), but dephosphorylation of PER by PP2a is not mediated by TIM (Fang *et al.*, 2007). Nevertheless, PP2a-dependent dephosphorylation of PER would further delay PER phosphorylation and degradation (Sathyanarayanan *et al.*, 2004). As phosphorylation enhances PER's ability to repress transcription (Nawathean and Rosbash, 2004), TIM-dependent stabilization of phosphorylated PER likely plays a key role in effectively repressing CLK–CYC transcription. It is not clear how PER phosphorylation enhances transcriptional repression, but it does not appear to act by enabling PER–CLK binding (Yu *et al.*, 2009). Thus, delays in the synthesis, nuclear localization, and degradation of PER contribute to the determination of circadian period.

B. CLK phosphorylation

CLK is phosphorylated coincident with the entry of PER repression complexes into the nucleus (Yu *et al.*, 2006). CLK phosphorylation coincides with transcriptional repression (Kim and Edery, 2006; Menet *et al.*, 2010; Yu *et al.*, 2006, 2009), but whether CLK phosphorylation is required for repression is not known (Fig. 5.3C). PER carries DBT into the nucleus (Kloss *et al.*, 2001), and DBT is required for CLK phosphorylation (Kim and Edery, 2006; Yu *et al.*, 2006, 2009). However, DBT does not phosphorylate CLK directly, though DBT (whether catalytically active or inactive) must be present in the PER repression complex to mediate CLK phosphorylation by one or more other kinases (Yu *et al.*, 2009). The kinase(s) responsible for most CLK phosphorylation has not been identified, but one kinase implicated in CLK phosphorylation is NMO (Fig. 5.3D). Loss of *nmo* function increases CLK levels and shortens circadian period, whereas increasing *nmo* function decreases CLK levels and lengthens circadian period (Yu *et al.*, 2011), consistent with period changes associated with increased or decreased CLK copy number (Kadener *et al.*, 2008). These results suggest that NMO promotes CLK degradation to slow the pace of the circadian cycle (Yu *et al.*, 2011), but whether NMO lowers CLK levels directly by phosphorylating CLK or indirectly by phosphorylating PER is not known.

Once PER is degraded, the levels of hypophosphorylated CLK increase along with CLK–CYC transcription. As overall CLK levels are essentially constant (Yu *et al.*, 2006), the increase in hypophosphorylated CLK levels could be due to the replacement of degraded hyperphosphorylated CLK with newly synthesized CLK, the dephosphorylation of hyperphosphorylated CLK, or both. Although *Clk* mRNA levels peak around dawn (Bae *et al.*, 1998), *Clk* mRNA levels only vary approximately threefold over the circadian cycle, thus increased synthesis is less likely to account for the almost complete conversion of hyperphosphorylated to hypophosphorylated CLK over 3–6 h. Phosphatases that would dephosphorylate CLK have not been identified, though PP1 and PP2a function within the core oscillator and PP2a regulatory subunit mRNAs cycle in abundance (Fang *et al.*, 2007; Sathyanarayanan *et al.*, 2004). Dephosphorylation of CLK after PER degradation may impose another delay within the feedback loop that is necessary to convert CLK–CYC to a transcriptionally active form. This is likely the case for CLOCK in mammals, as CLK is phosphorylated upon interaction with PER–CRY repressor complexes and returned to a dephosphorylated form when CLOCK–BMAL1 activates transcription (Lee *et al.*, 2001; Spengler *et al.*, 2009; Yoshitane *et al.*, 2009).

IV. LIGHT ENTRAINMENT OF THE *DROSOPHILA* CIRCADIAN FEEDBACK LOOPS

In animals, environmental cycles of light, temperature, and/or social cues set the phase of (i.e., entrain) circadian oscillators so that overt rhythms in physiology, metabolism, and behavior occur at the appropriate time of day. The most potent and reliable environmental cue is light, which mediates entrainment via different mechanisms depending on tissue type and species. In mammals, light is detected by melanopsin in retinal ganglion cells in the eye, which transmit signals to the "master clock" in the suprachiasmatic nucleus (SCN) through the retinohypothalamic tract (reviewed in Golombek and Rosenstein, 2010). These signals entrain circadian oscillators in the SCN, which relay information about their new phase via humoral signals to entrain oscillators in peripheral tissues (reviewed in Dibner *et al.*, 2010). This two-stage entrainment process (one stage for the SCN and a second for peripheral oscillators) differs from that in *Drosophila* because almost all *Drosophila* oscillator cells either detect light

directly or receive light information from photoreceptor cells. The cell-autonomous nature of circadian light entrainment is best illustrated in flies that use a *per*
promoter-driven luciferase reporter gene (*per*-luc) to drive rhythms in bioluminescence in isolated wings, and antennae and probosci that can be entrained by
light (Plautz et al., 1997). Thus, there is no "master clock" in *Drosophila* that
entrains peripheral oscillators (and no closed circulatory system to efficiently
send entrainment signals through even if there were a master clock).

Nevertheless, like the SCN clock, pacemaker cells in the *Drosophila*
brain control locomotor activity rhythms, which can be monitored with great
precision to detect light-induced shifts in oscillator phase. Light alters the phase
of activity differently depending on when it is detected during the circadian cycle
(Pittendrigh and Minis, 1964). A light pulse applied during the early night delays
the phase of activity by up to 4 h, a light pulse applied during the day (or
subjective day if the flies are in kept in constant darkness) has little effect on
activity phase, and a light pulse applied during the late night advances the phase
of activity up to 3 h (Myers et al., 1996; Saunders et al., 1994). Light induces
phase shifts by delaying or advancing the phase of the core feedback loop. This
section not only focuses primarily on the molecular mechanism that governs
light-induced phase shifts but will also discuss how light integrates with other
environmental cues to shift circadian phase.

A. TIM degradation, CRY, and other circadian photoreceptors

In 1996, three groups discovered that light induces the degradation of TIM in fly
heads (Hunter-Ensor et al., 1996; Myers et al., 1996; Zeng et al., 1996). Light
drastically reduces TIM levels within 30 min, thereby destabilizing PER and
shifting the phase of the core feedback loop. Although light's effect on TIM
levels is unidirectional, how reduced levels of TIM are interpreted by the clock
depends on when the light pulse is applied. Light-induced degradation of TIM
during the early evening produces a phase delay because *tim* mRNA levels are
high, and new synthesis can replenish TIM levels within a few hours. During late
night, light-induced degradation of TIM produces a phase advance because *tim*
mRNA levels are low, thus premature loss of TIM resets the core loop to its
normal state near dawn. Little or no phase shifts are seen during the actual or
subjective day because TIM levels are normally very low and cannot be further
reduced. Thus, light-dependent degradation of TIM initiates a phase shift, and
the levels of *tim* mRNA and protein determine the direction of the phase shift
through their effect on the core loop. Light induces TIM degradation in the
proteasome (Naidoo et al., 1999). Upon light exposure, TIM is phosphorylated
by a tyrosine kinase based on pharmacological experiments (Naidoo et al., 1999),
but this kinase has not been identified.

Though TIM is rapidly degraded after light exposure, TIM is not itself a photoreceptor. Loss of external eyes and ocelli or eliminating opsin-based photoreception via vitamin A depletion or phototransduction mutants reduces light sensitivity of the clock but does not abolish entrainment of locomotor activity rhythms by light (Blaschke et al., 1996; Hu et al., 1978; Ohata et al., 1998; Pearn et al., 1996), which suggests that other photoreceptors function to entrain circadian oscillators in brain pacemaker cells. A screen for mutants that disrupted rhythms in bioluminescence produced by per-luc identified CRYPTO-CHROME (CRY) (Emery et al., 1998; Stanewsky et al., 1998), which is an ortholog of blue light photoreceptor cryptochromes in plants (Ahmad and Cashmore, 1993; Lin et al., 1998). The severely hypomorphic cry^b mutant abolishes bioluminescence rhythms in whole flies during and after entrainment in 12-h light:12-h dark (LD) cycles, but not after entrainment to temperature cycles, suggesting a defect in the light entrainment pathway (Stanewsky et al., 1998). Despite the loss of bioluminescence rhythms, cry^b flies are behaviorally rhythmic during and after LD entrainment, indicating that pacemaker cells continue to receive light information. However, cry^b flies cannot be phase shifted by brief (e.g., 10 min) pulses of light (Stanewsky et al., 1998) and remain rhythmic in constant light after entrainment in LD cycles (Emery et al., 2000a), in contrast to wild-type flies that become arrhythmic in constant light (Konopka et al., 1989). Defects in behavioral rhythms due to cry^b can be rescued by expressing CRY in brain pacemaker neurons, demonstrating that CRY functions cell autonomously (Emery et al., 2000b). Taken as a whole, these studies demonstrate that CRY is a cell-autonomous photoreceptor that resets the phase of behavioral rhythms in Drosophila to short pulses of light but is not required for light entrainment to LD cycles.

Loss of phototransduction in external eyes (i.e., compound eyes and ocelli) by removing the NORP-A phospholipase C together with CRY further weakens, but does not abolish, light entrainment of behavioral rhythms (Stanewsky et al., 1998), indicating that other photoreceptors and/or phototransduction cascades function to entrain locomotor activity rhythms to light. The remaining photoreceptors capable of mediating light entrainment of locomotor activity are a pair of extraretinal eyes called the Hofbauer–Buchner (H–B) eyelet; eliminating all external photoreceptors, the H–B eyelet and CRY render flies unentrainable to light (Helfrich-Forster et al., 2001). Eliminating phototransduction in external eyes along with CRY abolishes core feedback loop function in all brain clock neurons except the LNvs and dorsal neuron 1s (DN1s), which implies that the H–B eyelet relays light information to this subset of clock neurons to mediate behavioral rhythms (Helfrich-Forster et al., 2001). Thus, behavioral rhythms can be entrained by independent light input pathways in external eyes, the H–B eyelet, and CRY, but light resetting to short pulses is mediated by CRY.

As TIM is degraded in response to short pulses of light, photodetection by CRY must initiate a cell-autonomous pathway leading to TIM degradation (Fig. 5.4). Indeed, CRY binds directly to TIM in a light-dependent manner, which irreversibly commits TIM to degradation in the proteasome (Busza *et al.*, 2004; Ceriani *et al.*, 1999; Dissel *et al.*, 2004; Naidoo *et al.*, 1999). CRY is also degraded in the proteasome upon activation by light, but CRY degradation occurs more slowly than TIM degradation (Busza *et al.*, 2004; Dissel *et al.*, 2004; Lin *et al.*, 2001). Although light-induced TIM and CRY degradation was thought to be controlled independently (Busza *et al.*, 2004; Sathyanarayanan *et al.*, 2008), more recent evidence suggests that they are linked (Peschel *et al.*, 2009). Light-dependent degradation of both TIM and CRY is mediated by the F-box protein JETLAG (JET) (Koh *et al.*, 2006; Peschel *et al.*, 2006, 2009). However, TIM and CRY are degraded sequentially due to a higher affinity of JET for TIM than CRY (Peschel *et al.*, 2009). Thus, when CRY is activated by light, it becomes a substrate for JET binding, and binding of activated CRY to TIM makes TIM an even higher affinity substrate for JET than CRY. Light-induced degradation of TIM requires the COP9 signalosome (Knowles *et al.*, 2009), which apparently acts downstream of JET to promote TIM (and maybe CRY) degradation in the proteasome. Mutants in *cry*, *jet*, and the COP9 signalosome are all rhythmic in constant light (Emery *et al.*, 2000a; Knowles *et al.*, 2009; Koh *et al.*, 2006; Peschel *et al.*, 2006), which suggests that these genes function in the same pathway to mediate light-dependent resetting of circadian oscillators in pacemaker cells that control rhythmic activity.

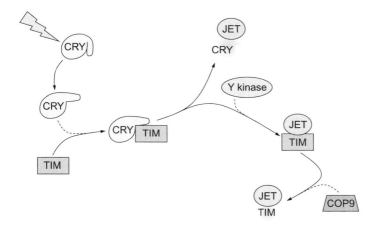

Figure 5.4. Light-induced phase resetting mechanism. All gene, regulatory element, and protein names are as defined in the text. All symbols are as defined in Figs. 5.1 and 5.3. Y kinase, tyrosine kinase. See text for detailed description. (See Color Insert.)

In contrast to CRY-mediated phase resetting to light, the molecular and cellular mechanisms by which the H–B eyelet and external photoreceptors in the compound eye and ocelli have only just begun to be characterized. The H–B eyelet projects to the large subset of ventrolateral neurons (lLNvs), which are necessary for light-induced phase resetting at dawn and increase their firing rate when stimulated by light (Shang et al., 2008; Sheeba et al., 2008). The lLNvs project throughout the optic lobe, including the vicinity of the small subset of ventrolateral neurons (sLNvs) (Helfrich-Forster et al., 2007), suggesting that light information from the H–B eyelet may be sent to lLNvs, and then on to the sLNvs to mediate phase resetting near dawn (Shang et al., 2008; Sheeba et al., 2010). How compound eyes and ocelli contribute to phase resetting is not known.

B. Resetting behavioral rhythms—Cellular and sensory integration

As TIM is rapidly degraded in fly heads after light exposure, it has been assumed that light leads to TIM degradation uniformly in all oscillator cells including the LNv pacemaker neurons. Although TIM is degraded in all brain oscillator neurons including the LNvs when phase advancing light pulses are applied at ZT21, TIM is not degraded in LNvs (but is degraded in dorsal brain oscillator neurons) when phase delaying light pulses are applied at ZT15 (Tang et al., 2010). Overexpressing JET in the LNvs of a jet null mutant enables light-induced TIM degradation at ZT15 but does not produce a phase delay in behavior (Tang et al., 2010), thus TIM degradation in LNvs is neither necessary nor sufficient for phase delays. If CRY expression is knocked down in only LNvs, light-induced phase delays and advances are both blocked, indicating that LNvs are necessary for behavioral phase shifts even though TIM is not degraded by a light pulse at ZT15 in these cells (Tang et al., 2010).

These results suggest that light-sensitive dorsal brain oscillator neurons signal through LNvs to promote light-induced phase delays. A good candidate for dorsal neurons that initiate light-induced phase delays is the DN1s because a subset of these neurons are CRY positive (Benito et al., 2008; Yoshii et al., 2008), light pulses at ZT15 consistently induce TIM degradation in these neurons (Tang et al., 2010), and LNvs signal to DN1s via PDF while DN1s may signal LNvs via projections to LNv cell bodies (Helfrich-Forster et al., 2007; Kaneko and Hall, 2000; Zhang et al., 2010a). Whether DN1s or other dorsal oscillator neurons mediate light-dependent phase delays via LNvs, it is clear that the standard cell-autonomous CRY-dependent degradation of TIM paradigm for light-induced phase resetting does not explain how light induces phase delays in locomotor activity.

During LD conditions, Drosophila display morning and evening peaks in activity that anticipates the lights-on and lights-off transitions, respectively. These activity peaks are controlled by separate sets of brain oscillator neurons, where morning activity is controlled by LNvs that express the PDF neuropeptide and evening activity is controlled by LNds and the PDF negative 5th sLNv (Grima *et al.*, 2004; Stoleru *et al.*, 2004). The phase of these morning and evening activity peaks adjusts to seasonal differences in photoperiod and temperature. When photoperiod is long and temperature is high in the summer, flies are more active before dawn and after dusk, but when photoperiod is short and temperatures are low, flies are more active after dawn and before dusk (Majercak *et al.*, 1999). This behavioral plasticity is influenced by molecular mechanisms that act to adjust oscillator phase including temperature sensitive splicing of *per* intron 8 (Collins *et al.*, 2004; Low *et al.*, 2008; Majercak *et al.*, 1999, 2004), temperature sensitive splicing of the last *tim* intron (Boothroyd *et al.*, 2007), and light-induced transcription of *tim* only at low temperatures (Chen *et al.*, 2006). Communication between morning cells that serve as dominant clocks in the dark and evening cells that serve as dominant clocks in light (along with a subset of DN1s) is also important for controlling seasonal differences in activity (Cusumano *et al.*, 2009; Murad *et al.*, 2007; Stoleru *et al.*, 2007; Zhang *et al.*, 2009). Integration of light and temperature information was recently shown to occur in the DN1s, which alter the phase and amplitude of morning and evening activity peaks depending on temperature and light intensity (Zhang *et al.*, 2010a,b). How temperature and light information is integrated in DN1s to modulate morning and evening activity will no doubt be the subject of future studies.

 Social interactions among flies also alter locomotor activity rhythms. Flies entrained to a particular circadian phase influence the activity of flies entrained to a different phase (Levine *et al.*, 2002b), and courtship interactions between males and females shift activity to the night and away from dusk (Fujii *et al.*, 2007). In both these cases, olfactory cues are responsible for the activity changes (Fujii *et al.*, 2007; Krupp *et al.*, 2008; Levine *et al.*, 2002b), which indicates that olfactory signaling, which is also under circadian control (Krishnan *et al.*, 1999, 2008; Tanoue *et al.*, 2004, 2008), is likely integrated with light and temperature to determine the timing of activity. Several excellent reviews have been published that provide a more in-depth description of the discovery, function and modulation of morning and evening oscillators (Allada and Chung, 2010; Choi and Nitabach, 2010; Dubruille and Emery, 2008), and insight into the mechanisms by which temperature entrains circadian oscillators to effect seasonal adaptation (Chen *et al.*, 2007; Dubruille and Emery, 2008; Glaser and Stanewsky, 2007).

C. Other functions of *Drosophila* CRY

Bioluminescence rhythms in cry^b flies bearing *per*-luc can be entrained by temperature cycles, but not LD cycles (Stanewsky *et al.*, 1998). As the vast majority of *per*-luc expression is from peripheral oscillators, this result implies that CRY is also required for light entrainment of circadian oscillators in peripheral tissues. While this may be the case, CRY is also required for oscillator function *per se* in at least some peripheral tissues including the eye, the antenna, Malpighian tubules (the fly kidney equivalent), and forelegs (Collins *et al.*, 2006; Ivanchenko *et al.*, 2001; Krishnan *et al.*, 2001; Levine *et al.*, 2002a). The evidence that CRY is required for oscillator function differs depending on the peripheral tissue. In antenna, rhythmic output in the form of daily cycles in electroanten-nagram (EAG) responses to odors is severely impaired or abolished in cry^b mutants during and after entrainment to LD or temperature cycles (Krishnan *et al.*, 2001). This loss of EAG rhythms in cry^b flies is due to the disruption of rhythmic expression within the core feedback loop, as reported by *per*-luc and *tim*-luc in antennae cultured during and after LD entrainment and after temperature entrainment (Krishnan *et al.*, 2001; Levine *et al.*, 2002a). This result is reminiscent of the arrhythmicity seen in mice lacking mCRY1 and mCRY2, which repress CLOCK–BMAL1-mediated transcription (Griffin *et al.*, 1999; Kume *et al.*, 1999; Lee *et al.*, 2001; Preitner *et al.*, 2002).

These experiments show that CRY is required for oscillator function in antennae, but it is difficult to know whether CRY also serves to entrain antennae to light cycles, as EAGs and gene expression in cry^b flies were arrhythmic during LD. Loss of rhythmic *per*-luc and *tim*-luc expressions in forelegs from cry^b flies during and after LD entrainment implies that CRY also functions within the oscillator, in contrast to the situation in wing where *per*-luc and *tim*-luc expressions are rhythmic in cry^b flies during and after LD entrainment (Levine *et al.*, 2002a). In Malpighian tubules from cry^b flies, loss of TIM degradation to light pulses suggests that CRY is required for phase resetting, but cycling of PER and TIM levels during LD cycles suggests that entrainment occurs through some other mechanism (Ivanchenko *et al.*, 2001). Rhythms in *tim*-luc and PER and TIM protein cycling were abolished in cry^b mutants during DD, demonstrating that CRY is necessary for oscillator function (Ivanchenko *et al.*, 2001). The role CRY plays within fly peripheral oscillators was investigated in fly heads, where > 80% of clock gene expression is from peripheral oscillators in the eye (Glossop and Hardin, 2002). In cry^b flies, expression of CLK–CYC-activated genes was constantly near peak levels, suggesting that CRY represses CLK–CYC transcription (Collins *et al.*, 2006). Consistent with this result, CRY collaborates with PER to inhibit CLK–CYC transcription in eyes and S2 cells, but not in sLNvs (Collins *et al.*, 2006). If CRY is indeed required for rhythmic transcription within peripheral oscillators, it is difficult to understand how oscillator function persists in cry^b fly heads after temperature entrainment (Stanewsky *et al.*, 1998). Despite this inconsistency, CRY- and PER-dependent repression of CLK–CYC transcription is completely in line with mCRY1 and mCRY2 function in mammals, which repress CLOCK–BMAL1 transcription in collaboration with mPER2 (Griffin *et al.*, 1999; Kume *et al.*, 1999; Lee *et al.*, 2001; Preitner *et al.*, 2002).

How *Drosophila* CRY carries out independent functions in brain pacemaker cells and peripheral oscillators is not known, though other insects (e.g., Monarch butterflies, mosquito, moth) carry out photoreceptor and transcriptional repressor function using two independent *cry* genes (Yuan *et al.*, 2007; Zhu *et al.*, 2008).

V. SUMMARY AND CONCLUSIONS

In this chapter, I have covered several topics related to the interlocked feedback loops that keep circadian time in *Drosophila*. There are several take home points from this review. First, many feedback loop components have been identified in *Drosophila* that carry out specific roles in regulating time-dependent transcription including the CLK and CYC activators and PER and TIM repressors. Although the *per* and *Clk* feedback loops drive rhythmic transcription in opposite phases of the circadian cycle, the *per* loop controls the *Clk* loop and functions to keep circadian time. Second, time delays are built into multiple steps of the core feedback loop to achieve a period of approximately 24 h. These delays are controlled posttranscriptionally via regulated synthesis, accumulation, nuclear localization, degradation, and activity of transcriptional activators and repressors. Third, light shifts the *Drosophila* circadian oscillator to a new circadian phase by promoting TIM degradation. In most oscillator cells, CRY acts as a cell-autonomous photoreceptor that not only binds TIM to induce JET-dependent degradation of both TIM and CRY in the proteasome after light exposure but also functions as a transcriptional repressor to support core feedback loop function in many peripheral oscillator cells. Retinal photoreceptors, extraretinal photoreceptors, and CRY all contribute to the entrainment of locomotor activity pacemaker cells to light–dark cycles, but other environmental factors such as temperature and olfactory cues also integrate with light information to modulate fly activity according to the season and social context. Understanding how the circadian clock is entrained by environmental cues and maintains an approximately 24-h period under constant conditions in a simple model organism such as *Drosophila* will no doubt continue to provide important insights into the basic mechanisms that control circadian clock function in mammals.

Despite the progress that has been made in identifying components of the circadian feedback loops, major gaps in our understanding of how these components contribute to circadian timekeeping remain. The *Clk* feedback loop controls rhythmic transcription that peaks near dawn but is dependent on the core loop. Though rhythmic outputs including locomotor activity are dependent on *Pdp1ε/δ* within the *Clk* loop (Benito *et al.*, 2008, 2010; Zheng *et al.*, 2009), we do not know if this feedback loop contributes to circadian timekeeping. The *Neurospora* interlocked feedback loop contributes to oscillator stability (Cheng *et al.*, 2001), whereas the mammalian interlocked loop influences period length and phase shifting to light (Preitner et al., 2002). Manipulating components of the *Clk* loop such as *vri* and *Clk* activators will provide the tools to test whether the *Clk* loop contributes to timekeeping function. Identifying *Clk* activators would fill a gap in our understanding of the *Clk* loop, as we would be able to determine whether these activators both determine clock cell identity and maintain *Clk* transcription in adults. Likewise, building tools to inducibly manipulate *vri* expression in adult flies would enable experiments to test whether *vri* is required for circadian timekeeping and/or output rhythms. The traditional role for *cwo* as a repressor of CLK–CYC transcription is based mainly on *in vitro* and cell culture data and does not jibe with *in vivo* data suggesting that *cwo* contributes to activating CLK–CYC transcription (Kadener et al., 2007; Lim et al., 2007; Matsumoto et al., 2007; Richier et al., 2008). Better tools for CWO detection and biochemistry are needed to determine how this gene contributes to the circadian timekeeping mechanism.

A major question in circadian biology is how circadian feedback loops maintain periods of approximately 24 h. While it is clear that delays are incorporated into various steps of the feedback loop, we are now beginning to understand the mechanisms that control these delays. The delayed accumulation of PER repressor complexes in the cytoplasm (at least in the LNvs) involves regulated translation by *tyf* (Lim et al., 2011), but it is not as clear how much PER phosphorylation by DBT and PER stabilization by TIM and PP1 and PP2a phosphatases contribute to slowing the rate of PER accumulation in the cytoplasm (Fang et al., 2007; Kloss et al., 1998, 2001; Ko et al., 2002; Price et al., 1998; Sathyanarayanan et al., 2004). Determining whether additional kinases, phosphatases, E3 ubiquitin ligases, or other factors promote PER degradation in the cytoplasm, and whether PER translation is regulated in other oscillator tissues, will be essential for understanding the extent to which these processes contribute to delays in PER cytoplasmic accumulation. SGG, CK2, and an as-yet unidentified proline-directed kinase phosphorylate PER and/or TIM to promote their nuclear localization (Akten et al., 2003; Ko et al., 2010; Lin et al., 2002a; Martinek et al., 2001), thus enabling these proteins to bind CLK–CYC and inhibit transcription. Movement of PER–TIM repressor complexes into the nucleus coincides with CLK phosphorylation, transcriptional repression, and the removal of CLK–CYC from E-boxes (Kim et al., 2007; Menet et al., 2010; Yu et al., 2006, 2009). The kinases that phosphorylate CLK have not been identified, though it is possible that NMO phosphorylates CLK to promote CLK degradation (Yu et al., 2011). To understand how PER complexes repress transcription, it is important to determine which kinases phosphorylate CLK, where CLK is phosphorylated, and how such phosphorylation represses transcription and induces CLK–CYC release from E-boxes. Kinases also contribute to light-induced and clock-dependent TIM degradation (Naidoo et al., 1999), but the identity of these kinases and their target sites on TIM have not been determined, but are essential to understanding how the clock maintains an approximately 24-h period, as the loss of TIM promotes PER degradation. Great strides have been made to understand how PER degradation is regulated in the nucleus. NMO phosphorylation of PER at S596 primes further phosphorylation by DBT to alter PER conformation, thus delaying phosphorylation of other DBT target sites including S47, which triggers SLIMB binding and PER degradation (Chiu et al., 2011). Determining the temporal profile of DBT phosphorylation sites on PER and the requirements for phosphorylation at specific sites would provide an unprecedented understanding of a process that is essential for setting circadian period. Once PER is degraded, CLK is returned to a dephosphorylated state that enables E-box binding and transcriptional activation by CLK–CYC (Kim and Edery, 2006; Yu et al., 2006, 2009). Identifying phosphatases that act on CLK during this phase of the circadian cycle will be key to understanding the extent to which CLK dephosphorylation contributes to period determination.

Although the basic molecular events that control light-induced phase shifting by CRY have been characterized, it is apparent that CRY mediates phase delays in locomotor activity independent of TIM degradation. Where light is detected to induce phase delays and how CRY acts to shift the LNv clock represent major gaps in our understanding. A subset of DN1s send projections to LNvs (Helfrich-Forster et al., 2007; Kaneko and Hall, 2000), and certain DN1s also express CRY (Benito et al., 2008; Yoshii et al., 2008). If the CRY positive DN1s also project to LNvs, these cells could receive light and signal the LNvs just as the H–B eyelet receives light and signals the LNvs (Helfrich-Forster et al., 2001). CRY is required for light shifting in LNvs in the absence of TIM degradation (Tang et al., 2010), which suggests that CRY induces phase delays in LNvs by altering the level or activity of some other core feedback loop component. Determining how LNvs receive light signals during the early evening and shift the oscillator in a CRY-dependent manner is critical for a comprehensive understanding of behavioral phase resetting to light. However, CRY is not the only photoreceptor capable of entraining behavioral rhythms to LD cycles. H–B eyelet neurons may effect light-induced entrainment of behavior by signaling lLNvs, which then relay this information to the sLNvs (Shang et al., 2008; Sheeba et al., 2010). Additional studies are needed to solidify this pathway, define the cellular pathways through which photoreceptors in the compound eyes and ocelli effect light entrainment, and determine how light information from other photoreceptor cells effect shifts in the sLNv molecular oscillator. Lastly, recent work shows that light and temperature information is integrated in DN1s to modulate the pattern of locomotor activity rhythms during LD conditions (Zhang et al., 2010a,b). How temperature information is received by DN1s, integrated with CRY-dependent light signals, and sent to LNvs to modify behavioral activity are important challenges for future studies, as is the integration of olfactory-based social cues that also modulate behavioral activity.

Although interlocked transcriptional feedback loops function to keep circadian time in eukaryotes, this is not the case in cyanobacteria (see Chapter 3). The core timekeeping mechanism in cyanobacteria does not require transcription but instead consists of a ~24-h phosphorylation/dephosphorylation cycle mediated by the KaiA, KaiB, and KaiC proteins. When KaiA, KaiB, KaiC are incubated with ATP, KaiA stimulates KaiC autophosphorylation and KaiB blocks KaiA to stimulate KaiC dephosphorylation, which results in a KaiC phosphorylation cycle with a ~24-h period (Nakajima et al., 2005). Although this cycle necessarily occurs independent of transcription, the Kai oscillator drives rhythms in virtually all transcripts by periodically altering chromosome compaction (Smith and Williams, 2006; Woelfle and Johnson, 2006; Woelfle et al., 2007). Although the core KaiC phosphorylation oscillator functions under most conditions in vitro, transcriptional rhythms contribute to the stability and entrainment of the Kai oscillator in vivo (see review by Dong and Golden, 2008).

Recent studies show that circadian clocks in some eukaryotic organisms and cell types can operate in the absence of rhythmic transcription. Red blood cells lack nuclei and mitochondria, and thus no transcription can occur. However, the dimerization of peroxyredoxin enzymes, which function to reduce reactive oxygen species, cycles with a circadian rhythm that is temperature entrainable (O'Neill and Reddy, 2011). In fact, binding of reactive oxygen species is what catalyzes the formation of peroxyredoxin dimers, which are then converted back to monomers via reduction by thioredoxin (O'Neill and Reddy, 2011). A similar phenomenon has been observed in the unicellular eukaryotic algae O. tauri, which cease transcriptional activity in the dark, yet circadian oscillations in peroxyredoxin dimerization persist (O'Neill et al., 2011). Rhythms in peroxyredoxin dimerization also occur under conditions when the transcriptional feedback loops are operating in Ostreococcus and mouse embryonic fibroblasts, thus both circadian oscillators function simultaneously in the same cell (O'Neill and Reddy, 2011; O'Neill et al., 2011). These oscillators apparently interact at some level as mutant mouse embryonic fibroblasts that abolish transcriptional feedback loops lengthen the period of peroxyredoxin dimerization rhythms (O'Neill and Reddy, 2011). Given that peroxyredoxin dimerization rhythms are seen in mouse embryonic fibroblasts, rhythms in peroxyredoxin dimerization likely occur in Drosophila. Once the oscillator that drives peroxyredoxin dimerization rhythms is identified, it will be fascinating to determine the extent to which this oscillator influences transcription-based oscillators.

Acknowledgments

I want to thank Dr. Wangjie Yu for comments on the chapter and Isaac Edery for communicating unpublished results. This work was supported by NIH Grant NS052854.

References

Ahmad, M., and Cashmore, A. R. (1993). HY4 gene of A. thaliana encodes a protein with characteristics of a blue-light photoreceptor. *Nature* **366**, 162–166.

Akten, B., Jauch, E., Genova, G. K., Kim, E. Y., Edery, I., Raabe, T., and Jackson, F. R. (2003). A role for CK2 in the *Drosophila* circadian oscillator. *Nat. Neurosci.* **6**, 251–257.

Allada, R., and Chung, B. Y. (2010). Circadian organization of behavior and physiology in Drosophila. *Annu. Rev. Physiol.* **72**, 605–624.

Allada, R., White, N. E., So, W. V., Hall, J. C., and Rosbash, M. (1998). A mutant *Drosophila* homolog of mammalian Clock disrupts circadian rhythms and transcription of *period* and *timeless*. *Cell* **93**, 791–804.

Bae, K., Lee, C., Sidote, D., Chuang, K. Y., and Edery, I. (1998). Circadian regulation of a *Drosophila* homolog of the mammalian Clock gene: PER and TIM function as positive regulators. *Mol. Cell. Biol.* **18**, 6142–6151.

Bae, K., Lee, C., Hardin, P. E., and Edery, I. (2000). dCLOCK is present in limiting amounts and likely mediates daily interactions between the dCLOCK-CYC transcription factor and the PER-TIM complex. *J. Neurosci.* **20**, 1746–1753.

Bargiello, T. A., and Young, M. W. (1984). Molecular genetics of a biological clock in Drosophila. *Proc. Natl. Acad. Sci. USA* **81**, 2142–2146.

Bargiello, T. A., Jackson, F. R., and Young, M. W. (1984). Restoration of circadian behavioural rhythms by gene transfer in Drosophila. *Nature* **312**, 752–754.

Baylies, M. K., Bargiello, T. A., Jackson, F. R., and Young, M. W. (1987). Changes in abundance or structure of the per gene product can alter periodicity of the Drosophila clock. *Nature* **326**, 390–392.

Baylies, M. K., Vosshall, L. B., Sehgal, A., and Young, M. W. (1992). New short period mutations of the Drosophila clock gene per. *Neuron* **9**, 575–581.

Bell-Pedersen, D., Cassone, V. M., Earnest, D. J., Golden, S. S., Hardin, P. E., Thomas, T. L., and Zoran, M. J. (2005). Circadian rhythms from multiple oscillators: Lessons from diverse organisms. *Nat. Rev. Genet.* **6**, 544–556.

Benito, J., Zheng, H., and Hardin, P. E. (2007). PDP1epsilon functions downstream of the circadian oscillator to mediate behavioral rhythms. *J. Neurosci.* **27**, 2539–2547.

Benito, J., Houl, J. H., Roman, G. W., and Hardin, P. E. (2008). The blue-light photoreceptor CRYPTOCHROME is expressed in a subset of circadian oscillator neurons in the Drosophila CNS. *J. Biol. Rhythms* **23**, 296–307.

Benito, J., Hoxha, V., Lama, C., Lazareva, A. A., Ferveur, J. F., Hardin, P. E., and Dauwalder, B. (2010). The circadian output gene takeout is regulated by Pdp1epsilon. *Proc. Natl. Acad. Sci. USA* **107**, 2544–2549.

Blaschke, I., Lang, P., Hofbauer, A., Engelmann, W., and Helfrich-Forster, C. (1996). Preliminary action spectra suggest that the clock cells of Drosophila are synchronized to the external LD-cycle by the compound eyes plus extraretinal photoreceptors. In "" (N. Elsner and H.-U. Schnitzler, eds.). *24th Gottingen Neurobiology Conference*. Thieme, Gottingen, Germany, Vol. I.

Blau, J., and Young, M. W. (1999). Cycling *vrille* expression is required for a functional *Drosophila* clock. *Cell* **99**, 661–671.

Boothroyd, C. E., Wijnen, H., Naef, F., Saez, L., and Young, M. W. (2007). Integration of light and temperature in the regulation of circadian gene expression in Drosophila. *PLoS Genet.* **3**, e54.

Busza, A., Emery-Le, M., Rosbash, M., and Emery, P. (2004). Roles of the two Drosophila CRYPTOCHROME structural domains in circadian photoreception. *Science* **304**, 1503–1506.

Ceriani, M. F., Darlington, T. K., Staknis, D., Mas, P., Petti, A. A., Weitz, C. J., and Kay, S. A. (1999). Light-dependent sequestration of TIMELESS by CRYPTOCHROME. *Science* **285**, 553–556.

Ceriani, M. F., Hogenesch, J. B., Yanovsky, M., Panda, S., Straume, M., and Kay, S. A. (2002). Genome-wide expression analysis in *Drosophila* reveals genes controlling circadian behavior. *J. Neurosci.* **22**, 9305–9319.

Chen, W. F., Majercak, J., and Edery, I. (2006). Clock-gated photic stimulation of timeless expression at cold temperatures and seasonal adaptation in Drosophila. *J. Biol. Rhythms* **21**, 256–271.

Chen, W. F., Low, K. H., Lim, C., and Edery, I. (2007). Thermosensitive splicing of a clock gene and seasonal adaptation. *Cold Spring Harb. Symp. Quant. Biol.* **72**, 599–606.

Cheng, P., Yang, Y., and Liu, Y. (2001). Interlocked feedback loops contribute to the robustness of the Neurospora circadian clock. *Proc. Natl. Acad. Sci. USA* **98**, 7408–7413.

Chiu, J. C., Vanselow, J. T., Kramer, A., and Edery, I. (2008). The phospho-occupancy of an atypical SLIMB-binding site on PERIOD that is phosphorylated by DOUBLETIME controls the pace of the clock. *Genes Dev.* **22**, 1758–1772.

Chiu, J., Ko, H. W., and Edery, I. (2011). NEMO/NLK primes phosphorylation of a time-delay phospho-cluster on PERIOD revealing a novel mechanism for how circadian clock speed is set. *Cell* **145**, 357–370.

Choi, C., and Nitabach, M. N. (2010). Circadian biology: Environmental regulation of a multi-oscillator network. *Curr. Biol.* **20**, R322–R324.

Claridge-Chang, A., Wijnen, H., Naef, F., Boothroyd, C., Rajewsky, N., and Young, M. W. (2001). Circadian regulation of gene expression systems in the *Drosophila* head. *Neuron* **32**, 657–671.

Collins, B. H., Rosato, E., and Kyriacou, C. P. (2004). Seasonal behavior in Drosophila melanogaster requires the photoreceptors, the circadian clock, and phospholipase C. *Proc. Natl. Acad. Sci. USA* **101**, 1945–1950.

Collins, B., Mazzoni, E. O., Stanewsky, R., and Blau, J. (2006). Drosophila CRYPTOCHROME is a circadian transcriptional repressor. *Curr. Biol.* **16**, 441–449.

Curtin, K. D., Huang, Z. J., and Rosbash, M. (1995). Temporally regulated nuclear entry of the *Drosophila period* protein contributes to the circadian clock. *Neuron* **14**, 365–372.

Cusumano, P., Klarsfeld, A., Chelot, E., Picot, M., Richier, B., and Rouyer, F. (2009). PDF-modulated visual inputs and cryptochrome define diurnal behavior in Drosophila. *Nat. Neurosci.* **12**, 1431–1437.

Cyran, S. A., Buchsbaum, A. M., Reddy, K. L., Lin, M. C., Glossop, N. R., Hardin, P. E., Young, M. W., Storti, R. V., and Blau, J. (2003). *vrille, Pdp1*, and *dClock* form a second feedback loop in the *Drosophila* circadian clock. *Cell* **112**, 329–341.

Darlington, T. K., Wager-Smith, K., Ceriani, M. F., Staknis, D., Gekakis, N., Steeves, T. D., Weitz, C. J., Takahashi, J. S., and Kay, S. A. (1998). Closing the circadian loop: CLOCK-induced transcription of its own inhibitors *per* and *tim*. *Science* **280**, 1599–1603.

Dauwalder, B., Tsujimoto, S., Moss, J., and Mattox, W. (2002). The *Drosophila takeout* gene is regulated by the somatic sex-determination pathway and affects male courtship behavior. *Genes Dev.* **16**, 2879–2892.

Dibner, C., Schibler, U., and Albrecht, U. (2010). The mammalian circadian timing system: Organization and coordination of central and peripheral clocks. *Annu. Rev. Physiol.* **72**, 517–549.

Dissel, S., Codd, V., Fedic, R., Garner, K. J., Costa, R., Kyriacou, C. P., and Rosato, E. (2004). A constitutively active cryptochrome in Drosophila melanogaster. *Nat. Neurosci.* **7**, 834–840.

Dong, G., and Golden, S. S. (2008). How a cyanobacterium tells time. *Curr. Opin. Microbiol.* **11**, 541–546.

Dubruille, R., and Emery, P. (2008). A plastic clock: How circadian rhythms respond to environmental cues in Drosophila. *Mol. Neurobiol.* **38**, 129–145.

Dunlap, J. C. (1999). Molecular bases for circadian clocks. *Cell* **96**, 271–290.

Edery, I., Zwiebel, L. J., Dembinska, M. E., and Rosbash, M. (1994). Temporal phosphorylation of the *Drosophila period* protein. *Proc. Natl. Acad. Sci. USA* **91**, 2260–2264.

Emery, P., So, W. V., Kaneko, M., Hall, J. C., and Rosbash, M. (1998). CRY, a *Drosophila* clock and light-regulated cryptochrome, is a major contributor to circadian rhythm resetting and photosensitivity. *Cell* **95,** 669–679.

Emery, P., Stanewsky, R., Hall, J. C., and Rosbash, M. (2000a). A unique circadian-rhythm photoreceptor. *Nature* **404,** 456–457.

Emery, P., Stanewsky, R., Helfrich-Forster, C., Emery-Le, M., Hall, J. C., and Rosbash, M. (2000b). *Drosophila* CRY is a deep brain circadian photoreceptor. *Neuron* **26,** 493–504.

Fang, Y., Sathyanarayanan, S., and Sehgal, A. (2007). Post-translational regulation of the Drosophila circadian clock requires protein phosphatase 1 (PP1). *Genes Dev.* **21,** 1506–1518.

Frisch, B., Hardin, P. E., Hamblen-Coyle, M. J., Rosbash, M., and Hall, J. C. (1994). A promoterless *period* gene mediates behavioral rhythmicity and cyclical *per* expression in a restricted subset of the *Drosophila* nervous system. *Neuron* **12,** 555–570.

Fujii, S., Krishnan, P., Hardin, P., and Amrein, H. (2007). Nocturnal male sex drive in Drosophila. *Curr. Biol.* **17,** 244–251.

Gekakis, N., Saez, L., Delahaye-Brown, A. M., Myers, M. P., Sehgal, A., Young, M. W., and Weitz, C. J. (1995). Isolation of *timeless* by PER protein interaction: Defective interaction between *timeless* protein and long-period mutant PERL. *Science* **270,** 811–815.

George, H., and Terracol, R. (1997). The *vrille* gene of *Drosophila* is a maternal enhancer of *decapentaplegic* and encodes a new member of the bZIP family of transcription factors. *Genetics* **146,** 1345–1363.

Glaser, F. T., and Stanewsky, R. (2007). Synchronization of the Drosophila circadian clock by temperature cycles. *Cold Spring Harb. Symp. Quant. Biol.* **72,** 233–242.

Glossop, N. R., and Hardin, P. E. (2002). Central and peripheral circadian oscillator mechanisms in flies and mammals. *J. Cell Sci.* **115,** 3369–3377.

Glossop, N. R., Lyons, L. C., and Hardin, P. E. (1999). Interlocked feedback loops within the *Drosophila* circadian oscillator. *Science* **286,** 766–768.

Glossop, N. R., Houl, J. H., Zheng, H., Ng, F. S., Dudek, S. M., and Hardin, P. E. (2003). VRILLE feeds back to control circadian transcription of *Clock* in the *Drosophila* circadian oscillator. *Neuron* **37,** 249–261.

Golombek, D. A., and Rosenstein, R. E. (2010). Physiology of circadian entrainment. *Physiol. Rev.* **90,** 1063–1102.

Griffin, E. A., Jr., Staknis, D., and Weitz, C. J. (1999). Light-independent role of CRY1 and CRY2 in the mammalian circadian clock. *Science* **286,** 768–771.

Grima, B., Lamouroux, A., Chelot, E., Papin, C., Limbourg-Bouchon, B., and Rouyer, F. (2002). The F-box protein slimb controls the levels of clock proteins period and timeless. *Nature* **420,** 178–182.

Grima, B., Chelot, E., Xia, R., and Rouyer, F. (2004). Morning and evening peaks of activity rely on different clock neurons of the Drosophila brain. *Nature* **431,** 869–873.

Hao, H., Allen, D. L., and Hardin, P. E. (1997). A circadian enhancer mediates PER-dependent mRNA cycling in *Drosophila melanogaster*. *Mol. Cell. Biol.* **17,** 3687–3693.

Hardin, P. E. (2004). Transcription regulation within the circadian clock: The E-box and beyond. *J. Biol. Rhythms* **19,** 348–360.

Hardin, P. E. (2005). The circadian timekeeping system of Drosophila. *Curr. Biol.* **15,** R714–R722.

Hardin, P. E. (2006). Essential and expendable features of the circadian timekeeping mechanism. *Curr. Opin. Neurobiol.* **16,** 686–692.

Hardin, P. E., Hall, J. C., and Rosbash, M. (1990). Feedback of the *Drosophila period* gene product on circadian cycling of its messenger RNA levels. *Nature* **343,** 536–540.

Hardin, P. E., Hall, J. C., and Rosbash, M. (1992). Circadian oscillations in *period* gene mRNA levels are transcriptionally regulated. *Proc. Natl. Acad. Sci. USA* **89,** 11711–11715.

Helfrich-Forster, C., Winter, C., Hofbauer, A., Hall, J. C., and Stanewsky, R. (2001). The circadian clock of fruit flies is blind after elimination of all known photoreceptors. *Neuron* **30,** 249–261.

Helfrich-Forster, C., Shafer, O. T., Wulbeck, C., Grieshaber, E., Rieger, D., and Taghert, P. (2007). Development and morphology of the clock-gene-expressing lateral neurons of Drosophila melanogaster. *J. Comp. Neurol.* **500,** 47–70.

Houl, J. H., Ng, F., Taylor, P., and Hardin, P. E. (2008). CLOCK expression identifies developing circadian oscillator neurons in the brains of Drosophila embryos. *BMC Neurosci.* **9,** 119.

Hu, K. G., Reichart, H., and Stark, W. S. (1978). Electrophysiological characterization of Drosophila ocelli. *J. Comp. Physiol.* **126,** 15–24.

Huang, Z. J., Edery, I., and Rosbash, M. (1993). PAS is a dimerization domain common to Drosophila period and several transcription factors. *Nature* **364,** 259–262.

Hunter-Ensor, M., Ousley, A., and Sehgal, A. (1996). Regulation of the *Drosophila* protein *timeless* suggests a mechanism for resetting the circadian clock by light. *Cell* **84,** 677–685.

Ivanchenko, M., Stanewsky, R., and Giebultowicz, J. M. (2001). Circadian photoreception in *Drosophila*: Functions of *cryptochrome* in peripheral and central clocks. *J. Biol. Rhythms* **16,** 205–215.

Kadener, S., Stoleru, D., McDonald, M., Nawathean, P., and Rosbash, M. (2007). Clockwork Orange is a transcriptional repressor and a new Drosophila circadian pacemaker component. *Genes Dev.* **21,** 1675–1686.

Kadener, S., Menet, J. S., Schoer, R., and Rosbash, M. (2008). Circadian transcription contributes to core period determination in Drosophila. *PLoS Biol.* **6,** e119.

Kadener, S., Menet, J. S., Sugino, K., Horwich, M. D., Weissbein, U., Nawathean, P., Vagin, V. V., Zamore, P. D., Nelson, S. B., and Rosbash, M. (2009). A role for microRNAs in the Drosophila circadian clock. *Genes Dev.* **23,** 2179–2191.

Kaneko, M., and Hall, J. C. (2000). Neuroanatomy of cells expressing clock genes in Drosophila: Transgenic manipulation of the period and timeless genes to mark the perikarya of circadian pacemaker neurons and their projections. *J. Comp. Neurol.* **422,** 66–94.

Keegan, K. P., Pradhan, S., Wang, J. P., and Allada, R. (2007). Meta-analysis of Drosophila circadian microarray studies identifies a novel set of rhythmically expressed genes. *PLoS Comput. Biol.* **3,** e208.

Kim, E. Y., and Edery, I. (2006). Balance between DBT/CKIepsilon kinase and protein phosphatase activities regulate phosphorylation and stability of Drosophila CLOCK protein. *Proc. Natl. Acad. Sci. USA* **103,** 6178–6183.

Kim, E. Y., Bae, K., Ng, F. S., Glossop, N. R., Hardin, P. E., and Edery, I. (2002). *Drosophila* CLOCK protein is under posttranscriptional control and influences light-induced activity. *Neuron* **34,** 69–81.

Kim, E. Y., Ko, H. W., Yu, W., Hardin, P. E., and Edery, I. (2007). A DOUBLETIME kinase binding domain on the Drosophila PERIOD protein is essential for its hyperphosphorylation, transcriptional repression, and circadian clock function. *Mol. Cell. Biol.* **27,** 5014–5028.

King, D. P., Vitaterna, M. H., Chang, A. M., Dove, W. F., Pinto, L. H., Turek, F. W., and Takahashi, J. S. (1997a). The mouse Clock mutation behaves as an antimorph and maps within the W19H deletion, distal of Kit. *Genetics* **146,** 1049–1060.

King, D. P., Zhao, Y., Sangoram, A. M., Wilsbacher, L. D., Tanaka, M., Antoch, M. P., Steeves, T. D., Vitaterna, M. H., Kornhauser, J. M., Lowrey, P. L., Turek, F. W., and Takahashi, J. S. (1997b). Positional cloning of the mouse circadian clock gene. *Cell* **89,** 641–653.

Kivimae, S., Saez, L., and Young, M. W. (2008). Activating PER repressor through a DBT-directed phosphorylation switch. *PLoS Biol.* **6,** e183.

Kloss, B., Price, J. L., Saez, L., Blau, J., Rothenfluh, A., Wesley, C. S., and Young, M. W. (1998). The *Drosophila* clock gene *double-time* encodes a protein closely related to human casein kinase Iepsilon. *Cell* **94,** 97–107.

Kloss, B., Rothenfluh, A., Young, M. W., and Saez, L. (2001). Phosphorylation of *period* is influenced by cycling physical associations of *double-time*, *period*, and *timeless* in the *Drosophila* clock. *Neuron* **30,** 699–706.

Knowles, A., Koh, K., Wu, J. T., Chien, C. T., Chamovitz, D. A., and Blau, J. (2009). The COP9 signalosome is required for light-dependent timeless degradation and Drosophila clock resetting. *J. Neurosci.* **29**, 1152–1162.

Ko, H. W., Jiang, J., and Edery, I. (2002). Role for Slimb in the degradation of Drosophila Period protein phosphorylated by Doubletime. *Nature* **420**, 673–678.

Ko, H. W., Kim, E. Y., Chiu, J., Vanselow, J. T., Kramer, A., and Edery, I. (2010). A hierarchical phosphorylation cascade that regulates the timing of PERIOD nuclear entry reveals novel roles for proline-directed kinases and GSK-3beta/SGG in circadian clocks. *J. Neurosci.* **30**, 12664–12675.

Koh, K., Zheng, X., and Sehgal, A. (2006). JETLAG resets the Drosophila circadian clock by promoting light-induced degradation of TIMELESS. *Science* **312**, 1809–1812.

Konopka, R. J., and Benzer, S. (1971). Clock mutants of Drosophila melanogaster. *Proc. Natl. Acad. Sci. USA* **68**, 2112–2116.

Konopka, R. J., Pittendrigh, C. S., and Orr, D. (1989). Reciprocal behaviour associated with altered homeostasis and photosensitivity of Drosophila clock mutants. *J. Neurogenet.* **6**, 1–10.

Konopka, R. J., Hamblen-Coyle, M. J., Jamison, C. F., and Hall, J. C. (1994). An ultrashort clock mutation at the period locus of Drosophila melanogaster that reveals some new features of the fly's circadian system. *J. Biol. Rhythms* **9**, 189–216.

Krishnan, B., Dryer, S. E., and Hardin, P. E. (1999). Circadian rhythms in olfactory responses of Drosophila melanogaster. *Nature* **400**, 375–378.

Krishnan, B., Levine, J. D., Lynch, M. K., Dowse, H. B., Funes, P., Hall, J. C., Hardin, P. E., and Dryer, S. E. (2001). A new role for cryptochrome in a Drosophila circadian oscillator. *Nature* **411**, 313–317.

Krishnan, P., Chatterjee, A., Tanoue, S., and Hardin, P. E. (2008). Spike amplitude of single-unit responses in antennal sensillae is controlled by the Drosophila circadian clock. *Curr. Biol.* **18**, 803–807.

Krupp, J. J., Kent, C., Billeter, J. C., Azanchi, R., So, A. K., Schonfeld, J. A., Smith, B. P., Lucas, C., and Levine, J. D. (2008). Social experience modifies pheromone expression and mating behavior in male Drosophila melanogaster. *Curr. Biol.* **18**, 1373–1383.

Kume, K., Zylka, M. J., Sriram, S., Shearman, L. P., Weaver, D. R., Jin, X., Maywood, E. S., Hastings, M. H., and Reppert, S. M. (1999). mCRY1 and mCRY2 are essential components of the negative limb of the circadian clock feedback loop. *Cell* **98**, 193–205.

Lee, C., Parikh, V., Itsukaichi, T., Bae, K., and Edery, I. (1996). Resetting the Drosophila clock by photic regulation of PER and a PER-TIM complex. *Science* **271**, 1740–1744.

Lee, C., Bae, K., and Edery, I. (1998). The Drosophila CLOCK protein undergoes daily rhythms in abundance, phosphorylation, and interactions with the PER-TIM complex. *Neuron* **21**, 857–867.

Lee, C., Bae, K., and Edery, I. (1999). PER and TIM inhibit the DNA binding activity of a Drosophila CLOCK-CYC/dBMAL1 heterodimer without disrupting formation of the heterodimer: A basis for circadian transcription. *Mol. Cell. Biol.* **19**, 5316–5325.

Lee, C., Etchegaray, J. P., Cagampang, F. R., Loudon, A. S., and Reppert, S. M. (2001). Posttranslational mechanisms regulate the mammalian circadian clock. *Cell* **107**, 855–867.

Levine, J. D., Funes, P., Dowse, H. B., and Hall, J. C. (2002a). Advanced analysis of a cryptochrome mutation's effects on the robustness and phase of molecular cycles in isolated peripheral tissues of Drosophila. *BMC Neurosci.* **3**, 5.

Levine, J. D., Funes, P., Dowse, H. B., and Hall, J. C. (2002b). Resetting the circadian clock by social experience in Drosophila melanogaster. *Science* **298**, 2010–2012.

Lim, C., Chung, B. Y., Pitman, J. L., McGill, J. J., Pradhan, S., Lee, J., Keegan, K. P., Choe, J., and Allada, R. (2007). Clockwork orange encodes a transcriptional repressor important for circadian-clock amplitude in Drosophila. *Curr. Biol.* **17**, 1082–1089.

Lim, C., Lee, J., Choi, C., Kilman, V. L., Kim, J., Park, S. M., Jang, S. K., Allada, R., and Choe, J. (2011). The novel gene twenty-four defines a critical translational step in the Drosophila clock. *Nature* **470**, 399–403.

Lin, C., Yang, H., Guo, H., Mockler, T., Chen, J., and Cashmore, A. R. (1998). Enhancement of blue-light sensitivity of Arabidopsis seedlings by a blue light receptor cryptochrome 2. *Proc. Natl. Acad. Sci. USA* **95,** 2686–2690.

Lin, F. J., Song, W., Meyer-Bernstein, E., Naidoo, N., and Sehgal, A. (2001). Photic signaling by cryptochrome in the *Drosophila* circadian system. *Mol. Cell. Biol.* **21,** 7287–7294.

Lin, J. M., Kilman, V. L., Keegan, K., Paddock, B., Emery-Le, M., Rosbash, M., and Allada, R. (2002a). A role for casein kinase 2alpha in the *Drosophila* circadian clock. *Nature* **420,** 816–820.

Lin, Y., Han, M., Shimada, B., Wang, L., Gibler, T. M., Amarakone, A., Awad, T. A., Stormo, G. D., Van Gelder, R. N., and Taghert, P. H. (2002b). Influence of the *period*-dependent circadian clock on diurnal, circadian, and aperiodic gene expression in *Drosophila melanogaster. Proc. Natl. Acad. Sci. USA* **99,** 9562–9567.

Lin, J. M., Schroeder, A., and Allada, R. (2005). In vivo circadian function of casein kinase 2 phosphorylation sites in Drosophila PERIOD. *J. Neurosci.* **25,** 11175–11183.

Low, K. H., Lim, C., Ko, H. W., and Edery, I. (2008). Natural variation in the splice site strength of a clock gene and species-specific thermal adaptation. *Neuron* **60,** 1054–1067.

Majercak, J., Sidote, D., Hardin, P. E., and Edery, I. (1999). How a circadian clock adapts to seasonal decreases in temperature and day length. *Neuron* **24,** 219–230.

Majercak, J., Chen, W. F., and Edery, I. (2004). Splicing of the period gene 3′-terminal intron is regulated by light, circadian clock factors, and phospholipase C. *Mol. Cell. Biol.* **24,** 3359–3372.

Martinek, S., Inonog, S., Manoukian, A. S., and Young, M. W. (2001). A role for the segment polarity gene *shaggy*/GSK-3 in the *Drosophila* circadian clock. *Cell* **105,** 769–779.

Matsumoto, A., Ukai-Tadenuma, M., Yamada, R. G., Houl, J., Uno, K. D., Kasukawa, T., Dauwalder, B., Itoh, T. Q., Takahashi, K., Ueda, R., Hardin, P. E., Tanimura, T., *et al.* (2007). A functional genomics strategy reveals clockwork orange as a transcriptional regulator in the Drosophila circadian clock. *Genes Dev.* **21,** 1687–1700.

McDonald, M. J., and Rosbash, M. (2001). Microarray analysis and organization of circadian gene expression in *Drosophila. Cell* **107,** 567–578.

Menet, J. S., Abruzzi, K. C., Desrochers, J., Rodriguez, J., and Rosbash, M. (2010). Dynamic PER repression mechanisms in the Drosophila circadian clock: From on-DNA to off-DNA. *Genes Dev.* **24,** 358–367.

Murad, A., Emery-Le, M., and Emery, P. (2007). A subset of dorsal neurons modulates circadian behavior and light responses in Drosophila. *Neuron* **53,** 689–701.

Myers, M. P., Wager-Smith, K., Rothenfluh-Hilfiker, A., and Young, M. W. (1996). Light-induced degradation of TIMELESS and entrainment of the *Drosophila* circadian clock. *Science* **271,** 1736–1740.

Naidoo, N., Song, W., Hunter-Ensor, M., and Sehgal, A. (1999). A role for the proteasome in the light response of the *timeless* clock protein. *Science* **285,** 1737–1741.

Nakajima, M., Imai, K., Ito, H., Nishiwaki, T., Murayama, Y., Iwasaki, H., Oyama, T., and Kondo, T. (2005). Reconstitution of circadian oscillation of cyanobacterial KaiC phosphorylation in vitro. *Science* **308,** 414–415.

Nambu, J. R., Lewis, J. O., Wharton, K. A., Jr., and Crews, S. T. (1991). The Drosophila single-minded gene encodes a helix-loop-helix protein that acts as a master regulator of CNS midline development. *Cell* **67,** 1157–1167.

Nawathean, P., and Rosbash, M. (2004). The doubletime and CKII kinases collaborate to potentiate *Drosophila* PER transcriptional repressor activity. *Mol. Cell* **13,** 213–223.

O'Neill, J. S., and Reddy, A. B. (2011). Circadian clocks in human red blood cells. *Nature* **469,** 498–503.

O'Neill, J. S., van Ooijen, G., Dixon, L. E., Troein, C., Corellou, F., Bouget, F. Y., Reddy, A. B., and Millar, A. J. (2011). Circadian rhythms persist without transcription in a eukaryote. *Nature* **469,** 554–558.

Ohata, K., Nishiyama, H., and Tsukahara, Y. (1998). Action spectrum of the circadian clock photoreceptor in Drosophila melanogaster. *In* "Biological Clocks: Mechanisms and Applications" (Y. Touitou, ed.), pp. 167–171. Elsevier, Amsterdam.

Pearn, M. T., Randall, L. L., Shortridge, R. D., Burg, M. G., and Pak, W. L. (1996). Molecular, biochemical, and electrophysiological characterization of Drosophila norpA mutants. *J. Biol. Chem.* **271,** 4937–4945.

Peschel, N., Veleri, S., and Stanewsky, R. (2006). Veela defines a molecular link between Crypto-chrome and Timeless in the light-input pathway to Drosophila's circadian clock. *Proc. Natl. Acad. Sci. USA* **103,** 17313–17318.

Peschel, N., Chen, K. F., Szabo, G., and Stanewsky, R. (2009). Light-dependent interactions between the Drosophila circadian clock factors cryptochrome, jetlag, and timeless. *Curr. Biol.* **19,** 241–247.

Pittendrigh, C. S., and Minis, D. H. (1964). The entrainment of circadian oscillations by light and their role as photoperiodic clocks. *Am. Nat.* **98,** 261–294.

Plautz, J. D., Kaneko, M., Hall, J. C., and Kay, S. A. (1997). Independent photoreceptive circadian clocks throughout *Drosophila. Science* **278,** 1632–1635.

Preitner, N., Damiola, F., Lopez-Molina, L., Zakany, J., Duboule, D., Albrecht, U., and Schibler, U. (2002). The orphan nuclear receptor REV-ERBalpha controls circadian transcription within the positive limb of the mammalian circadian oscillator. *Cell* **110,** 251–260.

Price, J. L., Dembinska, M. E., Young, M. W., and Rosbash, M. (1995). Suppression of PERIOD protein abundance and circadian cycling by the *Drosophila* clock mutation *timeless. EMBO J.* **14,** 4044–4049.

Price, J. L., Blau, J., Rothenfluh, A., Abodeely, M., Kloss, B., and Young, M. W. (1998). *double-time* is a novel *Drosophila* clock gene that regulates PERIOD protein accumulation. *Cell* **94,** 83–95.

Reddy, P., Zehring, W. A., Wheeler, D. A., Pirrotta, V., Hadfield, C., Hall, J. C., and Rosbash, M. (1984). Molecular analysis of the period locus in Drosophila melanogaster and identification of a transcript involved in biological rhythms. *Cell* **38,** 701–710.

Renn, S. C., Park, J. H., Rosbash, M., Hall, J. C., and Taghert, P. H. (1999). A *pdf* neuropeptide gene mutation and ablation of PDF neurons each cause severe abnormalities of behavioral circadian rhythms in *Drosophila. Cell* **99,** 791–802.

Richier, B., Michard-Vanhee, C., Lamouroux, A., Papin, C., and Rouyer, F. (2008). The clockwork orange Drosophila protein functions as both an activator and a repressor of clock gene expression. *J. Biol. Rhythms* **23,** 103–116.

Rutila, J. E., Edery, I., Hall, J. C., and Rosbash, M. (1992). The analysis of new short-period circadian rhythm mutants suggests features of D. melanogaster period gene function. *J. Neurogenet.* **8,** 101–113.

Rutila, J. E., Suri, V., Le, M., So, W. V., Rosbash, M., and Hall, J. C. (1998). CYCLE is a second bHLH-PAS clock protein essential for circadian rhythmicity and transcription of *Drosophila period* and *timeless. Cell* **93,** 805–814.

Sathyanarayanan, S., Zheng, X., Xiao, R., and Sehgal, A. (2004). Posttranslational regulation of Drosophila PERIOD protein by protein phosphatase 2A. *Cell* **116,** 603–615.

Sathyanarayanan, S., Zheng, X., Kumar, S., Chen, C. H., Chen, D., Hay, B., and Sehgal, A. (2008). Identification of novel genes involved in light-dependent CRY degradation through a genome-wide RNAi screen. *Genes Dev.* **22,** 1522–1533.

Saunders, D. S., Gillanders, L., and Lewis, R. D. (1994). Light-pulse locomotor phase response curves for the activity rhythm in period mutants of Drosophila melanogaster. *J. Insect Physiol.* **40,** 957–968.

Sawyer, L. A., Hennessy, J. M., Peixoto, A. A., Rosato, E., Parkinson, H., Costa, R., and Kyriacou, C. P. (1997). Natural variation in a Drosophila clock gene and temperature compensa-tion. *Science* **278,** 2117–2120.

Sehgal, A., Price, J. L., Man, B., and Young, M. W. (1994). Loss of circadian behavioral rhythms and per RNA oscillations in the Drosophila mutant timeless. *Science* **263,** 1603–1606.

Sehgal, A., Rothenfluh-Hilfiker, A., Hunter-Ensor, M., Chen, Y., Myers, M. P., and Young, M. W. (1995). Rhythmic expression of *timeless*: A basis for promoting circadian cycles in *period* gene autoregulation. *Science* **270,** 808–810.

Shang, Y., Griffith, L. C., and Rosbash, M. (2008). Light-arousal and circadian photoreception circuits intersect at the large PDF cells of the Drosophila brain. *Proc. Natl. Acad. Sci. USA* **105,** 19587–19594.

Sheeba, V., Fogle, K. J., Kaneko, M., Rashid, S., Chou, Y. T., Sharma, V. K., and Holmes, T. C. (2008). Large ventral lateral neurons modulate arousal and sleep in Drosophila. *Curr. Biol.* **18,** 1537–1545.

Sheeba, V., Fogle, K. J., and Holmes, T. C. (2010). Persistence of morning anticipation behavior and high amplitude morning startle response following functional loss of small ventral lateral neurons in Drosophila. *PLoS One* **5,** e11628.

Siwicki, K. K., Eastman, C., Petersen, G., Rosbash, M., and Hall, J. C. (1988). Antibodies to the period gene product of Drosophila reveal diverse tissue distribution and rhythmic changes in the visual system. *Neuron* **1,** 141–150.

Smith, R. M., and Williams, S. B. (2006). Circadian rhythms in gene transcription imparted by chromosome compaction in the cyanobacterium Synechococcus elongatus. *Proc. Natl. Acad. Sci. USA* **103,** 8564–8569.

Spengler, M. L., Kuropatwinski, K. K., Schumer, M., and Antoch, M. P. (2009). A serine cluster mediates BMAL1-dependent CLOCK phosphorylation and degradation. *Cell Cycle* **8,** 4138–4146.

Stanewsky, R., Kaneko, M., Emery, P., Beretta, B., Wager-Smith, K., Kay, S. A., Rosbash, M., and Hall, J. C. (1998). The cry^b mutation identifies cryptochrome as a circadian photoreceptor in *Drosophila*. *Cell* **95,** 681–692.

Stoleru, D., Peng, Y., Agosto, J., and Rosbash, M. (2004). Coupled oscillators control morning and evening locomotor behaviour of Drosophila. *Nature* **431,** 862–868.

Stoleru, D., Nawathean, P., Fernandez Mde, L., Menet, J. S., Ceriani, M. F., and Rosbash, M. (2007). The Drosophila circadian network is a seasonal timer. *Cell* **129,** 207–219.

Tang, C. H., Hinteregger, E., Shang, Y., and Rosbash, M. (2010). Light-mediated TIM degradation within Drosophila pacemaker neurons (s-LNvs) is neither necessary nor sufficient for delay zone phase shifts. *Neuron* **66,** 378–385.

Tanoue, S., Krishnan, P., Krishnan, B., Dryer, S. E., and Hardin, P. E. (2004). Circadian clocks in antennal neurons are necessary and sufficient for olfaction rhythms in Drosophila. *Curr. Biol.* **14,** 638–649.

Tanoue, S., Krishnan, P., Chatterjee, A., and Hardin, P. E. (2008). G protein-coupled receptor kinase 2 is required for rhythmic olfactory responses in Drosophila. *Curr. Biol.* **18,** 787–794.

Ueda, H. R., Matsumoto, A., Kawamura, M., Iino, M., Tanimura, T., and Hashimoto, S. (2002). Genome-wide transcriptional orchestration of circadian rhythms in *Drosophila*. *J. Biol. Chem.* **277,** 14048–14052.

Vitaterna, M. H., King, D. P., Chang, A. M., Kornhauser, J. M., Lowrey, P. L., McDonald, J. D., Dove, W. F., Pinto, L. H., Turek, F. W., and Takahashi, J. S. (1994). Mutagenesis and mapping of a mouse gene, Clock, essential for circadian behavior. *Science* **264,** 719–725.

Wijnen, H., Naef, F., Boothroyd, C., Claridge-Chang, A., and Young, M. W. (2006). Control of daily transcript oscillations in Drosophila by light and the circadian clock. *PLoS Genet.* **2,** e39.

Woelfle, M. A., and Johnson, C. H. (2006). No promoter left behind: Global circadian gene expression in cyanobacteria. *J. Biol. Rhythms* **21,** 419–431.

Woelfle, M. A., Xu, Y., Qin, X., and Johnson, C. H. (2007). Circadian rhythms of superhelical status of DNA in cyanobacteria. *Proc. Natl. Acad. Sci. USA* **104,** 18819–18824.

Yang, M., Lee, J. E., Padgett, R. W., and Edery, I. (2008). Circadian regulation of a limited set of conserved microRNAs in Drosophila. *BMC Genomics* **9,** 83.

Yoshii, T., Todo, T., Wulbeck, C., Stanewsky, R., and Helfrich-Forster, C. (2008). Cryptochrome operates in the compound eyes and a subset of Drosophila's clock neurons. *J. Comp. Neurol.* **508,** 952–966.

Yoshitane, H., Takao, T., Satomi, Y., Du, N. H., Okano, T., and Fukada, Y. (2009). Roles of CLOCK phosphorylation in suppression of E-box-dependent transcription. *Mol. Cell. Biol.* **29,** 3675–3686.

Young, M. W., and Kay, S. A. (2001). Time zones: A comparative genetics of circadian clocks. *Nat. Rev. Genet.* **2,** 702–715.

Yu, W., and Hardin, P. E. (2006). Circadian oscillators of Drosophila and mammals. *J. Cell Sci.* **119,** 4793–4795.

Yu, Q., Jacquier, A. C., Citri, Y., Hamblen, M., Hall, J. C., and Rosbash, M. (1987). Molecular mapping of point mutations in the period gene that stop or speed up biological clocks in Drosophila melanogaster. *Proc. Natl. Acad. Sci. USA* **84,** 784–788.

Yu, W., Zheng, H., Houl, J. H., Dauwalder, B., and Hardin, P. E. (2006). PER-dependent rhythms in CLK phosphorylation and E-box binding regulate circadian transcription. *Genes Dev.* **20,** 723–733.

Yu, W., Zheng, H., Price, J. L., and Hardin, P. E. (2009). DOUBLETIME plays a noncatalytic role to mediate CLOCK phosphorylation and repress CLOCK-dependent transcription within the Drosophila circadian clock. *Mol. Cell. Biol.* **29,** 1452–1458.

Yu, W., Houl, J. H., and Hardin, P. E. (2011). NEMO kinase contributes to core period determination by slowing the pace of the Drosophila circadian oscillator. *Curr. Biol.* **21,** 756–761.

Yuan, Q., Lin, F., Zheng, X., and Sehgal, A. (2005). Serotonin modulates circadian entrainment in Drosophila. *Neuron* **47,** 115–127.

Yuan, Q., Metterville, D., Briscoe, A. D., and Reppert, S. M. (2007). Insect cryptochromes: Gene duplication and loss define diverse ways to construct insect circadian clocks. *Mol. Biol. Evol.* **24,** 948–955.

Zehring, W. A., Wheeler, D. A., Reddy, P., Konopka, R. J., Kyriacou, C. P., Rosbash, M., and Hall, J. C. (1984). P-element transformation with period locus DNA restores rhythmicity to mutant, arrhythmic Drosophila melanogaster. *Cell* **39,** 369–376.

Zeng, H., Hardin, P. E., and Rosbash, M. (1994). Constitutive overexpression of the Drosophila period protein inhibits period mRNA cycling. *EMBO J.* **13,** 3590–3598.

Zeng, H., Qian, Z., Myers, M. P., and Rosbash, M. (1996). A light-entrainment mechanism for the Drosophila circadian clock. *Nature* **380,** 129–135.

Zhang, L., Lear, B. C., Seluzicki, A., and Allada, R. (2009). The CRYPTOCHROME photoreceptor gates PDF neuropeptide signaling to set circadian network hierarchy in Drosophila. *Curr. Biol.* **19,** 2050–2055.

Zhang, L., Chung, B. Y., Lear, B. C., Kilman, V. L., Liu, Y., Mahesh, G., Meissner, R. A., Hardin, P. E., and Allada, R. (2010a). DN1(p) circadian neurons coordinate acute light and PDF inputs to produce robust daily behavior in Drosophila. *Curr. Biol.* **20,** 591–599.

Zhang, Y., Liu, Y., Bilodeau-Wentworth, D., Hardin, P. E., and Emery, P. (2010b). Light and temperature control the contribution of specific DN1 neurons to Drosophila circadian behavior. *Curr. Biol.* **20,** 600–605.

Zheng, X., and Sehgal, A. (2008). Probing the relative importance of molecular oscillations in the circadian clock. *Genetics* **178,** 1147–1155.

Zheng, X., Koh, K., Sowcik, M., Smith, C. J., Chen, D., Wu, M. N., and Sehgal, A. (2009). An isoform-specific mutant reveals a role of PDP1 epsilon in the circadian oscillator. *J. Neurosci.* **29,** 10920–10927.

Zhu, H., Sauman, I., Yuan, Q., Casselman, A., Emery-Le, M., Emery, P., and Reppert, S. M. (2008). Cryptochromes define a novel circadian clock mechanism in monarch butterflies that may underlie sun compass navigation. *PLoS Biol.* **6,** e4.

6

Genetics of Circadian Rhythms in Mammalian Model Organisms

Phillip L. Lowrey* and Joseph S. Takahashi[†,1]
*Department of Biology, Rider University, Lawrenceville, New Jersey, USA
†Department of Neuroscience, Howard Hughes Medical Institute,
University of Texas Southwestern Medical Center, Dallas, Texas, USA
[1]Corresponding author

Advances in Genetics, Vol. 74
Copyright 2011, Elsevier Inc. All rights reserved.

0065-2660/11 $35.00
DOI: 10.1016/B978-0-12-387690-4.00006-4

ABSTRACT

The mammalian circadian system is a complex hierarchical temporal network which is organized around an ensemble of uniquely coupled cells comprising the principal circadian pacemaker in the suprachiasmatic nucleus of the hypothalamus. This central pacemaker is entrained each day by the environmental light/dark cycle and transmits synchronizing cues to cell-autonomous oscillators in tissues throughout the body. Within cells of the central pacemaker and the peripheral tissues, the underlying molecular mechanism by which oscillations in gene expression occur involves interconnected feedback loops of transcription and translation. Over the past 10 years, we have learned much regarding the genetics of this system, including how it is particularly resilient when challenged by single-gene mutations, how accessory transcriptional loops enhance the robustness of oscillations, how epigenetic mechanisms contribute to the control of circadian gene expression, and how, from coupled neuronal networks, emergent clock properties arise. Here, we will explore the genetics of the mammalian circadian system from cell-autonomous molecular oscillations, to interactions among central and peripheral oscillators and ultimately, to the daily rhythms of behavior observed in the animal. © 2011, Elsevier Inc.

I. INTRODUCTION

The rising and setting of the sun each day causes predictable environmental changes to which most organisms on earth have adapted by evolving endogenous biological timing systems with a period of approximately 24 h (Young and Kay, 2001). These circadian ($\sim$24 h) clocks anticipate environmental cycles and control daily rhythms in biochemistry, physiology, and behavior. Across phyla, all circadian clocks share several fundamental properties: they are synchronized (entrained) each day to external cues, they are self-sustained and produce oscillations that persist in the absence of any external cues, they are temperature compensated such that temperature changes in the physiological range do not alter their endogenous period, and of particular relevance to this review, they are cell-autonomous and genetically determined. In all of the major model organisms in which circadian rhythms have been studied, there has emerged a central organizing principle of the molecular clockwork: within cells, a set of clock genes and their protein products together participate in autoregulatory feedback loops of transcription and translation to produce an oscillation with a period of about 24 h (Lowrey and Takahashi, 2004; Takahashi et al., 2008).

Recent work, however, has prompted a reappraisal of the transcription/translation model as the sole generative mechanism of the molecular circadian oscillator in mammals. For example, it is now clear that oscillations of some mammalian core clock components are dispensable for circadian function (Fan et al., 2007; Liu et al., 2008), and there is some evidence, albeit preliminary, for circadian rhythms in the absence of transcription in some mammalian cells (O'Neill and Reddy, 2011). Perhaps more importantly, however, limitations of the conventional perturbation analysis methods that helped elucidate the transcription/translation model have become apparent. No longer is it sufficient to knock out a clock gene in a mouse and then assess the consequences on behavior (locomotor activity) or gene expression (changes in RNA and protein levels in cells) alone. We now appreciate that the mammalian circadian clock is a more complex hierarchical system than originally imagined, and thus understanding it requires analysis at many levels.

New technologies and clock models have revealed higher-order genetic properties of the mammalian clock system in which the elimination of one component may be compensated for by other components in ways that are more complex than simple redundancy, and they have demonstrated the important roles of accessory feedback loops and gene networks in conferring stability and robustness on the system (Baggs et al., 2009; Ueda et al., 2005; Ukai-Tadenuma et al., 2008). Further, novel approaches have elucidated the importance of networks of coupled cells from which emergent circadian clock properties arise and even buffer the system against the effects of mutations (Abraham et al., 2010; Buhr et al., 2010; Ko et al., 2010; Liu et al., 2007b). These, and other advances, are making clearer the fundamental properties of each level of organization of the mammalian circadian system from cell-autonomous molecular oscillations to tissue-specific properties, to the interaction of central and peripheral oscillators, and ultimately, to the overt daily rhythms of behavior observed in the animal.

Here, we present some of the key findings in the field of mammalian circadian biology over the past 10 years and introduce many of the new technologies that are revolutionizing our understanding of the clock system. Our emphasis will be primarily on work from the principal model organism used to study mammalian biology—the mouse. Indeed, for no other mammalian model is there the extensive repertoire of experimental resources and techniques as for the mouse (Adams and van der Weyden, 2008; Blake et al., 2010; Fox et al., 2007; Hedrich and Bullock, 2004; Nagy et al., 2003; Silver, 1995). We will not, however, explore in depth the intriguing link between the mammalian circadian clock and metabolism, first proposed by McKnight and colleagues a decade ago (Rutter et al., 2002), and now well established, as it is beyond the scope of this review. Instead, we refer the reader to several recent comprehensive treatments of this specific topic (Asher and Schibler, 2011; Bass and Takahashi, 2010; Green et al., 2008; Maury et al., 2010).

II. THE BEGINNING OF MAMMALIAN CLOCK GENETICS

A. Serendipitous discovery of the Syrian hamster *tau* mutant

Before discussing the current state of mammalian clock genetics and the details of the molecular clockwork in mammals, we would first like to reflect back briefly on the period from approximately 1985 to 2000 when the study of mammalian clock genetics began. Indeed, it was in 1985 that Martin Ralph, at the time a graduate student in the laboratory of Michael Menaker (then at the University of Oregon), identified a single outbred Syrian hamster (*Mesocricetus auratus*) with an unusually early onset of locomotor activity. Following transfer from a light/dark (LD) cycle to constant darkness (DD), this animal exhibited an endogenous free-running period of 22 h compared to 23.5 h, the shortest previously reported circadian period for this species. Recognizing the implications of possibly discovering the first mammalian circadian mutant, Ralph had the foresight to cross this animal to wild-type hamsters and analyze the behavioral rhythms in the offspring. The free-running periods for the resulting F_1 progeny (1:1 ratio; 22 h:24 h) confirmed that the aberrant phenotype was heritable. Intercrosses produced an F_2 generation with a 1:2:1 Mendelian ratio of 20 h:22 h:24 h periods. Thus, this spontaneous mutation designated *tau* (after the circadian symbol for period length), segregated in a semidominant manner, and seemed to involve a single autosomal locus (Table 6.1; Ralph and Menaker, 1988).

As the first mammalian circadian mutation, *tau* figured prominently in many studies addressing behavioral and physiological aspects of mammalian circadian biology. Perhaps the most important result obtained from the *tau* model was the definitive demonstration through transplantation experiments that the suprachiasmatic nucleus (SCN) of the hypothalamus harbors the central circadian pacemaker in mammals. Adult hamsters rendered behaviorally arrhythmic by SCN lesioning exhibited restored rhythmicity following transplantation of donor SCN tissue into the third ventricle. Not only was host rhythmicity rescued but also the restored rhythms had periods reflecting the genotype of the donor animal (Ralph et al., 1990). Further, when firing rate rhythms of individual SCN neurons on fixed microelectrode plates were recorded, cells from *tau* animals helped show that the circadian period of the whole tissue/animal is determined by averaging widely dispersed periods of

Table 6.1. Behavioral Phenotypes of Mutations in Mouse Clock and Clock-Related Genes

Gene(s)	Protein product(s)	Mutant allele(s)	Mutant phenotype(s)	References
Clock	bHLH-PAS transcription factor	Clock$^{Δ19/Δ19}$	4-h longer period/ arrhythmic	Vitaterna et al. (1994)
Npas2 (Mop4)	bHLH-PAS transcription factor	Clock$^{-/-}$ Npas2$^{-/-}$	0.4-h shorter period 0.2-h shorter period	Debruyne et al. (2006) Dudley et al. (2003)
Clock/Npas2	bHLH-PAS transcription factors	Clock$^{-/-}$/Npas2$^{-/-}$	Arrhythmic	DeBruyne et al. (2007a)
Bmal1 (Arntl, Mop3)	bHLH-PAS transcription factor	Bmal1$^{-/-}$	Arrhythmic	Bunger et al. (2000)
Cry1	Flavoprotein	Cry1$^{-/-}$	1-h shorter period	van der Horst et al. (1999), Vitaterna et al. (1999)
Cry2	Flavoprotein	Cry2$^{-/-}$	1-h longer period	Thresher et al. (1998), van der Horst et al. (1999)
Cry1/Cry2	Flavoproteins	Cry1$^{-/-}$/Cry2$^{-/-}$	Arrhythmic	van der Horst et al. (1999), Vitaterna et al. (1999)
Per1	PAS protein	Per1$^{-/-}$ Per1^{brdm1} Per1^{ldc}	0.7-h shorter period 1-h shorter period 0.5-h shorter period/ arrhythmic	Cermakian et al., (2001) Zheng et al. (2001) Bae et al. (2001)
Per2	PAS protein	Per2^{brdm1}	1.5-h shorter period/ arrhythmic	Zheng et al. (1999)
Per3	PAS protein	Per2^{ldc} Per3$^{-/-}$	Arrhythmic 0–0.5-h shorter period	Bae et al. (2001) Shearman et al. (2000a)
Per1/Per2	PAS proteins	Per1^{brdm1}/Per2^{brdm1} Per1^{ldc}/Per2^{ldc}	Arrhythmic Arrhythmic	Zheng et al. (2001) Bae et al. (2001)
Per1/Cry1	PAS protein/ flavoprotein	Per1^{brdm1}/Cry1$^{-/-}$	Normal behavior	Oster et al. (2003)

(Continues)

Table 6.1. (*Continued*)

Gene(s)	Protein product(s)	Mutant allele(s)	Mutant phenotype(s)	References
Per1/Cry2	PAS protein/flavoprotein	Per1bmal1/Cry2−/−	<6 months, 1.5-h longer period; >6 months, arrhythmic	Oster et al. (2003)
Per2/Cry1	PAS protein/flavoprotein	Per2bmal1/Cry1−/−	Arrhythmic	Oster et al. (2002)
Per2/Cry2	PAS protein/flavoprotein	Per2bmal1/Cry2−/−	0–0.4-h shorter period	Oster et al. (2002)
Rev-erbα (Nr1d1)	Nuclear receptor	Rev-erbα−/−	0.5-h shorter period; disrupted entrainment	Preitner et al. (2002)
Rev-erbβ (Nr1d2)	Nuclear receptor	–	–	–
Rorα (Rora)	Nuclear receptor	Rorα−/− (staggerer)	0.5-h shorter period; disrupted entrainment	Sato et al. (2004)
Rorβ (Rorb)	Nuclear receptor	Rorβ−/−	0.5-h longer period	Masana et al. (2007)
Rorγ (Rorc)	Nuclear receptor	Rorγ−/−	Normal behavior	Liu et al. (2008)
Dec1 (Bhlhe40, Stra13, Sharp-2)	bHLH transcription factor	Stra13−/−	0.15-h longer period	Nakashima et al. (2008)
Dec2 (Bhlhe41, Sharp-1)	bHLH transcription factor	Sharp-1−/−	Delayed resetting	Rossner et al. (2008)
CK1δ (Csnk1d)	Casein kinase 1	CK1δ+/−	0–0.5-h longer period	Xu et al. (2005), Etchegaray et al. (2009)
CK1ε (Csnk1e)	Casein kinase 1	CK1ε−/−	0.2–0.4-h longer period	Meng et al. (2008), Etchegaray et al. (2009)
		aCK1εtau	4-h shorter period	Lowrey et al. (2000), Meng et al. (2008)
CK1α (Csnk1a1)	Casein kinase 1	–	–	–
Fbxl3	F-box protein	Fbxl3Ovtm	2-h longer period	Siepka et al. (2007)
		Fbxl3Afh	3 hr longer period	Godinho et al. (2007)

Gene symbol	Description	Mutant	Phenotype	Reference
Bmal2 (Amtl2, Mop9, Clif)	bHLH-PAS transcription factor	—	—	—
Pgc1a (Ppargc1a)	Transcriptional coactivator	Pgc1a⁻/⁻	0.3-h longer period	Liu et al. (2007c)
Mtnr1a (Mel1a)	G protein-coupled receptor	Mtnr1a⁻/⁻	Normal behavior	Liu et al. (1997a)
Mtnr1b (Mel1b)	G protein-coupled receptor	Mtnr1b⁻/⁻	Normal behavior	Jin et al. (2003)
Opn4	Melanopsin, opsin 4	Opn4⁻/⁻	Attenuated photic responses	Panda et al. (2002b), Ruby et al. (2002)
Dbp	PAR bZIP transcription factor	Dbp⁻/⁻	0.5-h shorter period	Lopez-Molina et al. (1997)
Vipr2	G protein-coupled receptor	Vipr2⁻/⁻	Disrupted locomotor rhythm	Harmar et al. (2002), Cutler et al. (2003)
Vip	Peptide hormone	Vip⁻/⁻	1-h shorter period/arrhythmic	Colwell et al. (2003)
Prok2 (PK2)	Secreted protein	Prok2⁻/⁻	Reduced locomotor activity	Li et al. (2006)
Nocturnin (Ccrn4l)	Deadenylase	Noc⁻/⁻	Normal behavior	Green et al. (2007)

Gene symbols listed here are the predominate forms used in the scientific literature; alternate forms are given in parentheses. When the predominant symbol differs from standard mouse gene nomenclature, the standard form is given in parentheses. For genes with more than two symbols in parentheses, the form adhering to standard mouse gene nomenclature is underlined.

[a]First identified as a mutation in Syrian hamster (*Mesocricetus auratus*).

individual SCN clock cells. This was the first demonstration that the *tau* mutation affects circadian function in a cell-autonomous manner (Liu *et al.*, 1997b). Two additional seminal findings using the *tau* model include the first report that a diffusible signal can drive circadian rhythms in a mammal (Silver *et al.*, 1996), and that SCN-independent circadian oscillators reside in the mammalian retina (Tosini and Menaker, 1996, 1998).

Despite the importance to the field of mammalian circadian biology of the *tau* model, from a genetic standpoint, it is unfortunate that the mutation occurred in the Syrian hamster rather than in the mouse, a mammalian model for which even in 1985 more comprehensive genetic resources were available. Efforts to develop dense genetic maps and physical mapping reagents for the mouse were well underway at the inception of the Human Genome Project and its inclusion of the mouse as a model genome sequencing project (Muller and Grossniklaus, 2010). The Syrian hamster, however, became one of many "orphan genomes" not included in the publically funded mapping effort (Jacob, 1996). And so the quest to clone and characterize the *tau* mutation would not come to fruition until 12 years after Ralph and Menaker's 1988 report of the mutant. Our laboratory, using a comparative genomics approach called positional syntenic cloning, demonstrated that the *tau* mutation results from a single nucleotide change in the gene encoding casein kinase I epsilon (*CK1ε*; Lowrey *et al.*, 2000).

B. Forward genetics and the *Clock* mutation

With the realization that the *tau* mutant hamster, while advantageous for physiological studies, was not immediately a genetically tractable model, our laboratory, in collaboration with William F. Dove, Lawrence H. Pinto, and Fred W. Turek, initiated a dominant circadian behavioral screen of first generation offspring of male C57BL/6J mice mutagenized with the alkylating agent *N*-ethyl-*N*-nitrosourea (ENU). We adopted this forward genetics approach in mice encouraged by the successful mutagenesis screens in *Drosophila* for circadian defects some 20 years earlier by Seymour Benzer's group (Konopka and Benzer, 1971). Of the 304 animals tested, we recovered a mouse with a single-gene, semidominant mutation, *Clock* (for circadian locomotor output cycles kaput), that significantly lengthened its free-running period (24.8 h) compared to wild-type C57BL/6J mice (23.3–23.8 h). Homozygous *Clock* mutant animals exhibited an extremely long initial free-running period in DD of 26–29 h, followed by a complete loss of rhythmicity after 2 weeks in DD (Table 6.1). We mapped the mutation to mouse chromosome 5 (Vitaterna *et al.*, 1994) and subsequently identified the gene through a combination of positional cloning and transgenic rescue of the mutant phenotype (Antoch *et al.*, 1997; King *et al.*, 1997b). Sequence analysis revealed that the *Clock* mutation is caused by an A → T transversion in a splice donor site in the intron between exons 18 and 19 of the gene, resulting in a transcript missing exon 19 ($Clock^{\Delta19}$). This deletion disrupts the transactivation domain of the basic helix-loop-helix (bHLH)-PAS (*Period-Arnt-Single-minded*) transcription factor encoded by *Clock*. Additional genetic approaches revealed that *Clock* is an antimorph—a specific type of dominant negative mutation (King *et al.*, 1997a).

Identification of the *Clock* mutation was proof of principle that, as in *Drosophila*, forward genetic screens for behavioral defects in mice are feasible (Bacon *et al.*, 2004; Clark *et al.*, 2004; Siepka and Takahashi, 2005; Takahashi *et al.*, 1994). Identifying mammalian clock genes by recovering mutants was, however, not the only approach during this "birth" of mammalian clock genetics in the 1990s. Several of what proved to be mammalian core clock genes were cloned by homology to known genes in other organisms, or by identification of paralogs in the same organism. These include three mouse orthologs of the *Drosophila period* gene (*Per1*, *Per2*, and *Per3*), two mouse *Cryptochrome* orthologs (*Cry1* and *Cry2*), and brain and muscle ARNT-like protein 1 (*Bmal1* or *Mop3*), another bHLH-PAS protein, all reviewed comprehensively elsewhere (Lowrey and Takahashi, 2000, 2004; Reppert and Weaver, 2002; Young and Kay, 2001). Indeed, the 1990s witnessed the description of a molecular model of the mammalian core circadian oscillator based on a transcription/translation feedback loop with striking similarity to models proposed for other, phylogenetically divergent organisms, including *Drosophila*, *Arabidopsis*, and *Neurospora* (Dunlap, 1999). This rapid elucidation of the basic mechanism by which mammalian cells keep time was aptly characterized by one colleague as a "clockwork explosion" (Reppert, 1998).

III. OVERVIEW OF THE MAMMALIAN CLOCK SYSTEM

The mammalian circadian system is organized around three major physiological components: an input pathway by which environmental cues (most importantly light) are transmitted to the central or "master" pacemaker, the central pacemaker itself, and finally, a set of output pathways by which the central pacemaker regulates circadian rhythms throughout the body (Lowrey and Takahashi, 2004; Quintero *et al.*, 2003; Takahashi *et al.*, 2001). Light entrainment of the circadian system relies on the eye (Foster *et al.*, 1991; Nelson and Zucker, 1981) where, within the retina, the rods and cones and a recently discovered subset ($\sim 1\%$) of intrinsically photosensitive retinal ganglion cells (ipRGCs) reside (Do and Yau, 2010). The ipRGCs respond to light stimulation independently of the rod–cone system (Berson *et al.*, 2002; Hattar *et al.*, 2002) and are directly photosensitive owing to their expression of the photopigment melanopsin (Dacey *et al.*, 2005; Fu *et al.*, 2005; Gooley *et al.*, 2001; Melyan *et al.*, 2005; Panda *et al.*, 2005; Provencio *et al.*, 1998; Qiu *et al.*, 2005). Mice lacking either the rod–cone system

(Freedman *et al.*, 1999) or melanopsin ($Opn4^{-/-}$) (Lucas *et al.*, 2003; Panda *et al.*, 2002b; Ruby *et al.*, 2002) exhibit normal entrainment to light. Loss of both the rod–cone system and melanopsin, however, renders mice unable to entrain to photic stimuli (Hattar *et al.*, 2003; Panda *et al.*, 2003). Because the ipRGCs, via which all retinal input to the SCN is transmitted, receive synaptic input containing nonvisual information from the rods and cones, selective destruction of the ipRGCs in mice also prevents circadian photoentrainment (Goz *et al.*, 2008; Hatori *et al.*, 2008).

Photic information received by the retina is transmitted via the retino-hypothalamic tract (RHT) which is formed from the axons of the ipRGCs, to the bilaterally paired suprachiasmatic nuclei of the hypothalamus, the location of the central pacemaker in mammals (Gooley *et al.*, 2001; Moore and Eichler, 1972; Moore and Lenn, 1972; Moore *et al.*, 1995; Stephan and Zucker, 1972). Light stimulation of the ipRGCs causes their axon terminals to release glutamate and pituitary adenylate cyclase-activating polypeptide onto postsynaptic SCN neurons (Ebling, 1996; Hannibal, 2002; Hannibal *et al.*, 2004; Michel *et al.*, 2006; Morin and Allen, 2006). Glutamate-induced calcium influx activates several protein kinase pathways in SCN neurons which ultimately lead to phosphorylation of Ca^{2+}/cAMP-response element binding protein (CREB) (Golombek and Rosenstein, 2010). Within the promoters of many core clock genes, reside Ca^{2+}/cAMP-response elements (CREs) to which phospho-CREB homodimers bind to activate transcription (Zhang *et al.*, 2005). Two particularly important CREB clock targets with respect to photic entrainment are the *Per1* and *Per2* genes, both of which contain CREs in their promoters (Travnickova-Bendova *et al.*, 2002) and are rapidly induced in SCN neurons by nocturnal light exposure (Albrecht *et al.*, 1997; Shearman *et al.*, 1997; Shigeyoshi *et al.*, 1997). Circadian rhythms of locomotor activity in mice are phase advanced or phase delayed depending on the time at night during which a photic stimulus occurs (Golombek and Rosenstein, 2010). The resulting increase in PER protein presumably affects the molecular clock in SCN neurons by opposing the action of the positive effectors of the core clock feedback loop discussed later (Yan and Silver, 2004).

Each SCN of the mouse contains approximately 10,000 neurons (Abrahamson and Moore, 2001). When dissociated from SCN tissue (Herzog *et al.*, 1998; Honma *et al.*, 1998; Liu *et al.*, 1997b; Welsh *et al.*, 1995) or when grown as immortalized cells (Earnest *et al.*, 1999), these neurons can independently generate self-sustained circadian rhythms. Intact SCN neurons, however, couple to form a network that expresses synchronized rhythms (Welsh *et al.*, 2010). Ongoing work seeks to clarify the nature of the coupling mechanisms that give rise to the unique SCN network, yet it is clear that neurotransmitters, neuropeptides, gap junctions, and chemical synaptic mechanisms are involved (Welsh *et al.*, 2010). For example, the presence of vasoactive intestinal polypeptide (VIP) and its G protein-coupled receptor, VPAC$_2$, is important for maintaining circadian rhythmicity of gene expression in SCN cells, and for normal expression of rhythmic behavior in mice. Disrupting the genes for VIP ($Vip^{-/-}$) (Colwell et al., 2003) or its receptor VPAC$_2$ ($Vipr2^{-/-}$) (Cutler et al., 2003; Harmar et al., 2002) leads to severely compromised circadian rhythms in behavior, neuronal firing, and gene expression, owing to intercellular desynchronization among SCN neurons (Table 6.1) (Aton et al., 2005; Brown et al., 2007; Hughes et al., 2008; Maywood et al., 2006). In normal mice, circadian rhythms generated within the SCN network are much more robust than those produced by individual neurons (Yamaguchi et al., 2003), even in the presence of clock gene mutations (Liu et al., 2007b). Indeed, recent work has shown that when the autonomous circadian oscillation of individual SCN neurons is eliminated by core clock gene mutation, molecular noise and intercellular coupling are sufficient to elicit stochastic, quasi-circadian oscillations as an emergent property of the SCN network (Ko et al., 2010).

Photic input to the SCN via the RHT is transduced into neural and humoral output signals that synchronize other rhythms in the body, including those of temperature, hormone secretion, and rest/wake (Aston-Jones et al., 2001; Brown et al., 2002; Buijs and Kalsbeek, 2001). Synchronizing signals reach peripheral tissues by both autonomic neural connections (Buijs and Kalsbeek, 2001; Vujovic et al., 2008) and the release of hormones such as glucocorticoids (Balsalobre et al., 2000; Le Minh et al., 2001). In the absence of the SCN, circadian rhythms in most peripheral tissues damp out after a few days from desynchronization among the cells in the tissue, yet at the single-cell level, circadian rhythms persist (Balsalobre et al., 1998; Nagoshi et al., 2004; Welsh et al., 2004; Yamazaki et al., 2000; Yoo et al., 2004). The circadian rhythm of locomotor activity commonly monitored in mice and other rodents to determine endogenous circadian period relies on diffusible signals released from the SCN (Silver et al., 1996), several candidate molecules for which have been identified including transforming growth factor α (Kramer et al., 2001, 2005), cardiotrophin-like cytokine (Kraves and Weitz, 2006), prokineticin 2 (PK2) (Cheng et al., 2002; Li et al., 2006), and potentially others (Hatcher et al., 2008).

It is important to note that there are oscillators in some mammalian brain regions and tissues that, in the absence of the SCN, can drive local physiological rhythms. Two such well-characterized regions include the retina (Tosini and Menaker, 1996) and the olfactory bulb (Granados-Fuentes et al., 2004a,b, 2006). Further, two extra-SCN pacemakers, the food-entrainable oscillator and the methamphetamine-sensitive circadian oscillator, can drive circadian behavioral and endocrine rhythms in the absence of the SCN or functional canonical clock genes (Honma and Honma, 2009; Honma et al., 2008; Mohawk et al., 2009; Pezuk et al., 2010; Storch and Weitz, 2009).

IV. THE MAMMALIAN CIRCADIAN MOLECULAR OSCILLATOR

The mammalian circadian molecular oscillator model proposed following the discovery of the core clock genes described earlier encompasses our current understanding of the circadian control of gene expression in cells throughout the body. Core circadian clock genes are genes whose protein products are necessary components for the generation and regulation of circadian rhythms, that is, proteins which form the primary molecular circadian oscillatory mechanism within individual cells. In this model, positive and negative core clock components form a feedback loop with a time constant of about 24 h per cycle (Fig. 6.1). This loop begins during the day when two bHLH-PAS transcription factors, CLOCK and BMAL1, heterodimerize and initiate transcription from genes containing E-box (5'-CACGTG-3') or E'-box (5'-CACGTT-3') cis-regulatory elements, including the Per and Cry genes (Bunger et al., 2000; Gekakis et al., 1998; King et al., 1997b; Kume et al., 1999; Yoo et al., 2005; Zheng et al., 2001). PER and CRY proteins heterodimerize and, along with other proteins such as CK1ε, form a complex in the cytoplasm that translocates to the nucleus where they accumulate and subsequently repress transcription of their own (and other) genes by directly inhibiting CLOCK/BMAL1 (Griffin et al., 1999; Kume et al., 1999; Lee et al., 2001; Sato et al., 2006). Thus, the CLOCK/BMAL1 heterodimer forms the positive, or transactivating component in this loop, while the PER/CRY complex acts as the negative, or transinhibiting component (Fig. 6.1). Following several posttranscriptional and posttranslational steps discussed later, the PER/CRY complex is targeted for degradation via the proteasomal pathway, thereby relieving inhibition such that CLOCK/BMAL1 can initiate a new cycle of transcription. This relatively straightforward feedback loop forms what has become known as the mammalian "core" oscillator mechanism.

The general molecular mechanism just described governs circadian output rhythms in all cells throughout the body, although there are tissue-specific differences. For example, in the forebrain, neuronal PAS domain protein 2 (NPAS2) appears to be the more relevant BMAL1 partner (Reick et al., 2001). Thus, the CLOCK(NPAS2)/BMAL1 complex initiates the rhythmic transcription of clock-controlled genes in tissues throughout the body. Microarray studies have shown that approximately 10–15% of all mammalian transcripts exhibit a circadian oscillation from one cell type/tissue to another (Akhtar et al., 2002; Duffield et al., 2002; Miller et al., 2007; Oishi et al., 2003; Panda et al., 2002a; Storch et al., 2002). These studies, however, may underrepresent the true number of genes under circadian control. Powerful statistical tests are required to identify cycling transcripts from noisy microarray data sets (Doherty and Kay, 2010). Development of new nonparametric statistical algorithms promises to provide more accurate measurements of period, phase, and amplitude than traditional analysis methods (Hughes et al., 2010). Hence, continued   improvements   in   data   analysis   methods   should   allow   the

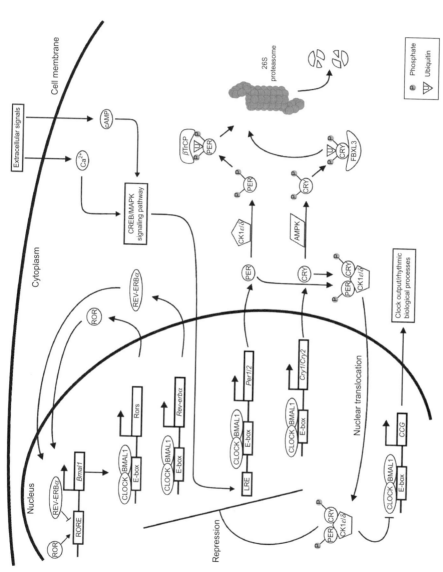

Figure 6.1. Model of the mammalian cell-autonomous oscillator as described in the text. *Abbreviations:* CCG, clock-controlled gene; P, phosphate; U, ubiquitin.

identification of rhythmic transcripts in noisy, low-amplitude data and provide a more precise estimate of the number of genes under circadian control in various mammalian tissues.

Following the discovery of several other *bona fide* clock genes, it soon became evident that accessory regulatory loops interconnect with the core loop just described and not only add robustness and stability to the clock mechanism but also provide additional layers of control and link to a myriad of other pathways in the cell. The first of these accessory loops involves members of the large nuclear receptor family. The mouse *Bmal1* promoter contains two cognate RevErbA/ROR-binding elements (ROREs) via which the nuclear receptors RORα (retinoic acid receptor-related orphan receptor α), RORβ, or RORγ activate (Akashi and Takumi, 2005; Guillaumond *et al.*, 2005; Sato *et al.*, 2004), and REV-ERBα (reverse orientation c-erbA α) or REV-ERBβ repress (Guillaumond *et al.*, 2005; Preitner *et al.*, 2002; Triqueneaux *et al.*, 2004), *Bmal1* expression, although the REV-ERBs are likely more important in this process (Fig. 6.1; Liu *et al.*, 2008). Interestingly, the expression of the aforementioned nuclear receptors is circadian and relies on CLOCK/BMAL1-mediated activation through E-boxes in their promoters, although the *Ror* and *Rev-erb* transcripts cycle antiphase to each other (Yang *et al.*, 2006). The RORα-mediated activation of *Bmal1* transcription is enhanced by peroxisome proliferator-activated receptor-γ (PPARγ) coactivator 1α (PGC1α), the expression of which cycles in liver and muscle (Liu *et al.*, 2007c). Members of the PAR bZIP transcription factor family, including the activators DBP (D-box binding protein), TEF (thyrotroph embryonic factor), HLF (hepatic leukemia factor), and the repressor E4BP4 (E4 promoter-binding protein 4), act via D-box elements in target genes to form a second accessory feedback loop (Mitsui *et al.*, 2001; Ohno *et al.*, 2007; Ueda *et al.*, 2005).

V. BEHAVIORAL, MOLECULAR, AND CELL/TISSUE EFFECTS OF CIRCADIAN CLOCK GENE MUTATIONS

Naturally occurring, chemically induced, or targeted mutations exist for all of the core clock genes (Table 6.1). These mutations have helped define the role of each component of the molecular oscillator (Ko and Takahashi, 2006; Lowrey and Takahashi, 2004; Takahashi *et al.*, 2008). At times, however, results from disruption of clock components have been unexpected.

A. Behavioral and molecular effects

One of these surprises occurred with the generation of a mouse *Clock* knockout model. Interestingly, unlike *Clock^{Δ19}* animals, *Clock^{−/−}* mice continue to express circadian rhythms of locomotor activity in DD, albeit with a slightly shorter period compared to wild-type animals, and experience only modest alterations in circadian gene expression in the SCN (Table 6.1; Debruyne et al., 2006). Subsequent work has revealed that in the SCN, NPAS2 can compensate for CLOCK by heterodimerizing with BMAL1 to activate transcription from E-box-containing target genes (DeBruyne et al., 2007a), but that molecular circadian rhythms in peripheral tissues are dependent on the presence of CLOCK (DeBruyne et al., 2007b). $Npas2^{-/-}$ animals, however, experience only subtle changes in circadian locomotor activity and gene expression (DeBruyne et al., 2007a; Dudley et al., 2003), suggesting that CLOCK has a more important role in the molecular oscillator. Hence, although there is partial functional redundancy between CLOCK and NPAS2, it is clearly tissue specific (DeBruyne et al., 2007b; Reick et al., 2001). As expected, CLOCK/NPAS2 double knockout animals are completely arrhythmic in DD (DeBruyne et al., 2007a).

Knockout of *Bmal1* in mice results in behavioral arrhythmicity in DD and disrupted molecular rhythms of gene expression even though its paralog, BMAL2 (MOP9), is also expressed in the SCN and can form a transcriptionally competent complex with CLOCK (Table 6.1; Bunger et al., 2000; Dardente et al., 2007; Hogenesch et al., 2000). Indeed, *Bmal1* is the only core clock gene for which loss of function causes an immediate loss of circadian locomotor behavior in DD. Constitutive expression of *Bmal1* in $Bmal1^{-/-}$ mice or cells restores circadian rhythmicity (Liu et al., 2008; McDearmon et al., 2006), demonstrating that cycling *Bmal1* mRNA is not necessary for circadian rhythm generation. That BMAL2 cannot rescue the $Bmal1^{-/-}$ phenotype most likely results from the dependence of *Bmal2* expression on CLOCK/BMAL1-mediated activation. Hence, disrupting *Bmal1* is likely functionally equivalent to creating a double *Bmal1/Bmal2* null animal (Shi et al., 2010).

Per1 null mutations have been independently generated by three groups (Table 6.1). The mutant progeny from these lines exhibit subtle differences in circadian behavior. Homozygous $Per1^{ldc}$ mice have about a 0.5-h shorter free-running period in DD than wild-type controls and experience a gradual loss of rhythmicity after 2 weeks in DD (Bae et al., 2001). $Per1^{Brdm1}$ mice express a free-running period approximately 1 h shorter than wild-type animals and maintain rhythmicity (Zheng et al., 2001). This result is consistent with the circadian behavior of the $Per1^{-/-}$ mutant line generated by a third group which exhibits a 0.7-h shorter free-running period with no loss of rhythmicity (Cermakian et al., 2001). The phenotypic disparities observed among the three *Per1* null studies may result from differences in targeting approaches or genetic backgrounds. Two independent null mutations in *Per2* have been reported (Table 6.1). Both mutant lines exhibit a loss of behavioral rhythmicity in DD, yet the $Per2^{Brdm1}$ line expresses a 1.5-h shorter period for several days before experiencing arrhythmicity (Zheng et al., 1999). Most animals of the $Per2^{ldc}$ line exhibit immediate arrhythmicity upon exposure to DD (Bae et al., 2001). As expected from the results just presented, double *Per1/Per2* knockout animals experience behavioral and molecular arrhythmicity in constant conditions (Bae *et al.*, 2001; Zheng *et al.*, 2001). Because loss of *Per3* has no effect on circadian rhythms either in *Per1/Per3* or *Per2/Per3* double mutant mice, *Per3* is not an essential component of the circadian core clock mechanism (Bae *et al.*, 2001; Shearman *et al.*, 2000a).

Targeted disruption of either of the two *Cry* genes results in opposite effects on circadian behavior—$Cry1^{-/-}$ animals have a 1-h shorter and $Cry2^{-/-}$ animals have a 1-h longer free-running period in DD compared to wild-type animals (Thresher *et al.*, 1998; van der Horst *et al.*, 1999; Vitaterna *et al.*, 1999) (Table 6.1). Similar to *Per1/Per2* double knockouts, *Cry1/Cry2* double knockout animals experience a complete loss of behavioral and molecular rhythmicity when transferred to DD (Okamura *et al.*, 1999; Thresher *et al.*, 1998; van der Horst *et al.*, 1999; Vitaterna *et al.*, 1999). *Per/Cry* compound knockouts also exhibit interesting behavioral phenotypes (Table 6.1). $Per1^{Brdm1}/Cry1^{-/-}$ mice have normal circadian rhythms of behavior and gene expression (Oster *et al.*, 2003). Deletion of *Per1* rescues the short-period phenotype observed in $Cry1^{-/-}$ mutants, revealing that *Per1* is a nonallelic suppressor of *Cry1*. $Per1^{Brdm1}/Cry2^{-/-}$ mice have more complex phenotypes. Mutants up to 6 months of age express behavioral rhythms 1.5 h longer than wild-type controls and have normal rhythms of *Per2* expression. After approximately 6 months of age, $Per1^{Brdm1}/Cry2^{-/-}$ animals exhibit disrupted entrainment to LD cycles and subsequently experience arrhythmicity upon release into DD (Oster *et al.*, 2003). In addition, the older animals have blunted *Per2* rhythms in the SCN revealing an age-sensitive effect in this compound mutant. $Per2^{Brdm1}/Cry1^{-/-}$ mutants experience immediate behavioral and molecular arrhythmicity in DD (Oster *et al.*, 2002). $Per2^{Brdm1}/Cry2^{-/-}$ animals, however, maintain behavioral and molecular circadian rhythmicity in DD with a slightly shorter free-running period compared to wild-type controls. Thus, inactivation of *Cry2* in *Per2* null animals restores circadian rhythmicity. As a result, *Cry2* is a nonallelic suppressor of *Per2* (Oster *et al.*, 2002).

The double *Per1/Per2* and *Cry1/Cry2* knockout results make sense given that the PER/CRY complex is necessary to directly inhibit CLOCK/BMAL1-mediated transcription (Griffin *et al.*, 1999; Kume *et al.*, 1999; Lee *et al.*, 2001; Sato *et al.*, 2006; Shearman *et al.*, 2000b). Until recently, transient transfection assays pointed to a more prominent role for CRY in inhibiting CLOCK/BMAL1 (Dardente *et al.*, 2007), yet new evidence suggests that PER may be the more important of the negative effectors as its constitutive expression disrupts the circadian clock in fibroblasts and hepatocytes. Further, constitutive PER2 expression in the SCN of transgenic mice results in the loss of circadian rhythms of locomotor behavior in a conditional and reversible manner (Chen *et al.*, 2009). Finally, biochemical evidence demonstrates that PER2 directly binds to the CLOCK/BMAL1 complex in a rhythmic way and that it brings CRY into contact with CLOCK/BMAL1. Rhythmic expression of PER, in turn, drives the rhythmic inhibition of CLOCK/BMAL1, and it is PER that is the rate-limiting component of the inhibitor complex (Chen et al., 2009). This is substantiated by independent work demonstrating that PER2 is also a more potent inhibitor of CLOCK/BMAL2-mediated transactivation than is CRY (Sasaki et al., 2009).

B. Cell/tissue effects

The analysis of behavioral (locomotor activity) and molecular (RNA/protein) rhythms in mice with mutations in core circadian clock genes just described is insufficient to provide a comprehensive view of molecular clock function. For example, most of the above studies do not take into account differences in central versus peripheral oscillators, potential intercellular interactions in producing the observed phenotypes, or reveal unique properties of individual cellular oscillators. New methods allowing continuous monitoring of circadian rhythms in cultured tissues and individual cells in real time for periods of 20 days or more via bioluminescent technology have revealed many clock properties not evident from behavioral and molecular analyses alone.

By crossing clock gene knockout mice to the $Per2^{Luciferase}$ ($Per2^{Luc}$) knockin reporter mouse line in which a luciferase gene is fused to the $3'$-end of the endogenous $Per2$ gene (Yoo et al., 2004), one group has measured the effects of clock gene perturbations at the level of tissues, populations of cultured cells, and single dissociated cells from both the SCN and the peripheral tissues (Liu et al., 2007b). In SCN tissue explants, disruption of $Per1$, $Per3$, $Cry1$, or $Cry2$ individually has no effect on the maintenance of circadian rhythmicity, and the observed period for each mutant SCN tissue reflects the free-running behavioral period of the corresponding mutant animal model (Tables 6.1 and 6.2). In peripheral tissue explants (e.g., liver, lung, cornea), unlike SCN explants, $Cry1^{-/-}$ and $Per1^{-/-}$ are required for robust, sustained circadian rhythmicity (Table 6.2), a property of peripheral tissues not evident from the previously described behavioral and molecular studies (Table 6.1). $Cry2^{-/-}$ and $Per3^{-/-}$ mutant peripheral tissues maintain rhythmicity with slightly longer and shorter periods, respectively, compared to wild-type controls, again consistent with behavioral results (Table 6.1). In dissociated fibroblast cells in culture, $Per1$, $Per2$, and $Cry1$ are required to maintain circadian oscillations. Thus it seems that, in fibroblast cultures at least, $Per1$ and $Per2$ are not functionally redundant (Liu et al., 2007b).

When single fibroblast cells are imaged for circadian rhythms of bioluminescence, again $Per1$ and $Cry1$ prove necessary to sustain circadian oscillations, confirming the results observed in fibroblast cultures (Liu et al., 2007b). Single $Cry2^{-/-}$ fibroblast cells are rhythmic with a slightly longer period compared

Table 6.2. Cell and Tissue Phenotypes of Mutations in Mouse Clock and Clock-Related Genes

Gene(s)	Mutant allele(s)	Cellular phenotype(s)	Tissue phenotype(s)	References
Clock	$Clock^{-/-}$	—	SCN: WT; lung, liver: AR	DeBruyne et al. (2007a,b)
Npas2 (Mop4)	$Npas2^{-/-}$	—	SCN, lung, liver: WT	DeBruyne et al. (2007b)
Bmal1 (Arntl, Mop3)	$Bmal1^{-/-}$	Fibroblasts, SCN neurons: AR	SCN: variable/stochastic; pituitary, liver, lung, cornea: AR	Liu et al. (2008), Ko et al. (2010)
Cry1	$Cry1^{-/-}$	Fibroblasts, SCN neurons: AR	SCN: short; lung, liver, cornea: AR	Liu et al. (2007b)
Cry2	$Cry2^{-/-}$	Fibroblasts, SCN neurons: long	SCN, lung, liver, cornea: long	Liu et al. (2007b)
Cry1/Cry2	$Cry1^{-/-}/Cry2^{-/-}$	Fibroblasts, SCN neurons: AR	SCN, lung, liver, cornea: AR	Yagita et al., (2001), Liu et al. (2007b)
Per1	$Per1^{ldc}$	Fibroblasts, SCN neurons: AR	SCN: WT; lung: AR	Liu et al. (2007b)
Per2	$Per2^{ldc}$	Fibroblasts: AR	—	Liu et al. (2007b)
Per3	$Per3^{-/-}$	Fibroblasts: short	SCN, lung: short	Liu et al. (2007b)
Rev-erbα (Nr1d1)	$Rev\text{-}erb\alpha^{-/-}$	Fibroblasts: WT	—	Liu et al. (2008)
Rorα (Rora)	$Ror\alpha^{-/-}$ (staggerer)	Fibroblasts: WT	—	Liu et al. (2008)
Rorγ (Rorc)	$Ror\gamma^{-/-}$	Fibroblasts: WT	Lung, liver: WT	Liu et al. (2008)
CK1δ (Csnk1d)	$CK1\delta^{-/-}$	Fibroblasts: long	[a]SCN, liver: long	Etchegaray et al. (2009, 2010), Lee et al. (2009)
CK1ε (Csnk1e)	$CK1\varepsilon^{-/-}$	Fibroblasts: WT	[a]SCN, liver: WT	Etchegaray et al. (2009, 2010)
	$CK1\varepsilon^{tau}$	Fibroblasts: short	SCN, pituitary, lung, kidney: short	Meng et al. (2008)

Gene symbols are as in Table 6.1. Abbreviations: AR, arrhythmic; WT, wild type.
[a]Neonatal tissue. Table modified from Baggs et al. (2009).

to individual wild-type cells, consistent with the behavioral phenotype of *Cry2* null mice (Tables 6.1 and 6.2). To measure rhythms of bioluminescence from single SCN neurons, they must first be uncoupled by mechanical dissociation into single cells (Herzog *et al.*, 1998; Welsh *et al.*, 1995). Similar to the result in single fibroblasts, single $Cry2^{-/-}$ SCN neurons are rhythmic with a period longer than wild-type SCN neurons (Table 6.2). Single $Cry1^{-/-}$ and $Per1^{-/-}$ SCN neurons, in contrast to single $Cry2^{-/-}$ SCN neurons or $Cry1^{-/-}$ and $Per1^{-/-}$ SCN tissue explants, exhibit arrhythmicity (Table 6.2). This is an important result not apparent in earlier behavioral studies of *Per1* and *Cry1* null mutants and demonstrates that the robustness of circadian oscillations observed in $Cry1^{-/-}$ and $Per1^{-/-}$ SCN explant tissue is not a cell-autonomous property. Instead, the ability of $Cry1^{-/-}$ and $Per1^{-/-}$ SCN tissue to maintain circadian rhythmicity despite mutations in core clock components that, at the single-cell level, cause arrhythmicity depends on intercellular coupling among SCN neurons (Liu *et al.*, 2007b). This property of intercellular coupling, and not unique intracellular molecular mechanisms, is what distinguishes SCN neurons from cells of peripheral tissues.

As mentioned previously, *Bmal1* null mutant animals experience an immediate loss of circadian behavior upon transfer to constant conditions (Bunger *et al.*, 2000). Using bioluminescence monitoring methods with $Bmal1^{-/-}$ SCN and peripheral tissues similar to those just described, a recent study has elucidated interesting properties of the SCN network not discovered in previous behavioral and molecular investigations (Ko *et al.*, 2010). SCN explants from $Bmal1^{-/-}$ mice crossed to the $Per2^{Luc}$ reporter line exhibit rhythmic but highly variable (noisy) oscillations of PER2::LUC bioluminescence for more than 35 days in culture (Table 6.2). This is an unexpected result given that the behavioral and molecular rhythms of gene expression in $Bmal1^{-/-}$ animals are arrhythmic (Table 6.1). The authors of this study refer to these quasi-circadian rhythms generated by the $Bmal1^{-/-}$ SCN explants as stochastic. As expected, bioluminescence monitoring demonstrates that all peripheral tissues from $Bmal1^{-/-}$ animals are arrhythmic (Table 6.2). Analysis of single dispersed $Bmal1^{-/-}$ SCN neurons, however, reveals that they are arrhythmic and do not exhibit the stochastic rhythms of bioluminescence observed in $Bmal1^{-/-}$ SCN explants. This result seems not to depend from what subtype of SCN neuron recordings are made or from what region of the SCN the neurons are obtained—all dispersed $Bmal1^{-/-}$ SCN neurons exhibit arrhythmicity. Further, $Bmal1^{-/-}$ SCN slices treated with tetrodotoxin (TTX) experience an immediate cessation of stochastic rhythmicity, a result of loss of rhythmicity at the single-cell level. Upon removal of TTX from SCN slices, stochastic rhythmicity is restored, thereby confirming the importance of intercellular coupling in generating the observed PER2::LUC rhythms in the SCN slices. Taken together, these bioluminescence results demonstrate that the quasi-circadian rhythms in $Bmal1^{-/-}$ SCN explants are not a cell-autonomous property, but rather an emergent rhythmic property of the SCN intercellular network (Ko *et al.*, 2010).

VI. POSTTRANSLATIONAL MODIFICATION OF CLOCK PROTEINS

Posttranslational modifications of the core clock components play a crucial role in generating the delays necessary to establish the ~24 h rhythm of the mammalian circadian clock. Some of these modifications are absolutely essential to clock function, while others simply fine-tune the rhythm. Phosphorylation of clock proteins was the first posttranslational process observed in the mammalian molecular clock, and we understand more about this mechanism than any other. The list of identified posttranslational modifications of mammalian clock proteins has grown rapidly and now, in addition to phosphorylation, includes dephosphorylation, ubiquitination, sumoylation, and acetylation.

A. Phosphorylation

1. Casein kinase 1

Posttranslational modification as an important clock-related process in higher eukaryotes became apparent with the identification in *Drosophila* of *doubletime*, a gene encoding a fly casein kinase 1 (CK1) ortholog that phosphorylates PER (Kloss *et al.*, 1998; Price *et al.*, 1998). Subsequently, we identified *CK1ε* as the gene affected by the Syrian hamster *tau* mutation and showed that *in vitro* CK1ε*tau* is hypomorphic toward various substrates, including mammalian PER1 and PER2 proteins (Lowrey *et al.*, 2000). Others demonstrated that, *in vitro* and in cultured cells, CK1ε and the closely related family member, CK1δ, can phosphorylate PER (Akashi *et al.*, 2002; Camacho *et al.*, 2001; Keesler *et al.*, 2000; Vielhaber *et al.*, 2000). Further work revealed that CK1δ/ε-mediated phosphorylation regulates PER subcellular localization and its ability to repress CLOCK/BMAL1-mediated transcription and promotes its ubiquitin-degradation via the 26S proteasome (Eide *et al.*, 2005b; Ohsaki *et al.*, 2008; Shirogane *et al.*, 2005; Vanselow *et al.*, 2006; Vielhaber *et al.*, 2000).

   Although our work showed that CK1ε*tau* was a hypomorph in *in vitro* assays, a study using mathematical modeling and *in vivo* analyses has reported that the CK1ε*tau* anion binding site mutation causes loss of enzyme function toward canonical acidic CK1ε substrates but gain of function toward the noncanonical β-transducin repeat-containing protein (βTrCP) binding site on PER2 (Gallego *et al.*, 2006a). Indeed, biochemical evidence substantiates the model's prediction as do behavioral, neurophysiological, and cellular studies from a mouse model of the hamster *tau* mutation (Gallego *et al.*, 2006a; Meng *et al.*, 2008). Another group, however, has published findings that contradict this interpretation. They report that CK1ε*tau* is actually a partial loss of function mutation as the mutant kinase is unable to phosphorylate sites that promote nuclear localization of PER2, but that it can phosphorylate amino acids required for PER2 proteasomal degradation (Vanselow *et al.*, 2006). By mapping all of the CK1ε phosphorylation sites on PER2, they opine that the different sites differentially target PER2 to two cellular locations—nucleus or proteasome. Both interpretations agree, however, that the *tau* allele is a particularly interesting mutation biochemically as it differentially affects CK1ε activity toward specific substrates.

The *tau* mutation focused much attention on the role of CK1ε in the mammalian molecular clock, yet as mentioned above, CK1δ also phosphorylates PER proteins and targets them for degradation (Camacho *et al.*, 2001; Xu *et al.*, 2005), and both kinases associate with PER/CRY repressor complexes *in vivo* (Fig. 6.1; Lee *et al.*, 2001). Thus, to better define the role of these two CK1 enzymes, null mutants of both have been generated independently by different laboratories. The free-running period of the locomotor activity rhythm of $CK1\varepsilon^{-/-}$ mice is slightly, but significantly, longer than wild-type controls (Table 6.1; Etchegaray *et al.*, 2009; Meng *et al.*, 2008). Two groups have reported that $CK1\delta^{-/-}$ mice die during the perinatal period; thus the free-running behavioral period has been studied in $CK1\delta^{+/-}$ heterozygous animals. In one case, one copy of the $CK1\delta$ null allele results in no difference in free-running period compared to controls (Xu *et al.*, 2005), while another group's heterozygous null animal exhibits a slight increase in circadian period (Etchegaray *et al.*, 2009). In addition, compared to CK1ε-deficient tissue, mouse embryonic fibroblasts (MEFs) and liver tissue deficient in $CK1\delta$ have about a 1- to 2-h longer circadian period *in vitro* (Etchegaray *et al.*, 2009; Lee *et al.*, 2009). When monitored from neonatal SCN explants, PER2::LUC bioluminescence rhythms from $CK1\delta^{-/-}$ mice are longer compared to wild-type controls, yet there is no significant difference in PER2::LUC rhythms from $CK1\varepsilon^{-/-}$ SCN compared to controls (Etchegaray *et al.*, 2010).

Pharmacological approaches have also been used to study the roles of CK1ε and CK1δ in the mammalian clock with the general CK1δ/ε inhibitors CKI-7, IC261, and D4476, all of which lengthen circadian period in cultured cells (Eide *et al.*, 2005a; Hirota *et al.*, 2008; Reischl *et al.*, 2007; Vanselow *et al.*, 2006). An inhibitor specific for CK1ε (PF-4800567) has only a slight effect on the period of oscillating rat-1 fibroblasts stably transfected with a *Per2::luc* reporter compared to the dual CK1δ/ε inhibitor (PF-670462) which causes an increase in fibroblast circadian period (Walton *et al.*, 2009). Single injections into rats of the dual CK1δ/ε inhibitor induce large phase delays in circadian locomotor rhythms under free-running and entrained conditions (Badura *et al.*, 2007; Sprouse *et al.*, 2010). Daily treatment with PF-670462 significantly lengthens locomotor behavioral rhythms in a dose-dependent manner in wild-type, $CK1\varepsilon^{tau}$, and $CK1\varepsilon^{-/-}$ mice (Meng *et al.*, 2010). Selective inhibition of CK1ε with PF-4800567 has no significant effect on behavioral rhythms in wild-type or $CK1\varepsilon^{-/-}$ mice, yet it lengthens the free-running locomotor activity rhythm of $CK1\varepsilon^{tau}$ animals. How does inhibition of the CK1 enzymes affect molecular clock function? PF-670462 seems to work by stabilizing PER2 nuclear localization in

SCN neurons and peripheral tissues. This prolongs PER2-mediated negative feedback, thereby lengthening circadian period (Meng et al., 2010). Together with the CK1 knockout experiments, these results suggest that CK1δ has a more prominent role compared to CK1ε in the mammalian clockwork; that is, the two kinases seem not to be equally redundant. It is clear, however, that loss of both CK1ε and CK1δ causes arrhythmicity in MEFs (Lee et al., 2009), and that knockdown of both kinases in human U2OS (osteosarcoma) cells additively lengthens circadian period to more than 30 h (Isojima et al., 2009). It remains to be determined if this partial functional redundancy derives from unequal expression levels of the two kinases in cells throughout the body. Some evidence suggests that, at least in MEFs, CK1δ is twice as abundant as CK1ε (Lee et al., 2009).

One group has reported a surprising result in their work with CK1δ/ε in which a novel, noncatalytic clock-related role for these kinases is revealed (Lee et al., 2009). Overexpression of the CK1δ/ε-binding domain of PER2 (CKBD-P2) in MEFs severely disrupts PER2::LUC rhythms of biolumines-cence, presumably because the CKBD-P2 fragment interferes (competes) with the interaction between CK1δ/ε and PER. In addition, PER1 and PER2::LUC levels are lower in the CKBD-P2-expressing MEFs, while CK1δ/ε levels are higher than normal. This suggests that the CKBD-P2 enhances the stability of CK1δ/ε via physical interaction. Further, the low levels of PER observed in the CKBD-P2-expressing MEFs may result from the inhibition of CK1δ/ε-specific phosphorylation of PER, and from the reduced physical interaction between CK1δ/ε and PER which, under normal circumstances, may confer stability to PER (Lee et al., 2009). Overexpression of the CKBD region of PER3 in MEFs does not have an effect on the circadian rhythms in these cells, mirroring previous reports showing that PER3 does not interact with CK1δ/ε. That the stabilizing role of CK1δ/ε toward PER is noncatalytic is supported by experi-ments in which dominant-negative CK1δ/ε (K38R)-expressing MEFs do not experience reduced PER levels (Lee et al., 2009). Interestingly, a noncatalytic circadian role has also been reported recently for the Drosophila CK1 family member, DBT (Yu et al., 2009).

Finally, in a high-throughput screen of approximately 120,000 com-pounds using U2OS cells expressing luciferase under the control of the mouse Bmal1 promoter, another CK1 family member, CK1α, has been identified as a mammalian clock regulatory kinase (Hirota et al., 2010). A purine derivative identified in this screen, longdaysin, inhibits CK1δ, CK1α, and ERK2 (MAPK1) and prevents them from phosphorylating PER1, causing a dramatic 13-h lengthening of period in U2OS cells. Knockdown by siRNA of one of the aforementioned kinases alone is insufficient to recapitulate the 13-h peri-od-lengthening effect, yet combinatorial knockdown of all three kinases additively increases period and closely mirrors the effect of longdaysin treatment (Hirota *et al.*, 2010). Results from this interesting study suggest that multiple kinases participate in a network to enhance robustness of the molecular clock mechanism.

2. Glycogen synthase kinase-3β

Glycogen synthase kinase-3 (GSK-3) is a serine–threonine, phosphate-directed protein kinase of which there are two isoforms in mammals: GSK-3α and GSK-3β (Ali *et al.*, 2001). GSK-3 was initially characterized as a kinase involved in metabolism and energy storage, yet it has since been shown to play a role in many intracellular pathways (Doble and Woodgett, 2003). Knockout models for both isoforms have been generated, but Gsk-$3\beta^{-/-}$ mice experience embryonic lethality (Hoeflich *et al.*, 2000; MacAulay *et al.*, 2007). Interestingly, GSK-3 is sensitive to lithium. Presumably, Li^+ competes directly for binding to GSK-3 with Mg^{2+}, a required cofactor for GSK-3 function (Klein and Melton, 1996; Ryves and Harwood, 2001; Stambolic *et al.*, 1996). Several studies have documented the effects of lithium treatment on circadian rhythms in mammals, including a consistent effect of lengthening the free-running period of behavioral rhythms, notably those of locomotor activity and drinking (Iwahana *et al.*, 2004; LeSauter and Silver, 1993; Seggie *et al.*, 1982). This period-lengthening effect of lithium is also observed for the firing rate rhythms in isolated mouse SCN neurons (Abe *et al.*, 2000). Other work has shown that GSK-3β is rhythmically expressed in the SCN and liver of mice, and that it undergoes a daily cycle in phosphorylation *in vivo* as well as in serum-shocked NIH3T3 mouse fibroblasts (Harada *et al.*, 2005; Iitaka *et al.*, 2005). Lithium chloride treatment phase delays, while overexpression of GSK-3β phase advances clock gene expression in fibroblasts (Iitaka *et al.*, 2005). In addition, *Gsk-3α* RNAi knockdown in Gsk-$3\beta^{-/-}$ MEFs induces a phase delay in the *Per2* RNA rhythm, as does treatment of MEFs with kenpaullone, a GSK-3 antagonist (Kaladchibachi *et al.*, 2007). Surprisingly, however, a recent high-throughput screen has identified small molecule inhibitors of GSK-3β that shorten circadian period, a result confirmed by siRNA knockdown of GSK-3β (Hirota *et al.*, 2008). These effects of inhibition of GSK-3β on the molecular clock in cells and on clock-controlled behavior in mammals have prompted further investigation into the potential clock-related targets of this enzyme. In addition, work in *Drosophila* has shown that *shaggy*, the fly ortholog of mammalian GSK-3β, is an important component in determining circadian period length in that organism (Martinek *et al.*, 2001).

Mammalian targets of GSK-3β phosphorylation include the positive and negative components of the core circadian transcriptional/translational feedback loop, CLOCK/BMAL1 and PER/CRY, respectively. Phosphorylation by GSK-3β of PER2 promotes its nuclear localization (Iitaka et al., 2005). GSK-3β-mediated phosphorylation at Ser553 of CRY2 leads to its degradation by the proteasome (Harada et al., 2005), yet GSK-3β phosphorylation promotes the stabilization of REV-ERBα (Yin et al., 2006). Lithium treatment of cultured cells results in the rapid proteasomal degradation of REV-ERBα and concomitant derepression of Bmal1 transcription owing to inhibition of GSK-3β (Yin et al., 2006). Others have identified a phoshodegron region on the CLOCK protein that is phosphorylated by GSK-3β and which promotes degradation of CLOCK (Spengler et al., 2009). Finally, BMAL1, the major CLOCK dimerization partner, is phosphorylated by GSK-3β on Ser17 and Thr21. This modification targets BMAL1 for ubiquitination and subsequent degradation by the proteasome (Sahar et al., 2010).

3. Casein kinase 2

Casein kinase 2 (CK2), a serine/threonine kinase, is a tetramer composed of two α catalytic and two β regulatory subunits and, similar to CK1, prefers acidic substrates (Meggio and Pinna, 2003). A role for this kinase in the regulation of circadian rhythms was first reported in Arabidopsis, Drosophila, and Neurospora (Gallego and Virshup, 2007). Recently, participation of CK2 in the mammalian molecular clock mechanism has been reported by three groups. First, BMAL1 is a substrate for CK2α at Ser90, a residue that undergoes rhythmic phosphorylation (Tamaru et al., 2009). Nuclear localization of BMAL1 seems to depend on CK2α phosphorylation, as knockdown of CK2α results in cytoplasmic BMAL1 accumulation and a concomitant disruption of Per2 mRNA rhythms (Tamaru et al., 2009).

PER2 is also a CK2α substrate. One study has shown that CK2α both phosphorylates PER2 at Ser53 and enhances CK1ε-mediated PER2 destabilization (Tsuchiya et al., 2009). These two actions of CK2α on PER2 are independent, as the CK2α-mediated potentiation of CK1ε-dependent degradation of PER2 does not require phosphorylation of Ser53. Although it has not been demonstrated, it may be that CK2α acts to phosphorylate CK1ε, thereby upregulating its activity toward PER2, given that a catalytically inactive form of CK2α (CK2α-K68A) fails to enhance CK1ε-dependent PER2 degradation (Tsuchiya et al., 2009). Finally, in a recent RNAi screen for clock-related components, downregulation of either CK2α or CK2β lengthens circadian period, while knockdown of both subunits leads to arrhythmicity (Maier et al., 2009). Further, overexpression of CK2α has a period-shortening effect. This same study showed that CK2α binds PER2 and promotes phosphorylation near the N-terminus, although the proposed sites do not correspond to Ser53 mentioned before. Also, unlike the previous study, CK2α phosphorylation is proposed to stabilize PER2 rather than enhance its degradation. Inhibition of CK2α was shown to delay PER2 nuclear accumulation, suggesting that CK2α phosphorylation of PER2 may provide a signal for nuclear entry (Maier et al., 2009). These, and other questions, need to be addressed to better understand the role of CK2 in the mammalian clock mechanism.

4. Other kinases

BMAL1 has been identified as a substrate for several kinases including CK1ε and GSK-3β as described above. In addition, mitogen-activated protein kinase (MAPK) phosphorylates BMAL1 on several residues, including Thr534 which negatively regulates the transactivation potential of the CLOCK/BMAL1 complex at E-boxes (Sanada et al., 2002). Another study has shown recently that RACK1 (receptor for activated C kinase-1) recruits activated protein kinase Cα (PKCα) to the CLOCK/BMAL1 complex during the negative feedback phase of the circadian cycle in the nucleus where it phosphorylates BMAL1 and suppresses CLOCK/BMAL1 transcriptional activity (Robles et al., 2010). Knockdown of either RACK1 or PKCα in fibroblast cultures shortens circadian period. $Prkca^{-/-}$ mice have a normal circadian period, yet they experience impaired photic resetting of behavioral rhythms (Jakubcakova et al., 2007). This difference between the knockout phenotype and cell culture results highlights the need for further work.

CLOCK has been shown to be phosphorylated throughout the circadian cycle, yet it is hyperphosphorylated during the negative feedback phase (Yoshitane et al., 2009). Specific residues have been identified as phosphorylation sites on CLOCK, including Ser38, Ser42, and Ser427. The kinase or kinases responsible for phosphorylating these sites, however, remain undetermined. CLOCKΔ19, which lacks the binding site for CLOCK-interacting protein circadian (CIPC), a PER/CRY-independent negative regulator of CLOCK/BMAL1-mediated transcription, is less phosphorylated and more stable than wild-type CLOCK (Zhao et al., 2007). Others have shown that CLOCK is a substrate for phosphorylation by protein kinase G II (PKG-II) and by two protein kinase C isoforms (PKCα and PKCγ; Shim et al., 2007; Tischkau et al., 2004).

Finally, it will be important to assess the contribution of sequential phosphorylation events on clock proteins. For example, some kinases such as CK1δ/ε require priming phosphorylation of their target sites (Knippschild et al., 2005). Phosphorylation of human PER2 at the Ser662 familial advanced sleep phase syndrome (FASPS) site occurs by an unidentified kinase and is necessary for subsequent CK1δ/ε phosphorylation (Xu et al., 2007). Similarly, GSK-3β phosphorylation of CRY2 requires priming phosphorylation by DYRK1A (dual-specificity tyrosine-phosphorylated and -regulated kinase 1A;

Kurabayashi *et al.*, 2010). In many cases, the priming kinases for these so-called phosphate-directed kinases such as CK1δ/ε and GSK-3 remain to be identified for particular circadian substrates.

B. Dephosphorylation

Protein phosphatases, although fewer in number in the mammalian genome relative to kinases, also play an important role in the molecular clock (Gallego and Virshup, 2007). Reversible phosphorylation of clock-relevant substrates presumably confers on clock the flexibility to respond appropriately to various stimuli. Few studies, however, have explored the role of phosphatases on the mammalian clock mechanism. The major clock kinases, CK1ε and CK1δ, undergo autophosphorylation which downregulates their activity. At least eight autophosphorylation sites must be dephosphorylated by phosphatases to activate CK1δ/ε (Gietzen and Virshup, 1999). Additional evidence suggests that CK1δ/ε participates in a futile autophosphorylation/dephosphorylation cycle *in vivo* which acts to regulate kinase activity (Rivers *et al.*, 1998). One group has shown that it is protein phosphatase 5 (PP5) that dephosphorylates CK1δ/ε. Further, the same study demonstrated that the CRY proteins interact with and inhibit noncompetitively, PP5. As a result, the CRY proteins seem to indirectly regulate the activity of CK1δ/ε by inhibiting the phosphatase activity of PP5 (Partch *et al.*, 2006). Another study has found a role for PP1 in the clock mechanism (Gallego *et al.*, 2006b). PP1 can bind to and dephosphorylate CK1δ/ε-phosphorylated PER2, thereby negatively regulating its degradation by the proteasome. Overexpression of a dominant negative form of PP1, or use of PP1 inhibitors, results in accelerated degradation of PER2 (Gallego *et al.*, 2006b).

C. Ubiquitination

One of the major pathways for protein degradation in cells is the ubiquitin-dependent proteasomal mechanism. This system requires the attachment of multiple ubiquitin molecules to lysine residues on the target protein (Ciechanover *et al.*, 2000; Nandi *et al.*, 2006). Polyubiquitinated proteins are directed to the 26S proteasome, a large multicatalytic protease, where they are then degraded to small peptides (Nandi *et al.*, 2006). Attachment of ubiquitin to a protein is a three-step process. First, ubiquitin is adenylated by an activating enzyme (E1), then transferred to a conjugating enzyme (E2), and finally linked to the target protein by a ligase (E3; Ciechanover *et al.*, 2000; Wilkinson, 1999). The specificity of this system is determined by the E3 ligases which can be categorized into at least six subtypes. With respect to the mammalian clock mechanism, the SCF E3 subtype is the most relevant. SCF complexes are multimers and are named for their constituent protein components, Skp1, Cdc53, or Cullin, and any one of a number of proteins containing an F-box motif, each of which recognizes a particular target protein (Nandi et al., 2006). It is the F-box protein that confers specificity to each SCF complex (Cardozo and Pagano, 2004). The SCF complexes are constitutively active enzymes that recognize and ubiquitinate only phosphorylated substrates (Kornitzer and Ciechanover, 2000). Consequently, this system links protein phosphorylation to proteolytic degradation by the 26S proteasome (Cardozo and Pagano, 2004).

Turnover of the PER/CRY repressor complex is an important event in relieving inhibition of CLOCK/BMAL1 such that a new circadian day can begin. Insight into how this repressor complex is cleared has come from several studies. Recently, our laboratory and another independently recovered long-period behavioral mutants, *Overtime* (Siepka et al., 2007) and *After hours* (Godinho et al., 2007), respectively, through ENU mutagenesis screens of mice. Both mutations affect the same gene, *Fbxl3*, which encodes the F-box and leucine-rich repeat protein 3 (FBXL3). FBXL3 is involved in ubiquitination of the CRY proteins, which targets them for degradation (Busino et al., 2007). The proteasomal-mediated degradation of CRY1 and CRY2, however, appears to be differentially regulated. CRY1 is phosphorylated by AMPK (AMP-activated protein kinase) which promotes its FBXL13-dependent degradation (Fig. 6.1; Lamia et al., 2009). CRY2, however, is phosphorylated by GSK-3β on Ser553 (Harada et al., 2005), which first requires priming phosphorylation by DYRK1A on Ser557, both of which are at the C-terminal tail of CRY2 (Kurabayashi et al., 2010). This Ser557/Ser553 phosphorylation mechanism promotes proteasomal degradation of CRY2 by what is likely an FBXL13-independent mechanism. Indeed, it seems that Ser557/Ser553 phosphorylation and subsequent turnover of CRY2 slows its cytosolic accumulation rate and allows its timely nuclear translocation (Kurabayashi et al., 2010). The FBXL13-dependent degradation mechanism acts during the declining phase of negative feedback when CRY2 nuclear clearance occurs (Godinho et al., 2007). Taken together, these two mechanisms of CRY2 degradation suggest a model by which negative feedback is controlled.

As discussed earlier, phosphorylation of the PER proteins by CK1δ/ϵ promotes their polyubiquitination (Eide et al., 2005a; Ohsaki et al., 2008; Reischl et al., 2007; Shirogane et al., 2005). Ubiquitination of the PERs is mediated by the F-box proteins βTrCP1 and/or βTrCP2 (Fig. 6.1; Eide et al., 2005b; Ohsaki et al., 2008; Shirogane et al., 2005; Vanselow et al., 2006; Vielhaber et al., 2000). Following ubiquitination, the PER proteins are degraded by the 26S proteasome (Gallego and Virshup, 2007).

D. Sumoylation

The BMAL1 protein undergoes extensive posttranslational modification in cells, including phosphorylation, acetylation, ubiquitination, and sumoylation. Small ubiquitin-like modifier (SUMO) proteins are covalently attached at lysine residues of target proteins to modify their function (Wilkinson and Henley, 2010). Sumoylation has been shown to affect nuclear/cytosolic localization of proteins, progression through the cell cycle, protein stability, and transcriptional regulation (Gareau and Lima, 2010). In mice and humans, there are three *Sumo* paralogs (*Sumo1–3*). SUMO2 and SUMO3 are 95% identical and are often referred to as SUMO2/3. Two groups have independently demonstrated that BMAL1 is rhythmically polysumoylated at a conserved lysine residue (K259) in the PAS domain linker region by all three SUMO proteins and that this process is dependent on CLOCK, the BMAL1 dimerization partner (Cardone *et al.*, 2005; Lee *et al.*, 2008). Further, sumoylation localizes BMAL1 to the promyelocytic leukemia nuclear body, potentiates CLOCK/BMAL1 transactivation of clock-controlled genes, and promotes BMAL1 ubiquitin-dependent proteasomal degradation (Lee *et al.*, 2008).

E. Acetylation, deacetylation, and chromatin remodeling

Rhythmic changes in chromatin architecture participate in the activation and repression of transcription via posttranslational modifications at histone N-terminal tail regions (Imhof and Becker, 2001). Acetylation of lysine residues or phosphorylation of serine residues in histone tails induces a transcription permissive nucleosome conformation, whereas deacetylation, dephosphorylation, or methylation of histone lysine residues promote a transcription inhibitory nucleosome conformation. These, and other covalent modifications that occur at histones to alter the degree of chromatin condensation, have become known as the "histone code" (Jenuwein and Allis, 2001; Strahl and Allis, 2000). Chromatin remodeling as a possible circadian regulatory mechanism was first suggested by an experiment demonstrating that in mice, light pulses during the subjective night promote phosphorylation of serine 10 of histone H3 (Crosio *et al.*, 2000). Subsequent work has shown that at the *Per1* and *Per2* promoters, lysine 9 of histone H3 is rhythmically acetylated (Etchegaray *et al.*, 2003) and that CLOCK and NPAS2 may act to recruit histone acetyltransferases (HATs) to the *Per1* promoter (Curtis *et al.*, 2004; Etchegaray *et al.*, 2003).

Interestingly, CLOCK has been shown to exhibit intrinsic HAT activity toward lysine residues of histones H3 and H4 (Doi *et al.*, 2006). This suggests that CLOCK, while activating transcription with its partner, BMAL1, may rhythmically acetylate histones at clock-controlled genes and thereby participate in chromatin remodeling. CLOCK also acetylates nonhistone substrates, including BMAL1 at lysine 537 (Hirayama et al., 2007), as well as the glucocorticoid receptor (Nader et al., 2009). Additional work is needed to clarify the relationship of CLOCK HAT activity to the circadian control of gene expression. Recently, a role for CLOCK HAT activity in facilitating herpes simplex virus gene expression in infected mammalian cells has been reported (Kalamvoki and Roizman, 2010); hence this activity of CLOCK may not be restricted solely to circadian regulation.

Histone deacetylases (HDACs) act to remove acetyl groups from histone tails to promote transcriptional repression (Finkel et al., 2009). The NAD^+-dependent HDAC Sirtuin 1 (SIRT1) has been shown recently to be involved in the mammalian molecular clock mechanism. SIRT1 acts to deacetylate lysines 9 and 14 of histone H3 as well as lysine 16 of histone H4 leading to chromatin condensation and transcriptional repression (Nakahata et al., 2008; Vaquero et al., 2004). Further, SIRT1 seems to bind CLOCK–BMAL1 to form a complex that is recruited in a circadian manner to promoters of clock-controlled genes. Through its interaction with CLOCK–BMAL1, SIRT1 participates in the circadian expression of Bmal1, Rorc, Per2, and Cry1 (Asher et al., 2008). Interestingly, SIRT1 also acts directly on two core clock components—it reverses the CLOCK-mediated acetylation of lysine 537 of BMAL1 (Nakahata et al., 2008) and deacetylates and promotes the degradation of PER2 (Asher et al., 2008).

It is interesting to note here that because SIRT1 is controlled by the cellular NAD^+:NADH ratio, its activity is intimately tied to a cell's redox state. This finding fits nicely with earlier work showing that the transcriptional activity of CLOCK and NPAS2 is regulated by cellular redox state (Rutter et al., 2001). Expression of the gene encoding the rate-limiting enzyme in the NAD^+ salvage pathway, nicotinamide phosphoribosyltransferase (NAMPT), is circadian owing to E-box-mediated CLOCK/BMAL1 activation. The resulting daily oscillation in NAMPT activity produces rhythmic levels of NAD^+ in cells (Nakahata et al., 2009; Ramsey et al., 2009). Hence, a novel feedback loop connecting cellular metabolism and the circadian clock has been uncovered, whereby CLOCK/BMAL1 positively regulates NAD^+ levels, and thus SIRT1 activity through the circadian control of Nampt expression. SIRT1, in turn, negatively regulates CLOCK/BMAL1 activity by promoting transcriptional repression and participates in the oscillation of its own coenzyme, NAD^+. As mentioned earlier, this and several other connections between the circadian system and metabolic pathways represent an emerging area of investigation as reviewed comprehensively elsewhere (Asher and Schibler, 2011; Bass and Takahashi, 2010; Bellet and Sassone-Corsi, 2010; Green et al., 2008; Maury et al., 2010).

VII. POSTTRANSCRIPTIONAL CLOCK MECHANISMS

Despite significant progress in elucidating the role of posttranslational regulation of the molecular clock in mammals, only recently have the contributions of posttranscriptional regulatory processes to clock function been explored (Kojima et al., 2011; Staiger and Koster, 2011). Because many of the core clock genes, as well as clock-controlled genes, exhibit circadian oscillations in their transcript levels, it is important to determine what processes mediate daily mRNA turnover in mammalian cells. Further, in mice, only between 33% and 50% of the genes that encode rhythmic proteins also manifest rhythmic transcript levels, indicating that the other mRNAs are regulated at the posttranscriptional level (Deery et al., 2009; Reddy et al., 2006).

A. MicroRNAs

Of particular relevance to the elucidation of posttranscriptional clock mechanisms is recent work revealing circadian control of the expression of microRNAs (miRNAs) and their role in the cycling of clock gene transcripts in mammals. Transcribed from noncoding genomic regions, miRNAs are short, single-stranded RNA molecules 19–25 nucleotides in length that interact with the 3'untranslated regions (3'UTRs) of target transcripts to induce the cleavage/destabilization of, or to repress translation of, the target mRNA (Bartel, 2009; Bushati and Cohen, 2007; Guo et al., 2010; Rana, 2007). If one report is correct in positing that most mammalian mRNAs are conserved targets of miRNAs (Friedman et al., 2008), it will be important to understand how this regulatory mechanism affects the molecular clock.

Results from a genome-wide screen to identify CREB-regulated miRNAs (Impey et al., 2004) have encouraged analysis of miRNA expression specifically in the SCN of mice (Cheng et al., 2007). Two miRNAs, miR-219 and miR-132, exhibit circadian expression in the SCN and harbor CREs in their promoters, yet only in miR-219 has an E-box element (noncanonical) been identified. Coexpression of CLOCK and BMAL1 in PC12 cells induces expression of miR-219, but not of miR-132. Under constant conditions, both miRNAs exhibit a circadian oscillation solely in the SCN, peaking during the mid-subjective day. This rhythm is abolished in Cry1/Cry2 double mutant mice. Evidence that miR-132 downregulates translation of Per2 comes from treatment of mice with an antagomir to this miRNA. Antagomirs, modified oligoribonucleotides complementary to a specific miRNA, act to block miRNA function (Krutzfeldt et al., 2005). The miR-132 antagomir results in increased PER2 expression in mice. Finally, circadian period length and light-dependent clock resetting are altered by the antagomir-mediated silencing of miR-219 and miR-132, respectively (Cheng et al., 2007).

To further investigate the role of miR-132 in photic entrainment of the SCN clock, one group has generated a transgenic mouse model that conditionally expresses miR-132 in the SCN and forebrain (Alvarez-Saavedra *et al.*, 2011). Following photic stimulation, the miR-123 transgenic animals experience attenuation in light-induced resetting of behavioral rhythms, and a concomitant decline in PER1 and PER2 levels. Analysis of putative miR-132 targets in the SCN has revealed several that are involved in chromatin remodeling and translational control. Through posttranscriptional regulation of these genes, it is proposed that miR-132 is able to control chromatin remodeling and translation within SCN neurons, thus mediating clock entrainment by regulating *Per* expression. Indeed, the *Per1* and *Per2* promoters are bound to, and activated by, methyl CpG island binding protein 2 (MeCP2), whereas PABP-interacting protein 2A (PAIP2A) and B-cell translocation gene 2 (BTG2) have the opposite effect—they suppress PER translation by promoting *Per* mRNA decay (Alvarez-Saavedra *et al.*, 2011).

Another group has identified a role for the miR-192/194 cluster in the regulation of the *Per* family in HeLa and NIH3T3 cells (Nagel *et al.*, 2009). The 3'UTRs of all three *Per* genes contain putative target sites for miR-192, miR-194, or both. Evidence shows that miR-192/194 downregulates the *Per* genes by acting on mRNA. This downregulation causes a shortening of the circadian period (Nagel *et al.*, 2009).

In liver, the most abundant miRNA, miR-122, is transcribed in a rhythmic manner with pri-mir-122 peaking at zeitgeber time 0 (ZT0) and reaching a nadir at ZT12, although the levels of miR-122 remain constant throughout the day (Gatfield *et al.*, 2009). The expression of miR-122 is most likely regulated by REV-ERBα/β through two conserved ROREs in this gene's promoter as *Rev-erbα$^{-/-}$* mice experience attenuation in pri-mir-122 accumulation. Microarray analyses of gene expression in liver tissue of mice treated with a miR-122 antisense oligonucleotide have demonstrated several mRNAs that are under circadian regulation including peroxisome proliferator-activated receptor β/δ (PPARβ/δ) and the PPARα coactivator SMARCD1/BAF60a (SWI/SNF-related, matrix-associated, actin-dependent regulator of chromatin, subfamily d, member 1), both circadian regulators of metabolism (Li *et al.*, 2008; Yang *et al.*, 2006). Indeed, the 3'UTR of *Pparβ/δ* contains target sites for miR-122, and this protein is upregulated by two-to-threefold upon inactivation of miR-122 (Gatfield *et al.*, 2009). A similar mechanism of miR-122 regulation at the 3'UTR of *Smarcd1/Baf60a* is apparent. The identification of the circadian regulation of miR-122 expression in liver and its role in the circadian control of downstream targets involved in metabolism provide further insight into tissue-specific links between the circadian system and metabolic function.

B. RNA-binding proteins

In addition to the miRNA mechanisms just described, several RNA-binding proteins have been identified that regulate clock-related RNA transcripts by different mechanisms. Most RNAs contain *cis*-acting elements in their 3′UTRs to which *trans*-acting protein factors can bind and regulate splicing, transport, stability, and translation (Gratacós and Brewer, 2010; Moore, 2005). The first RNA-binding protein with a demonstrated effect on the mouse molecular clockwork is LARK, for which there are two forms: LARK1 (RBM4) and LARK2 (RBM4B; Kojima et al., 2007). Although transcripts for both *Lark* isoforms are found in the SCN, only the protein levels oscillate. Interestingly, the phase of the LARK protein oscillation is similar to that for PER1. Investigation of the *Per1* 3′UTR has revealed a specific *cis* element to which LARK binds and promotes translation. Knockdown of *Lark* transcript causes a shorter circadian period, while overexpression of LARK protein increases circadian period length. Thus, LARK seems to act on *Per1* posttranscriptionally by enhancing translation and conferring robustness on the PER1 protein oscillation (Kojima et al., 2007).

One of the more common *cis*-acting elements in mammalian 3′UTRs is the AU-rich element (ARE; Gratacós and Brewer, 2010). A protein which binds these AREs and is abundant in many tissues is polypyrimidine tract-binding protein 1 (PTBP1; also known as hnRNP I). Recently, circadian oscillation of the mouse *Per2* mRNA has been shown to be regulated by PTBP1 (Woo et al., 2009). PTBP1 binds to the 3′UTR of the *Per2* transcript and promotes its decay. Cytoplasmic PTBP1 levels increase at the time at which there is a concomitant rapid decline in *Per2* RNA levels. Upon knockdown of PTBP1 expression with siRNA, the peak amplitude of *Per2* expression increases. Because several other RNA-binding proteins have been detected at the *Per2* 3′UTR, it may be that the effect of PTBP1 on *Per2* RNA decay relies on additional proteins (Woo et al., 2009).

Another RNA-binding protein implicated in the regulation of a circadian gene is heterogeneous nuclear ribonucleoprotein D (HNRNPD, also known as AUF1). HNRNPD binds to specific sequences in target mRNA 3′UTRs and regulates transcript stability or the promotion of translation (Gratacós and Brewer, 2010). One group's examination of the *Cry1* transcript has revealed a 610-bp 3′UTR which contains an ARE that, when deleted, results in *Cry1* mRNA stability (Woo et al., 2010). HNRNPD has been identified as the protein responsible for binding to the *Cry1* ARE and promoting mRNA turnover. Similar to the findings for PTPB1 and *Per2*, the cytoplasmic levels of HNRNPD levels exhibit a maximum as *Cry1* mRNA levels are in decline. Knockdown of HNRNPD results in stabilized *Cry1* transcript and enhanced oscillation amplitude (Woo et al., 2010). Thus, cytoplasmic destabilization of both the *Per2* and *Cry1* transcripts share many similarities. It will be interesting to determine what posttranscriptional processes act on the other core clock gene transcripts.

Decay pathways for mRNAs in mammalian cells are either deadenylation-dependent or -independent (Gratacós and Brewer, 2010). Both PTBP1 and HNRNPD are involved in the deadenylation-dependent pathway. Deadenylation involves the removal of a transcript's poly(A) tail followed by breakdown of the RNA. It is interesting to note that a circadian deadenylase gene, *Nocturnin* (*Ccrn4l*), has been characterized in mammals where it is rhythmically expressed in multiple tissues, including the SCN (Wang et al., 2001). *Noc* circadian expression is damped in the liver of *Clock* mutant mice and is constitutively elevated in *Cry* double mutant animals (Oishi et al., 2003). Although $Noc^{-/-}$ mice exhibit normal locomotor activity rhythms and clock gene expression, they do have several aberrant metabolic phenotypes (Green et al., 2007). Thus, *Noc* is a clock-controlled gene—a clock output. Interestingly, it has recently been shown that circadian expression of *Noc* in mouse liver is controlled by miR-122 (Kojima et al., 2010).

VIII. EFFECTS OF TEMPERATURE ON THE MAMMALIAN CLOCK

A. Temperature as an entraining agent

Temperature is an important environmental entraining agent for many organisms, yet in homoeothermic vertebrates, including mammals, changes in ambient temperature either do not entrain circadian rhythms of locomotor activity or do so poorly (Aschoff and Tokura, 1986; Francis and Coleman, 1997; Hoffmann, 1969; Palkova et al., 1999). Mammals do, however, experience circadian rhythms in core body temperature with a fluctuation of 1–4 °C that are regulated by the SCN (Refinetti and Menaker, 1992). As mentioned previously, cells of most peripheral tissues throughout the mammalian body harbor cell-autonomous circadian oscillators (Balsalobre et al., 1998; Nagoshi et al., 2004; Welsh et al., 1995, 2004; Yoo et al., 2004), which are synchronized to external cues by rhythmic signals from the SCN (Buijs and Kalsbeek, 2001; Earnest et al., 1999; Silver et al., 1996). These observations raise the intriguing question as to what effect, if any, the normal circadian variation in core body temperature may have upon the cell-autonomous oscillators in peripheral tissues. Could body temperature entrain peripheral oscillators in mammals and, perhaps just as interesting, what properties of the SCN prevent it from being synchronized by environmental temperature cycles?

Studies have demonstrated that circadian rhythms of gene expression in cultures of rat-1 fibroblasts (Brown et al., 2002) and rat astrocytes (Prolo et al., 2005) can be entrained to temperature fluctuations of 4 and 1.5 °C, respectively. Although rhythmicity in both cell types damps within a few cycles upon cessation of the temperature rhythms, damping can be delayed by exposing fibroblasts to natural body temperature oscillations (Brown et al., 2002) or by coculturing astrocytes with SCN explants (Prolo et al., 2005). Further, when mice are exposed to inverted environmental temperature cycles of 37 °C during the day and 24 °C during the night, circadian rhythms of gene expression in the liver are reversed without affecting the central clock in the SCN (Brown et al., 2002). It should be mentioned that this phenomenon of decoupling peripheral circadian rhythms from the SCN has also been observed under paradigms of restricted feeding in mammals and that restricted feeding can alter body temperature rhythms (Damiola et al., 2000; Stokkan et al., 2001). Indeed, it has been proposed that food availability and environmental temperature are related, but independent, entraining cues (Brown et al., 2002). Thus, it is clear that in cultured nonneuronal mammalian cells, and in peripheral tissues in vivo in mammals, temperature cycles can entrain circadian rhythms of gene expression.

In contrast to peripheral oscillators, SCN rhythms in vivo are unaffected by environmental temperature changes that phase shift circadian rhythms in other brain regions (Brown et al., 2002). Recent work in our lab has extended this finding by demonstrating that in mice, the resistance of the SCN to entrainment by ambient temperature is not a cell-autonomous property of individual SCN neurons. Rather, this phenomenon is an emergent property of the SCN network and is therefore dependent upon intercellular coupling among neurons (Buhr et al., 2010). Compelling evidence for the role of intercellular coupling comes from experiments showing that blocking either voltage-gated Na^+ channels with TTX or L-type, but not T-type, Ca^{2+} channels with nimodipine renders the SCN susceptible to resetting by temperature pulses. Further, communication between the neurochemically distinct ventral and dorsal regions of the SCN is necessary as cultures of either of these regions alone are shifted in response to temperature changes (Buhr et al., 2010).

Work implying a role for heat shock factor 1 (HSF1) and other components of the heat shock response pathway in circadian gene expression in mammalian liver (Kornmann et al., 2007; Reinke et al., 2008) has encouraged investigation of the possible involvement of this pathway in temperature resetting in mammals. When SCN cultures are pulsed for 1 h with the heat shock pathway antagonist KNK437, no phase shifts occur. In contrast, the same treatment induces strong phase shifts in pituitary and lung cultures, indicating that in non-SCN tissues, KNK437 mimics the effect of 1-h cool

(33.5 °C) pulses in reducing HSF1-mediated transcription (Buhr et al., 2010). Phase shifts to warm (38.6 °C) pulses observed in lung and pituitary are blocked by both KNK437 and quercetin, another HSF1 inhibitor. Continuous inhibition of HSF1-mediated transcription via chronic KNK437 administration to SCN, lung, and pituitary cultures causes an increase in circadian period. This effect is consistent with period lengthening of the circadian rhythm of locomotor activity observed in $Hsf1^{-/-}$ mice (Reinke et al., 2008). Taken together, these results suggest a molecular mechanism that involves HSF1 in temperature resetting in mammalian peripheral tissues (Buhr et al., 2010).

It is important to mention here that two reports, in contrast to the work just presented, provide evidence that rat SCN slice cultures do exhibit entrainment to temperature cycles (Herzog and Huckfeldt, 2003; Ruby et al., 1999). Several differences in these studies may account for this discrepancy. First, the work of Ruby et al. (1999) relied on extracellular recordings from single neurons as glass electrodes advanced along tracks through SCN slices maintained in culture for up to 60 h. Herzog and Huckfeldt (2003) measured PER1::LUC rhythms from neonatal and juvenile rat SCN slices in cultures for up to 2 weeks and observed that these slice preparations were more sensitive to temperature pulses than adult tissues. In contrast, our studies used organotypic cultures of SCN and peripheral tissues from $Per2^{Luc}$ reporter mice and measured in real time the PER2::LUC bioluminescence rhythm continuously for several days (Buhr et al., 2010; Yoo et al., 2004). The in vivo study of Brown et al. (2002) examined mice exposed to environmental temperature changes over a several day period and measured Dbp expression in SCN sections by in situ hybridization. Thus, the SCN culture techniques and output rhythms measured in each study differed. A final possibility is that species-specific differences between mice and rats account for the differences observed in some of these studies. Further work is necessary to clarify these issues.

B. Temperature compensation

One of the hallmarks of circadian rhythms in all organisms, from cyanobacteria to mammals, is that they are temperature compensated—daily rhythms remain constant as temperature increases or decreases across a physiologically viable range (Pittendrigh, 1993). This clock property can be expressed quantitatively as the Q_{10}, or temperature coefficient—the change in the rate of a biological rhythm or biochemical reaction as a result of increasing the ambient temperature by 10 °C. The Q_{10} for most biochemical reactions is 2–3, yet circadian rhythms usually have a Q_{10} between 0.8 and 1.2 (Sweeney and Hastings, 1960). Efforts to elucidate a molecular mechanism underlying the temperature compensation property of circadian clocks have been unsuccessful until recently. The striking demonstration that the circadian rhythm of phosphorylation of the cyanobacterial clock protein, KaiC, can be reconstituted *in vitro* with just three Kai proteins and ATP has led to experiments revealing that this *in vitro* oscillator is also temperature compensated (Nakajima *et al.*, 2005; Tomita *et al.*, 2005). Hence, as the cyanobacterial example aptly illustrates, temperature compensation must be an inherent property of at least some of the biochemical reactions comprising the molecular mechanism of the circadian clock of any given species.

In mammals, temperature compensation has been demonstrated in cell culture for rat-1 (Izumo *et al.*, 2003) and mouse (Dibner *et al.*, 2009; Tsuchiya *et al.*, 2003) fibroblasts, and for neonatal rat SCN neurons (Herzog and Huckfeldt, 2003). As for whole neural tissues, hamster retina (Tosini and Menaker, 1998), and rat (Ruby *et al.*, 1999), mouse (Buhr *et al.*, 2010), and ground squirrel (Ruby and Heller, 1996) SCN are temperature compensated. Several mammalian peripheral tissues also exhibit this property (Reyes *et al.*, 2008), as do human red blood cells (O'Neill and Reddy, 2011). Work by our group has also shown that the temperature compensation observed in SCN tissue is a cell-autonomous property and does not rely on intercellular coupling, unlike the resistance to temperature resetting of the SCN which is an emergent property of the SCN network (Buhr *et al.*, 2010).

Genetic mutations affecting the period of the circadian clock in mammals can also affect temperature compensation as shown for isolated *tau* mutant hamster retinal cultures (Tosini and Menaker, 1998). This is particularly intriguing in light of recent work demonstrating both in cell culture and in cell-free *in vitro* reactions, that the enzymatic activity of CK1δ/ε toward circadian substrates such as PER2 is temperature insensitive, but temperature sensitive toward noncircadian substrates (Isojima *et al.*, 2009). Both wild type and CK1ε^{tau} exhibit temperature insensitive phosphorylation of a peptide containing the PER2 βTrCP-binding region, yet when a peptide substrate derived from the FASPS region of PER2 is tested, an increase in the temperature sensitivity of the CK1ε^{tau} mutant enzyme is revealed (Isojima *et al.*, 2009). An effect of autophosphorylation state on CK1δ/ε activity toward circadian-relevant substrates *in vitro* is also apparent. Thus, similar to the cyanobacterial system mentioned previously, the temperature insensitive phosphorylation of mammalian circadian substrates by CK1δ/ε can be reconstituted *in vitro* and this property of CK1δ/ε is dependent both on the substrate and on the phosphorylation state of the enzyme (Isojima *et al.*, 2009). Additional work will be necessary to replicate these results and to elucidate the biochemical mechanism underlying CK1δ/ε circadian-specific temperature compensation.

IX. UNRESOLVED ISSUES AND FUTURE DIRECTIONS

As mentioned at the beginning of this review, extensive work has shown that, across phyla, the primary molecular mechanism underlying cell-autonomous circadian oscillators is composed of autoregulatory feedback loops of transcription and translation. Hence, the existence of transcription-independent oscillations and a potential role for such oscillations in the function of the cellular clock in mammals and other organisms seem surprising. Indeed, it was the cell-free recapitulation in a test tube of the circadian rhythm of KaiC phosphorylation by colleagues working on the cyanobacterial circadian clock (Nakajima et al., 2005; Tomita et al., 2005) that generated recent interest in transcription-independent clock processes. Further work in cyanobacteria has shown that both circadian transcriptional/translational mechanisms and transcription-independent posttranslational mechanisms are necessary for a competent circadian clock in this organism (Kitayama et al., 2008).

Hints that transcription-independent processes in the mammalian circadian clock exist come from several studies. The surprising discovery that rhythmic transcription of the core clock genes *Bmal1*, *Cry1*, and *Cry2* (Fan et al., 2007; Liu et al., 2008) is not necessary for circadian clock function in mammalian cells, and that the mammalian cellular clock is particularly resilient to attenuation of transcription (Dibner et al., 2009), suggests that mechanisms other than the transcriptional/translational feedback loop are involved in the generation of oscillations. Perhaps the most striking demonstration of a transcription-independent circadian rhythm in mammals is the recent report of a daily oscillation in human red blood cells of the oxidation and subsequent monomer-to-dimer transition of peroxiredoxin (PRX), a protein that inactivates reactive oxygen species (O'Neill and Reddy, 2011). In liver cells, expression of PRX is circadian (Reddy et al., 2006), but this is not possible in mature erythrocytes which have no nucleus. Inhibitors of transcription and translation have no effect on the circadian rhythm of PRX oxidation in human red blood cells further suggesting that the observed PRX oxidation rhythm in erythrocytes is a transcription-independent process. Moreover, detection of another circadian rhythm in erythrocytes—the transition of hemoglobin between dimer and tetramer states—seems also to be transcription independent. A circadian rhythm of PRX oxidation has also been demonstrated by the same group in the green algae *Ostreococcus tauri*, a primitive eukaryote, even when gene expression is halted by exposing these cells to DD (O'Neill et al., 2011). Although these findings must be repeated and validated in other model organisms, they suggest an intriguing avenue for further work.

A major goal toward understanding any biological system is to successfully model that system mathematically. Accomplishing this requires that biologists and modelers work together to incorporate experimental results into models such that they may be used to make testable predictions. The ongoing development of models of the mammalian circadian clock has been particularly helpful recently as they take into account the "combinatorial complexity" of clock component interactions as well as stochastic properties (Yamada and Forger, 2010). Model-based predictions regarding the effects of mutations in core clock genes (Liu et al., 2007b), the role of posttranslational processes on clock components (Gallego et al., 2006a), and the electrical properties of the SCN (Belle et al., 2009) have been validated empirically. The necessity for both molecular noise and intercellular coupling to induce rhythms in a population of SCN neurons has also been predicted via modeling and subsequently confirmed by experiment (Ko et al., 2010). A future challenge to clock modelers will be to address the complex connections and interactions between the mammalian circadian clock and other systems, including the many emerging links between the clock and metabolism.

While there are many potentially fruitful paths of pursuit in the study of mammalian circadian clock genetics, we will conclude by mentioning a few here. Large-scale screening of small molecule libraries may yield promising targets for therapeutic intervention of circadian-related genetic disorders in humans including FASPS and seasonal affective disorder (Liu et al., 2007a). New approaches in synthetic biology in which artificial transcriptional circuits are used to define networks of oscillating genes or to interrogate the function of circadian-related cis-acting elements promise to further our understanding of clock transcriptional pathways (Kumaki et al., 2008; Ueda et al., 2005; Ukai-Tadenuma et al., 2008). Understanding circadian phenotypes that occur in mammals in the absence of underlying changes in DNA sequence—via epigenetic processes—will require ongoing work (Bellet and Sassone-Corsi, 2010; Ripperger and Merrow, 2011). Elucidation of the molecular and biochemical mechanisms of temperature insensitive phenomena in mammalian clock systems remains an important goal (Brown et al., 2002; Buhr et al., 2010; Isojima et al., 2009). In all of these endeavors, the mouse will remain the mammalian genetic model of choice.

X. CONCLUSION

Over the past 10 years, remarkable progress has been made in our understanding of the genetics of the mammalian circadian clock. The transcriptional/translational feedback loop model of the molecular oscillator within cells, for which there is evidence across phyla, has formed the foundation of our understanding of the molecular clockwork. This model, however, must be modified, given the new levels of hierarchy and complexity evident from recent work. It is necessary to study the mammalian clock at all levels from single cells to cell–cell interactions within a tissue, to tissue-level properties, and finally, at the level of behavior. Emergent clock properties arise from the interactions among cells—properties that cannot be studied at the single-cell level. Likewise, focusing on behavior or tissue-specific processes alone will overlook cell-autonomous clock properties. New advances in reporter technology, microarrays, mathematical modeling, perturbation analysis methods, and systems biology will continue to elucidate properties of the molecular clock. Hence, a lesson learned from the work presented herein is that with respect to the mammalian circadian clock, a systems-wide approach must be advocated. In the near future, we should look forward to a better understanding of how the mammalian clockwork is integrated with the other physiological systems of the body and perhaps be better able to develop therapies for human clock-related disorders.

Acknowledgments

This work was supported by NIH Grant 1R15GM086825-01 to P. L. L. and NIH Grants U01 MH61915, P50 MH074924, and R01 MH078024 to J. S. T. J. S. T. is an investigator in the Howard Hughes Medical Institute.

References

Abe, M., Herzog, E. D., and Block, G. D. (2000). Lithium lengthens the circadian period of individual suprachiasmatic nucleus neurons. *Neuroreport* **11**, 3261–3264.

Abraham, U., Granada, A. E., Westermark, P. O., Heine, M., Kramer, A., and Herzel, H. (2010). Coupling governs entrainment range of circadian clocks. *Mol. Syst. Biol.* **6**, 438.

Abrahamson, E. E., and Moore, R. Y. (2001). Suprachiasmatic nucleus in the mouse: Retinal innervation, intrinsic organization and efferent projections. *Brain Res.* **916**, 172–191.

Adams, D. J., and van der Weyden, L. (2008). Contemporary approaches for modifying the mouse genome. *Physiol. Genomics* **34**, 225–238.

Akashi, M., and Takumi, T. (2005). The orphan nuclear receptor RORα regulates circadian transcription of the mammalian core-clock Bmal1. *Nat. Struct. Mol. Biol.* **12**, 441–448.

Akashi, M., Tsuchiya, Y., Yoshino, T., and Nishida, E. (2002). Control of intracellular dynamics of mammalian period proteins by casein kinase I ε (CKIε) and CKIδ in cultured cells. *Mol. Cell. Biol.* **22**, 1693–1703.

Akhtar, R. A., Reddy, A. B., Maywood, E. S., Clayton, J. D., King, V. M., Smith, A. G., Gant, T. W., Hastings, M. H., and Kyriacou, C. P. (2002). Circadian cycling of the mouse liver transcriptome, as revealed by cDNA microarray, is driven by the suprachiasmatic nucleus. *Curr. Biol.* **12**, 540–550.

Albrecht, U., Sun, Z. S., Eichele, G., and Lee, C. C. (1997). A differential response of two putative mammalian circadian regulators, *mper1* and *mper2*, to light. *Cell* **91**, 1055–1064.

Ali, A., Hoeflich, K. P., and Woodgett, J. R. (2001). Glycogen synthase kinase-3: Properties, functions, and regulation. *Chem. Rev.* **101**, 2527–2540.

Alvarez-Saavedra, M., Antoun, G., Yanagiya, A., Oliva-Hernandez, R., Cornejo-Palma, D., Perez-Iratxeta, C., Sonenberg, N., and Cheng, H. Y. (2011). miRNA-132 orchestrates chromatin remodeling and translational control of the circadian clock. *Hum. Mol. Genet.* **20,** 731–751.

Antoch, M. P., Song, E. J., Chang, A. M., Vitaterna, M. H., Zhao, Y., Wilsbacher, L. D., Sangoram, A. M., King, D. P., Pinto, L. H., and Takahashi, J. S. (1997). Functional identification of the mouse circadian *Clock* gene by transgenic BAC rescue. *Cell* **89,** 655–667.

Aschoff, J., and Tokura, H. (1986). Circadian activity rhythms in squirrel monkeys: Entrainment by temperature cycles. *J. Biol. Rhythms* **1,** 91–99.

Asher, G., and Schibler, U. (2011). Crosstalk between components of circadian and metabolic cycles in mammals. *Cell Metab.* **13,** 125–137.

Asher, G., Gatfield, D., Stratmann, M., Reinke, H., Dibner, C., Kreppel, F., Mostoslavsky, R., Alt, F. W., and Schibler, U. (2008). SIRT1 regulates circadian clock gene expression through PER2 deacetylation. *Cell* **134,** 317–328.

Aston-Jones, G., Chen, S., Zhu, Y., and Oshinsky, M. L. (2001). A neural circuit for circadian regulation of arousal. *Nat. Neurosci.* **4,** 732–738.

Aton, S. J., Colwell, C. S., Harmar, A. J., Waschek, J., and Herzog, E. D. (2005). Vasoactive intestinal polypeptide mediates circadian rhythmicity and synchrony in mammalian clock neurons. *Nat. Neurosci.* **8,** 476–483.

Bacon, Y., Ooi, A., Kerr, S., Shaw-Andrews, L., Winchester, L., Breeds, S., Tymoska-Lalanne, Z., Clay, J., Greenfield, A. G., and Nolan, P. M. (2004). Screening for novel ENU-induced rhythm, entrainment and activity mutants. *Genes Brain Behav.* **3,** 196–205.

Badura, L., Swanson, T., Adamowicz, W., Adams, J., Cianfrogna, J., Fisher, K., Holland, J., Kleiman, R., Nelson, F., Reynolds, L., *et al.* (2007). An inhibitor of casein kinase Iε induces phase delays in circadian rhythms under free-running and entrained conditions. *J. Pharmacol. Exp. Ther.* **322,** 730–738.

Bae, K., Jin, X., Maywood, E. S., Hastings, M. H., Reppert, S. M., and Weaver, D. R. (2001). Differential functions of *mPer1*, *mPer2*, and *mPer3* in the SCN circadian clock. *Neuron* **30,** 525–536.

Baggs, J. E., Price, T. S., DiTacchio, L., Panda, S., Fitzgerald, G. A., and Hogenesch, J. B. (2009). Network features of the mammalian circadian clock. *PLoS Biol.* **7,** e52.

Balsalobre, A., Damiola, F., and Schibler, U. (1998). A serum shock induces circadian gene expression in mammalian tissue culture cells. *Cell* **93,** 929–937.

Balsalobre, A., Brown, S. A., Marcacci, L., Tronche, F., Kellendonk, C., Reichardt, H. M., Schutz, G., and Schibler, U. (2000). Resetting of circadian time in peripheral tissues by glucocorticoid signaling. *Science* **289,** 2344–2347.

Bartel, D. P. (2009). MicroRNAs: Target recognition and regulatory functions. *Cell* **136,** 215–233.

Bass, J., and Takahashi, J. S. (2010). Circadian integration of metabolism and energetics. *Science* **330,** 1349–1354.

Belle, M. D., Diekman, C. O., Forger, D. B., and Piggins, H. D. (2009). Daily electrical silencing in the mammalian circadian clock. *Science* **326,** 281–284.

Bellet, M. M., and Sassone-Corsi, P. (2010). Mammalian circadian clock and metabolism—The epigenetic link. *J. Cell Sci.* **123,** 3837–3848.

Berson, D. M., Dunn, F. A., and Takao, M. (2002). Phototransduction by retinal ganglion cells that set the circadian clock. *Science* **295,** 1070–1073.

Blake, J. A., Bult, C. J., Kadin, J. A., Richardson, J. E., and Eppig, J. T. (2010). The Mouse Genome Database (MGD): Premier model organism resource for mammalian genomics and genetics. *Nucleic Acids Res.* **39,** D842–D848.

Brown, S., Zumbrunn, G., Fleury-Olela, F., Preitner, N., and Schibler, U. (2002). Rhythms of mammalian body temperature can sustain peripheral circadian clocks. *Curr. Biol.* **12,** 1574–1583.

Brown, T. M., Colwell, C. S., Waschek, J. A., and Piggins, H. D. (2007). Disrupted neuronal activity rhythms in the suprachiasmatic nuclei of vasoactive intestinal polypeptide-deficient mice. *J. Neurophysiol.* **97,** 2553–2558.

Buhr, E. D., Yoo, S. H., and Takahashi, J. S. (2010). Temperature as a universal resetting cue for mammalian circadian oscillators. *Science* **330,** 379–385.

Buijs, R. M., and Kalsbeek, A. (2001). Hypothalamic integration of central and peripheral clocks. *Nat. Rev. Neurosci.* **2,** 521–526.

Bunger, M. K., Wilsbacher, L. D., Moran, S. M., Clendenin, C., Radcliffe, L. A., Hogenesch, J. B., Simon, M. C., Takahashi, J. S., and Bradfield, C. A. (2000). *Mop3* is an essential component of the master circadian pacemaker in mammals. *Cell* **103,** 1009–1017.

Bushati, N., and Cohen, S. M. (2007). microRNA functions. *Annu. Rev. Cell Dev. Biol.* **23,** 175–205.

Busino, L., Bassermann, F., Maiolica, A., Lee, C., Nolan, P. M., Godinho, S. I., Draetta, G. F., and Pagano, M. (2007). SCF[Fbxl3] controls the oscillation of the circadian clock by directing the degradation of cryptochrome proteins. *Science* **316,** 900–904.

Camacho, F., Cilio, M., Guo, Y., Virshup, D. M., Patel, K., Khorkova, O., Styren, S., Morse, B., Yao, Z., and Keesler, G. A. (2001). Human casein kinase Iδ phosphorylation of human circadian clock proteins *period* 1 and 2. *FEBS Lett.* **489,** 159–165.

Cardone, L., Hirayama, J., Giordano, F., Tamaru, T., Palvimo, J. J., and Sassone-Corsi, P. (2005). Circadian clock control by SUMOylation of BMAL1. *Science* **309,** 1390–1394.

Cardozo, T., and Pagano, M. (2004). The SCF ubiquitin ligase: Insights into a molecular machine. *Nat. Rev. Mol. Cell Biol.* **5,** 739–751.

Cermakian, N., Monaco, L., Pando, M. P., Dierich, A., and Sassone-Corsi, P. (2001). Altered behavioral rhythms and clock gene expression in mice with a targeted mutation in the *Period1* gene. *EMBO J.* **20,** 3967–3974.

Chen, R., Schirmer, A., Lee, Y., Lee, H., Kumar, V., Yoo, S. H., Takahashi, J. S., and Lee, C. (2009). Rhythmic PER abundance defines a critical nodal point for negative feedback within the circadian clock mechanism. *Mol. Cell* **36,** 417–430.

Cheng, M. Y., Bullock, C. M., Li, C., Lee, A. G., Bermak, J. C., Belluzzi, J., Weaver, D. R., Leslie, F. M., and Zhou, Q. Y. (2002). Prokineticin 2 transmits the behavioural circadian rhythm of the suprachiasmatic nucleus. *Nature* **417,** 405–410.

Cheng, H. Y., Papp, J. W., Varlamova, O., Dziema, H., Russell, B., Curfman, J. P., Nakazawa, T., Shimizu, K., Okamura, H., Impey, S., et al. (2007). microRNA modulation of circadian-clock period and entrainment. *Neuron* **54,** 813–829.

Ciechanover, A., Orian, A., and Schwartz, A. L. (2000). Ubiquitin-mediated proteolysis: Biological regulation via destruction. *Bioessays* **22,** 442–451.

Clark, A. T., Goldowitz, D., Takahashi, J. S., Vitaterna, M. H., Siepka, S. M., Peters, L. L., Frankel, W. N., Carlson, G. A., Rossant, J., Nadeau, J. H., et al. (2004). Implementing large-scale ENU mutagenesis screens in North America. *Genetica* **122,** 51–64.

Colwell, C. S., Michel, S., Itri, J., Rodriguez, W., Tam, J., Lelievre, V., Hu, Z., Liu, X., and Waschek, J. A. (2003). Disrupted circadian rhythms in VIP- and PHI-deficient mice. *Am. J. Physiol. Regul. Integr. Comp. Physiol.* **285,** R939–R949.

Crosio, C., Cermakian, N., Allis, C. D., and Sassone-Corsi, P. (2000). Light induces chromatin modification in cells of the mammalian circadian clock. *Nat. Neurosci.* **3,** 1241–1247.

Curtis, A. M., Seo, S. B., Westgate, E. J., Rudic, R. D., Smyth, E. M., Chakravarti, D., FitzGerald, G. A., and McNamara, P. (2004). Histone acetyltransferase-dependent chromatin remodeling and the vascular clock. *J. Biol. Chem.* **279,** 7091–7097.

Cutler, D. J., Haraura, M., Reed, H. E., Shen, S., Sheward, W. J., Morrison, C. F., Marston, H. M., Harmar, A. J., and Piggins, H. D. (2003). The mouse VPAC$_2$ receptor confers suprachiasmatic nuclei cellular rhythmicity and responsiveness to vasoactive intestinal polypeptide *in vitro*. *Eur. J. Neurosci.* **17,** 197–204.

Dacey, D. M., Liao, H. W., Peterson, B. B., Robinson, F. R., Smith, V. C., Pokorny, J., Yau, K. W., and Gamlin, P. D. (2005). Melanopsin-expressing ganglion cells in primate retina signal colour and irradiance and project to the LGN. *Nature* **433,** 749–754.

Damiola, F., Le Minh, N., Preitner, N., Kornmann, B., Fleury-Olela, F., and Schibler, U. (2000). Restricted feeding uncouples circadian oscillators in peripheral tissues from the central pacemaker in the suprachiasmatic nucleus. *Genes Dev.* **14,** 2950–2961.

Dardente, H., Fortier, E. E., Martineau, V., and Cermakian, N. (2007). Cryptochromes impair phosphorylation of transcriptional activators in the clock: A general mechanism for circadian repression. *Biochem. J.* **402,** 525–536.

Debruyne, J. P., Noton, E., Lambert, C. M., Maywood, E. S., Weaver, D. R., and Reppert, S. M. (2006). A clock shock: Mouse CLOCK is not required for circadian oscillator function. *Neuron* **50,** 465–477.

DeBruyne, J. P., Weaver, D. R., and Reppert, S. M. (2007a). CLOCK and NPAS2 have overlapping roles in the suprachiasmatic circadian clock. *Nat. Neurosci.* **10,** 543–545.

DeBruyne, J. P., Weaver, D. R., and Reppert, S. M. (2007b). Peripheral circadian oscillators require CLOCK. *Curr. Biol.* **17,** R538–R539.

Deery, M. J., Maywood, E. S., Chesham, J. E., Sladek, M., Karp, N. A., Green, E. W., Charles, P. D., Reddy, A. B., Kyriacou, C. P., Lilley, K. S., et al. (2009). Proteomic analysis reveals the role of synaptic vesicle cycling in sustaining the suprachiasmatic circadian clock. *Curr. Biol.* **19,** 2031–2036.

Dibner, C., Sage, D., Unser, M., Bauer, C., d'Eysmond, T., Naef, F., and Schibler, U. (2009). Circadian gene expression is resilient to large fluctuations in overall transcription rates. *EMBO J.* **28,** 123–134.

Do, M. T., and Yau, K. W. (2010). Intrinsically photosensitive retinal ganglion cells. *Physiol. Rev.* **90,** 1547–1581.

Doble, B. W., and Woodgett, J. R. (2003). GSK-3: Tricks of the trade for a multi-tasking kinase. *J. Cell Sci.* **116,** 1175–1186.

Doherty, C. J., and Kay, S. A. (2010). Circadian control of global gene expression patterns. *Annu. Rev. Genet.* **44,** 419–444.

Doi, M., Hirayama, J., and Sassone-Corsi, P. (2006). Circadian regulator CLOCK is a histone acetyltransferase. *Cell* **125,** 497–508.

Dudley, C. A., Erbel-Sieler, C., Estill, S. J., Reick, M., Franken, P., Pitts, S., and McKnight, S. L. (2003). Altered patterns of sleep and behavioral adaptability in NPAS2-deficient mice. *Science* **301,** 379–383.

Duffield, G. E., Best, J. D., Meurers, B. H., Bittner, A., Loros, J. J., and Dunlap, J. C. (2002). Circadian programs of transcriptional activation, signaling, and protein turnover revealed by microarray analysis of mammalian cells. *Curr. Biol.* **12,** 551–557.

Dunlap, J. C. (1999). Molecular bases for circadian clocks. *Cell* **96,** 271–290.

Earnest, D. J., Liang, F. Q., Ratcliff, M., and Cassone, V. M. (1999). Immortal time: Circadian clock properties of rat suprachiasmatic cell lines. *Science* **283,** 693–695.

Ebling, F. J. (1996). The role of glutamate in the photic regulation of the suprachiasmatic nucleus. *Prog. Neurobiol.* **50,** 109–132.

Eide, E. J., Kang, H., Crapo, S., Gallego, M., and Virshup, D. M. (2005a). Casein kinase I in the mammalian circadian clock. *Methods Enzymol.* **393,** 408–418.

Eide, E. J., Woolf, M. F., Kang, H., Woolf, P., Hurst, W., Camacho, F., Vielhaber, E. L., Giovanni, A., and Virshup, D. M. (2005b). Control of mammalian circadian rhythm by CKIɛ-regulated proteasome-mediated PER2 degradation. *Mol. Cell. Biol.* **25,** 2795–2807.

Etchegaray, J. P., Lee, C., Wade, P. A., and Reppert, S. M. (2003). Rhythmic histone acetylation underlies transcription in the mammalian circadian clock. *Nature* **421,** 177–182.

Etchegaray, J. P., Machida, K. K., Noton, E., Constance, C. M., Dallmann, R., Di Napoli, M. N., DeBruyne, J. P., Lambert, C. M., Yu, E. A., Reppert, S. M., et al. (2009). Casein kinase 1 delta regulates the pace of the mammalian circadian clock. Mol. Cell. Biol. **29**, 3853–3866.

Etchegaray, J. P., Yu, E. A., Indic, P., Dallmann, R., and Weaver, D. R. (2010). Casein kinase 1 delta (CK1δ) regulates period length of the mouse suprachiasmatic circadian clock in vitro. PLoS One **5**, e10303.

Fan, Y., Hida, A., Anderson, D. A., Izumo, M., and Johnson, C. H. (2007). Cycling of CRYPTO-CHROME proteins is not necessary for circadian-clock function in mammalian fibroblasts. Curr. Biol. **17**, 1091–1100.

Finkel, T., Deng, C. X., and Mostoslavsky, R. (2009). Recent progress in the biology and physiology of sirtuins. Nature **460**, 587–591.

Foster, R. G., Provencio, I., Hudson, D., Fiske, S., De Grip, W., and Menaker, M. (1991). Circadian photoreception in the retinally degenerate mouse (rd/rd). J. Comp. Physiol. A **A169**, 39–50.

Fox, J., Barthold, S., Davvison, M., Newcomer, C., Quimby, F., and Smith, A. (eds.) (2007). The Mouse in Biomedical Research. 2nd edn., Elsevier, Boston. 4 Vols.

Francis, A. J., and Coleman, G. J. (1997). Phase response curves to ambient temperature pulses in rats. Physiol. Behav. **62**, 1211–1217.

Freedman, M. S., Lucas, R. J., Soni, B., von Schantz, M., Munoz, M., David-Gray, Z., and Foster, R. (1999). Regulation of mammalian circadian behavior by non-rod, non-cone, ocular photoreceptors. Science **284**, 502–504.

Friedman, R. C., Farh, K. K. H., Burge, C. B., and Bartel, D. P. (2008). Most mammalian mRNAs are conserved targets of microRNAs. Genome Res. **19**, 92–105.

Fu, Y., Zhong, H., Wang, M. H., Luo, D. G., Liao, H. W., Maeda, H., Hattar, S., Frishman, L. J., and Yau, K. W. (2005). Intrinsically photosensitive retinal ganglion cells detect light with a vitamin A-based photopigment, melanopsin. Proc. Natl. Acad. Sci. USA **102**, 10339–10344.

Gallego, M., and Virshup, D. M. (2007). Post-translational modifications regulate the ticking of the circadian clock. Nat. Rev. Mol. Cell Biol. **8**, 139–148.

Gallego, M., Eide, E. J., Woolf, M. F., Virshup, D. M., and Forger, D. B. (2006a). An opposite role for tau in circadian rhythms revealed by mathematical modeling. Proc. Natl. Acad. Sci. USA **103**, 10618–10623.

Gallego, M., Kang, H., and Virshup, D. M. (2006b). Protein phosphatase 1 regulates the stability of the circadian protein PER2. Biochem. J. **399**, 169–175.

Gareau, J. R., and Lima, C. D. (2010). The SUMO pathway: Emerging mechanisms that shape specificity, conjugation and recognition. Nat. Rev. Mol. Cell Biol. **11**, 861–871.

Gatfield, D., Le Martelot, G., Vejnar, C. E., Gerlach, D., Schaad, O., Fleury-Olela, F., Ruskeepaa, A. L., Oresic, M., Esau, C. C., Zdobnov, E. M., et al. (2009). Integration of microRNA miR-122 in hepatic circadian gene expression. Genes Dev. **23**, 1313–1326.

Gekakis, N., Staknis, D., Nguyen, H. B., Davis, F. C., Wilsbacher, L. D., King, D. P., Takahashi, J. S., and Weitz, C. J. (1998). Role of the CLOCK protein in the mammalian circadian mechanism. Science **280**, 1564–1569.

Gietzen, K. F., and Virshup, D. M. (1999). Identification of inhibitory autophosphorylation sites in casein kinase Iε. J. Biol. Chem. **274**, 32063–32070.

Godinho, S. I., Maywood, E. S., Shaw, L., Tucci, V., Barnard, A. R., Busino, L., Pagano, M., Kendall, R., Quwailid, M. M., Romero, M. R., et al. (2007). The after-hours mutant reveals a role for Fbxl3 in determining mammalian circadian period. Science **316**, 897–900.

Golombek, D. A., and Rosenstein, R. E. (2010). Physiology of circadian entrainment. Physiol. Rev. **90**, 1063–1102.

Gooley, J. J., Lu, J., Chou, T. C., Scammell, T. E., and Saper, C. B. (2001). Melanopsin in cells of origin of the retinohypothalamic tract. Nat. Neurosci. **4**, 1165.

Goz, D., Studholme, K., Lappi, D. A., Rollag, M. D., Provencio, I., and Morin, L. P. (2008). Targeted destruction of photosensitive retinal ganglion cells with a saporin conjugate alters the effects of light on mouse circadian rhythms. *PLoS One* **3,** e3153.

Granados-Fuentes, D., Prolo, L. M., Abraham, U., and Herzog, E. D. (2004a). The suprachiasmatic nucleus entrains, but does not sustain, circadian rhythmicity in the olfactory bulb. *J. Neurosci.* **24,** 615–619.

Granados-Fuentes, D., Saxena, M. T., Prolo, L. M., Aton, S. J., and Herzog, E. D. (2004b). Olfactory bulb neurons express functional, entrainable circadian rhythms. *Eur. J. Neurosci.* **19,** 898–906.

Granados-Fuentes, D., Tseng, A., and Herzog, E. D. (2006). A circadian clock in the olfactory bulb controls olfactory responsivity. *J. Neurosci.* **26,** 12219–12225.

Gratacós, F. M., and Brewer, G. (2010). The role of AUF1 in regulated mRNA decay. *WIREs RNA* **1,** 457–473.

Green, C. B., Douris, N., Kojima, S., Strayer, C. A., Fogerty, J., Lourim, D., Keller, S. R., and Besharse, J. C. (2007). Loss of Nocturnin, a circadian deadenylase, confers resistance to hepatic steatosis and diet-induced obesity. *Proc. Natl. Acad. Sci. USA* **104,** 9888–9893.

Green, C. B., Takahashi, J. S., and Bass, J. (2008). The meter of metabolism. *Cell* **134,** 728–742.

Griffin, E. A., Jr., Staknis, D., and Weitz, C. J. (1999). Light-independent role of CRY1 and CRY2 in the mammalian circadian clock. *Science* **286,** 768–771.

Guillaumond, F., Dardente, H., Giguere, V., and Cermakian, N. (2005). Differential control of Bmal1 circadian transcription by REV-ERB and ROR nuclear receptors. *J. Biol. Rhythms* **20,** 391–403.

Guo, H., Ingolia, N. T., Weissman, J. S., and Bartel, D. P. (2010). Mammalian microRNAs predominantly act to decrease target mRNA levels. *Nature* **466,** 835–840.

Hannibal, J. (2002). Neurotransmitters of the retino-hypothalamic tract. *Cell Tissue Res.* **309,** 73–88.

Hannibal, J., Hindersson, P., Ostergaard, J., Georg, B., Heegaard, S., Larsen, P. J., and Fahrenkrug, J. (2004). Melanopsin is expressed in PACAP-containing retinal ganglion cells of the human retinohypothalamic tract. *Invest. Ophthalmol. Vis. Sci.* **45,** 4202–4209.

Harada, Y., Sakai, M., Kurabayashi, N., Hirota, T., and Fukada, Y. (2005). Ser-557-phosphorylated mCRY2 is degraded upon synergistic phosphorylation by glycogen synthase kinase-3β. *J. Biol. Chem.* **280,** 31714–31721.

Harmar, A. J., Marston, H. M., Shen, S., Spratt, C., West, K. M., Sheward, W. J., Morrison, C. F., Dorin, J. R., Piggins, H. D., Reubi, J. C., *et al.* (2002). The VPAC$_2$ receptor is essential for circadian function in the mouse suprachiasmatic nuclei. *Cell* **109,** 497–508.

Hatcher, N. G., Atkins, N., Jr., Annangudi, S. P., Forbes, A. J., Kelleher, N. L., Gillette, M. U., and Sweedler, J. V. (2008). Mass spectrometry-based discovery of circadian peptides. *Proc. Natl. Acad. Sci. USA* **105,** 12527–12532.

Hatori, M., Le, H., Vollmers, C., Keding, S. R., Tanaka, N., Buch, T., Waisman, A., Schmedt, C., Jegla, T., and Panda, S. (2008). Inducible ablation of melanopsin-expressing retinal ganglion cells reveals their central role in non-image forming visual responses. *PLoS One* **3,** e2451.

Hattar, S., Liao, H. W., Takao, M., Berson, D. M., and Yau, K. W. (2002). Melanopsin-containing retinal ganglion cells: Architecture, projections, and intrinsic photosensitivity. *Science* **295,** 1065–1070.

Hattar, S., Lucas, R. J., Mrosovsky, N., Thompson, S., Douglas, R. H., Hankins, M. W., Lem, J., Biel, M., Hofmann, F., Foster, R. G., *et al.* (2003). Melanopsin and rod-cone photoreceptive systems account for all major accessory visual functions in mice. *Nature* **424,** 75–81.

Hedrich, H. J., and Bullock, G. (eds.) (2004). The Laboratory Mouse. Elsevier Academic Press, San Diego, CA.

Herzog, E. D., and Huckfeldt, R. M. (2003). Circadian entrainment to temperature, but not light, in the isolated suprachiasmatic nucleus. *J. Neurophysiol.* **90,** 763–770.

Herzog, E. D., Takahashi, J. S., and Block, G. D. (1998). Clock controls circadian period in isolated suprachiasmatic nucleus neurons. *Nat. Neurosci.* **1,** 708–713.

Hirayama, J., Sahar, S., Grimaldi, B., Tamaru, T., Takamatsu, K., Nakahata, Y., and Sassone-Corsi, P. (2007). CLOCK-mediated acetylation of BMAL1 controls circadian function. *Nature* **450**, 1086–1090.

Hirota, T., Lewis, W. G., Liu, A. C., Lee, J. W., Schultz, P. G., and Kay, S. A. (2008). A chemical biology approach reveals period shortening of the mammalian circadian clock by specific inhibition of GSK-3β. *Proc. Natl. Acad. Sci. USA* **105**, 20746–20751.

Hirota, T., Lee, J. W., Lewis, W. G., Zhang, E. E., Breton, G., Liu, X., Garcia, M., Peters, E. C., Etchegaray, J. P., Traver, D., et al. (2010). High-throughput chemical screen identifies a novel potent modulator of cellular circadian rhythms and reveals CKIα as a clock regulatory kinase. *PLoS Biol.* **8**, e1000559.

Hoeflich, K. P., Luo, J., Rubie, E. A., Tsao, M. S., Jin, O., and Woodgett, J. R. (2000). Requirement for glycogen synthase kinase-3β in cell survival and NF-κB activation. *Nature* **406**, 86–90.

Hoffmann, K. (1969). Die relative Wirksamkeit von Zeitgebern. *Oecologia* **3**, 184–206.

Hogenesch, J. B., Gu, Y. Z., Moran, S. M., Shimomura, K., Radcliffe, L. A., Takahashi, J. S., and Bradfield, C. A. (2000). The basic helix-loop-helix-PAS protein MOP9 is a brain-specific heterodimeric partner of circadian and hypoxia factors. *J. Neurosci.* **20**, RC83.

Honma, K., and Honma, S. (2009). The SCN-independent clocks, methamphetamine and food restriction. *Eur. J. Neurosci.* **30**, 1707–1717.

Honma, S., Shirakawa, T., Katsuno, Y., Namihira, M., and Honma, K. (1998). Circadian periods of single suprachiasmatic neurons in rats. *Neurosci. Lett.* **250**, 157–160.

Honma, S., Yasuda, T., Yasui, A., van der Horst, G. T., and Honma, K. (2008). Circadian behavioral rhythms in Cry1/Cry2 double-deficient mice induced by methamphetamine. *J. Biol. Rhythms* **23**, 91–94.

Hughes, A. T., Guilding, C., Lennox, L., Samuels, R. E., McMahon, D. G., and Piggins, H. D. (2008). Live imaging of altered *period1* expression in the suprachiasmatic nuclei of Vipr2$^{-/-}$ mice. *J. Neurochem.* **106**, 1646–1657.

Hughes, M. E., Hogenesch, J. B., and Kornacker, K. (2010). JTK_CYCLE: An efficient nonparametric algorithm for detecting rhythmic components in genome-scale data sets. *J. Biol. Rhythms* **25**, 372–380.

Iitaka, C., Miyazaki, K., Akaike, T., and Ishida, N. (2005). A role for glycogen synthase kinase-3β in the mammalian circadian clock. *J. Biol. Chem.* **280**, 29397–29402.

Imhof, A., and Becker, P. B. (2001). Modifications of the histone N-terminal domains. Evidence for an "epigenetic code"? *Mol. Biotechnol.* **17**, 1–13.

Impey, S., McCorkle, S. R., Cha-Molstad, H., Dwyer, J. M., Yochum, G. S., Boss, J. M., McWeeney, S., Dunn, J. J., Mandel, G., and Goodman, R. H. (2004). Defining the CREB regulon: A genome-wide analysis of transcription factor regulatory regions. *Cell* **119**, 1041–1054.

Isojima, Y., Nakajima, M., Ukai, H., Fujishima, H., Yamada, R. G., Masumoto, K. H., Kiuchi, R., Ishida, M., Ukai-Tadenuma, M., Minami, Y., et al. (2009). CKIε/δ-dependent phosphorylation is a temperature-insensitive, period-determining process in the mammalian circadian clock. *Proc. Natl. Acad. Sci. USA* **106**, 15744–15749.

Iwahana, E., Akiyama, M., Miyakawa, K., Uchida, A., Kasahara, J., Fukunaga, K., Hamada, T., and Shibata, S. (2004). Effect of lithium on the circadian rhythms of locomotor activity and glycogen synthase kinase-3 protein expression in the mouse suprachiasmatic nuclei. *Eur. J. Neurosci.* **19**, 2281–2287.

Izumo, M., Johnson, C. H., and Yamazaki, S. (2003). Circadian gene expression in mammalian fibroblasts revealed by real-time luminescence reporting: Temperature compensation and damping. *Proc. Natl. Acad. Sci. USA* **100**, 16089–16094.

Jacob, H. J. (1996). A landmark for orphan genomes? *Nat. Genet.* **13**, 14–15.

Jakubcakova, V., Oster, H., Tamanini, F., Cadenas, C., Leitges, M., van der Horst, G. T., and Eichele, G. (2007). Light entrainment of the mammalian circadian clock by a PRKCA-dependent posttranslational mechanism. *Neuron* **54,** 831–843.

Jenuwein, T., and Allis, C. D. (2001). Translating the histone code. *Science* **293,** 1074–1080.

Jin, X., von Gall, C., Pieschl, R. L., Gribkoff, V. K., Stehle, J. H., Reppert, S. M., and Weaver, D. R. (2003). Targeted disruption of the mouse Mel_{1b} melatonin receptor. *Mol. Cell. Biol.* **23,** 1054–1060.

Kaladchibachi, S. A., Doble, B., Anthopoulos, N., Woodgett, J. R., and Manoukian, A. S. (2007). Glycogen synthase kinase 3, circadian rhythms, and bipolar disorder: A molecular link in the therapeutic action of lithium. *J. Circadian Rhythms* **5,** 3.

Kalamvoki, M., and Roizman, B. (2010). Circadian CLOCK histone acetyl transferase localizes at ND10 nuclear bodies and enables herpes simplex virus gene expression. *Proc. Natl. Acad. Sci. USA* **107,** 17721–17726.

Keesler, G. A., Camacho, F., Guo, Y., Virshup, D., Mondadori, C., and Yao, Z. (2000). Phosphorylation and destabilization of human period I clock protein by human casein kinase I epsilon. *Neuroreport* **11,** 951–955.

King, D. P., Vitaterna, M. H., Chang, A. M., Dove, W. F., Pinto, L. H., Turek, F. W., and Takahashi, J. S. (1997a). The mouse *Clock* mutation behaves as an antimorph and maps within the W^{19H} deletion, distal of *Kit*. *Genetics* **146,** 1049–1060.

King, D. P., Zhao, Y., Sangoram, A. M., Wilsbacher, L. D., Tanaka, M., Antoch, M. P., Steeves, T. D., Vitaterna, M. H., Kornhauser, J. M., Lowrey, P. L., *et al.* (1997b). Positional cloning of the mouse circadian *Clock* gene. *Cell* **89,** 641–653.

Kitayama, Y., Nishiwaki, T., Terauchi, K., and Kondo, T. (2008). Dual KaiC-based oscillations constitute the circadian system of cyanobacteria. *Genes Dev.* **22,** 1513–1521.

Klein, P. S., and Melton, D. A. (1996). A molecular mechanism for the effect of lithium on development. *Proc. Natl. Acad. Sci. USA* **93,** 8455–8459.

Kloss, B., Price, J. L., Saez, L., Blau, J., Rothenfluh, A., Wesley, C. S., and Young, M. W. (1998). The *Drosophila* clock gene *double-time* encodes a protein closely related to human casein kinase Iε. *Cell* **94,** 97–107.

Knippschild, U., Gocht, A., Wolff, S., Huber, N., Lohler, J., and Stoter, M. (2005). The casein kinase 1 family: Participation in multiple cellular processes in eukaryotes. *Cell. Signal.* **17,** 675–689.

Ko, C. H., and Takahashi, J. S. (2006). Molecular components of the mammalian circadian clock. *Hum. Mol. Genet.* **15 Spec No 2,** R271–R277.

Ko, C. H., Yamada, Y. R., Welsh, D. K., Buhr, E. D., Liu, A. C., Zhang, E. E., Ralph, M. R., Kay, S. A., Forger, D. B., and Takahashi, J. S. (2010). Emergence of noise-induced oscillations in the central circadian pacemaker. *PLoS Biol.* **8,** e1000513.

Kojima, S., Matsumoto, K., Hirose, M., Shimada, M., Nagano, M., Shigeyoshi, Y., Hoshino, S., Ui-Tei, K., Saigo, K., Green, C. B., *et al.* (2007). LARK activates posttranscriptional expression of an essential mammalian clock protein, PERIOD1. *Proc. Natl. Acad. Sci. USA* **104,** 1859–1864.

Kojima, S., Gatfield, D., Esau, C. C., and Green, C. B. (2010). MicroRNA-122 modulates the rhythmic expression profile of the circadian deadenylase *Nocturnin* in mouse liver. *PLoS One* **5,** e11264.

Kojima, S., Shingle, D. L., and Green, C. B. (2011). Post-transcriptional control of circadian rhythms. *J. Cell Sci.* **124,** 311–320.

Konopka, R. J., and Benzer, S. (1971). Clock mutants of *Drosophila melanogaster*. *Proc. Natl. Acad. Sci. USA* **68,** 2112–2116.

Kornitzer, D., and Ciechanover, A. (2000). Modes of regulation of ubiquitin-mediated protein degradation. *J. Cell. Physiol.* **182,** 1–11.

Kornmann, B., Schaad, O., Bujard, H., Takahashi, J. S., and Schibler, U. (2007). System-driven and oscillator-dependent circadian transcription in mice with a conditionally active liver clock. *PLoS Biol.* **5**, e34.

Kramer, A., Yang, F. C., Snodgrass, P., Li, X., Scammell, T. E., Davis, F. C., and Weitz, C. J. (2001). Regulation of daily locomotor activity and sleep by hypothalamic EGF receptor signaling. *Science* **294**, 2511–2515.

Kramer, A., Yang, F. C., Kraves, S., and Weitz, C. J. (2005). A screen for secreted factors of the suprachiasmatic nucleus. *Methods Enzymol.* **393**, 645–663.

Kraves, S., and Weitz, C. J. (2006). A role for cardiotrophin-like cytokine in the circadian control of mammalian locomotor activity. *Nat. Neurosci.* **9**, 212–219.

Krutzfeldt, J., Rajewsky, N., Braich, R., Rajeev, K. G., Tuschl, T., Manoharan, M., and Stoffel, M. (2005). Silencing of microRNAs in vivo with 'antagomirs'. *Nature* **438**, 685–689.

Kumaki, Y., Ukai-Tadenuma, M., Uno, K. D., Nishio, J., Masumoto, K. H., Nagano, M., Komori, T., Shigeyoshi, Y., Hogenesch, J. B., and Ueda, H. R. (2008). Analysis and synthesis of high-amplitude Cis-elements in the mammalian circadian clock. *Proc. Natl. Acad. Sci. USA* **105**, 14946–14951.

Kume, K., Zylka, M. J., Sriram, S., Shearman, L. P., Weaver, D. R., Jin, X., Maywood, E. S., Hastings, M. H., and Reppert, S. M. (1999). mCRY1 and mCRY2 are essential components of the negative limb of the circadian clock feedback loop. *Cell* **98**, 193–205.

Kurabayashi, N., Hirota, T., Sakai, M., Sanada, K., and Fukada, Y. (2010). DYRK1A and glycogen synthase kinase 3β, a dual-kinase mechanism directing proteasomal degradation of CRY2 for circadian timekeeping. *Mol. Cell. Biol.* **30**, 1757–1768.

Lamia, K. A., Sachdeva, U. M., DiTacchio, L., Williams, E. C., Alvarez, J. G., Egan, D. F., Vasquez, D. S., Juguilon, H., Panda, S., Shaw, R. J., *et al.* (2009). AMPK regulates the circadian clock by cryptochrome phosphorylation and degradation. *Science* **326**, 437–440.

Le Minh, N., Damiola, F., Tronche, F., Schutz, G., and Schibler, U. (2001). Glucocorticoid hormones inhibit food-induced phase-shifting of peripheral circadian oscillators. *EMBO J.* **20**, 7128–7136.

Lee, C., Etchegaray, J. P., Cagampang, F. R., Loudon, A. S., and Reppert, S. M. (2001). Posttranslational mechanisms regulate the mammalian circadian clock. *Cell* **107**, 855–867.

Lee, J., Lee, Y., Lee, M. J., Park, E., Kang, S. H., Chung, C. H., Lee, K. H., and Kim, K. (2008). Dual modification of BMAL1 by SUMO2/3 and ubiquitin promotes circadian activation of the CLOCK/BMAL1 complex. *Mol. Cell. Biol.* **28**, 6056–6065.

Lee, H., Chen, R., Lee, Y., Yoo, S., and Lee, C. (2009). Essential roles of CKIδ and CKIε in the mammalian circadian clock. *Proc. Natl. Acad. Sci. USA* **106**, 21359–21364.

LeSauter, J., and Silver, R. (1993). Lithium lengthens the period of circadian rhythms in lesioned hamsters bearing SCN grafts. *Biol. Psychiatry* **34**, 75–83.

Li, J. D., Hu, W. P., Boehmer, L., Cheng, M. Y., Lee, A. G., Jilek, A., Siegel, J. M., and Zhou, Q. Y. (2006). Attenuated circadian rhythms in mice lacking the prokineticin 2 gene. *J. Neurosci.* **26**, 11615–11623.

Li, S., Liu, C., Li, N., Hao, T., Han, T., Hill, D. E., Vidal, M., and Lin, J. D. (2008). Genome-wide coactivation analysis of PGC-1alpha identifies BAF60a as a regulator of hepatic lipid metabolism. *Cell Metab.* **8**, 105–117.

Liu, C., Weaver, D. R., Jin, X., Shearman, L. P., Pieschl, R. L., Gribkoff, V. K., and Reppert, S. M. (1997a). Molecular dissection of two distinct actions of melatonin on the suprachiasmatic circadian clock. *Neuron* **19**, 91–102.

Liu, C., Weaver, D. R., Strogatz, S. H., and Reppert, S. M. (1997b). Cellular construction of a circadian clock: Period determination in the suprachiasmatic nuclei. *Cell* **91**, 855–860.

Liu, A. C., Lewis, W. G., and Kay, S. A. (2007a). Mammalian circadian signaling networks and therapeutic targets. *Nat. Chem. Biol.* **3**, 630–639.

Liu, A. C., Welsh, D. K., Ko, C. H., Tran, H. G., Zhang, E. E., Priest, A. A., Buhr, E. D., Singer, O., Meeker, K., Verma, I. M., *et al.* (2007b). Intercellular coupling confers robustness against mutations in the SCN circadian clock network. *Cell* **129,** 605–616.

Liu, C., Li, S., Liu, T., Borjigin, J., and Lin, J. D. (2007c). Transcriptional coactivator PGC-1α integrates the mammalian clock and energy metabolism. *Nature* **447,** 477–481.

Liu, A. C., Tran, H. G., Zhang, E. E., Priest, A. A., Welsh, D. K., and Kay, S. A. (2008). Redundant function of REV-ERBα and β and non-essential role for Bmal1 cycling in transcriptional regulation of intracellular circadian rhythms. *PLoS Genet.* **4,** e1000023.

Lopez-Molina, L., Conquet, F., Dubois-Dauphin, M., and Schibler, U. (1997). The DBP gene is expressed according to a circadian rhythm in the suprachiasmatic nucleus and influences circadian behavior. *EMBO J.* **16,** 6762–6771.

Lowrey, P. L., and Takahashi, J. S. (2000). Genetics of the mammalian circadian system: Photic entrainment, circadian pacemaker mechanisms, and posttranslational regulation. *Annu. Rev. Genet.* **34,** 533–562.

Lowrey, P. L., and Takahashi, J. S. (2004). Mammalian circadian biology: Elucidating genome-wide levels of temporal organization. *Annu. Rev. Genom. Hum. Genet.* **5,** 407–441.

Lowrey, P. L., Shimomura, K., Antoch, M. P., Yamazaki, S., Zemenides, P. D., Ralph, M. R., Menaker, M., and Takahashi, J. S. (2000). Positional syntenic cloning and functional characterization of the mammalian circadian mutation *tau. Science* **288,** 483–492.

Lucas, R. J., Hattar, S., Takao, M., Berson, D. M., Foster, R. G., and Yau, K. W. (2003). Diminished pupillary light reflex at high irradiances in melanopsin-knockout mice. *Science* **299,** 245–247.

MacAulay, K., Doble, B. W., Patel, S., Hansotia, T., Sinclair, E. M., Drucker, D. J., Nagy, A., and Woodgett, J. R. (2007). Glycogen synthase kinase 3α-specific regulation of murine hepatic glycogen metabolism. *Cell Metab.* **6,** 329–337.

Maier, B., Wendt, S., Vanselow, J. T., Wallach, T., Reischl, S., Oehmke, S., Schlosser, A., and Kramer, A. (2009). A large-scale functional RNAi screen reveals a role for CK2 in the mammalian circadian clock. *Genes Dev.* **23,** 708–718.

Martinek, S., Inonog, S., Manoukian, A. S., and Young, M. W. (2001). A role for the segment polarity gene shaggy/GSK-3 in the Drosophila circadian clock. *Cell* **105,** 769–779.

Masana, M. I., Sumaya, I. C., Becker-Andre, M., and Dubocovich, M. L. (2007). Behavioral characterization and modulation of circadian rhythms by light and melatonin in C3H/HeN mice homozygous for the RORβ knockout. *Am. J. Physiol. Regul. Integr. Comp. Physiol.* **292,** R2357–R2367.

Maury, E., Ramsey, K. M., and Bass, J. (2010). Circadian rhythms and metabolic syndrome: From experimental genetics to human disease. *Circ. Res.* **106,** 447–462.

Maywood, E. S., Reddy, A. B., Wong, G. K., O'Neill, J. S., O'Brien, J. A., McMahon, D. G., Harmar, A. J., Okamura, H., and Hastings, M. H. (2006). Synchronization and maintenance of timekeeping in suprachiasmatic circadian clock cells by neuropeptidergic signaling. *Curr. Biol.* **16,** 599–605.

McDearmon, E. L., Patel, K. N., Ko, C. H., Walisser, J. A., Schook, A. C., Chong, J. L., Wilsbacher, L. D., Song, E. J., Hong, H. K., Bradfield, C. A., *et al.* (2006). Dissecting the functions of the mammalian clock protein BMAL1 by tissue-specific rescue in mice. *Science* **314,** 1304–1308.

Meggio, F., and Pinna, L. A. (2003). One-thousand-and-one substrates of protein kinase CK2? *FASEB J.* **17,** 349–368.

Melyan, Z., Tarttelin, E. E., Bellingham, J., Lucas, R. J., and Hankins, M. W. (2005). Addition of human melanopsin renders mammalian cells photoresponsive. *Nature* **433,** 741–745.

Meng, Q. J., Logunova, L., Maywood, E. S., Gallego, M., Lebiecki, J., Brown, T. M., Sladek, M., Semikhodskii, A. S., Glossop, N. R., Piggins, H. D., *et al.* (2008). Setting clock speed in mammals: The CK1ε *tau* mutation in mice accelerates circadian pacemakers by selectively destabilizing PERIOD proteins. *Neuron* **58**, 78–88.

Meng, Q. J., Maywood, E. S., Bechtold, D. A., Lu, W. Q., Li, J., Gibbs, J. E., Dupre, S. M., Chesham, J. E., Rajamohan, F., Knafels, J., *et al.* (2010). Entrainment of disrupted circadian behavior through inhibition of casein kinase 1 (CK1) enzymes. *Proc. Natl. Acad. Sci. USA* **107**, 15240–15245.

Michel, S., Itri, J., Han, J. H., Gniotczynski, K., and Colwell, C. S. (2006). Regulation of glutamatergic signalling by PACAP in the mammalian suprachiasmatic nucleus. *BMC Neurosci.* **7**, 15.

Miller, B. H., McDearmon, E. L., Panda, S., Hayes, K. R., Zhang, J., Andrews, J. L., Antoch, M. P., Walker, J. R., Esser, K. A., Hogenesch, J. B., *et al.* (2007). Circadian and CLOCK-controlled regulation of the mouse transcriptome and cell proliferation. *Proc. Natl. Acad. Sci. USA* **104**, 3342–3347.

Mitsui, S., Yamaguchi, S., Matsuo, T., Ishida, Y., and Okamura, H. (2001). Antagonistic role of E4BP4 and PAR proteins in the circadian oscillatory mechanism. *Genes Dev.* **15**, 995–1006.

Mohawk, J. A., Baer, M. L., and Menaker, M. (2009). The methamphetamine-sensitive circadian oscillator does not employ canonical clock genes. *Proc. Natl. Acad. Sci. USA* **106**, 3519–3524.

Moore, M. J. (2005). From birth to death: The complex lives of eukaryotic mRNAs. *Science* **309**, 1514–1518.

Moore, R. Y., and Eichler, V. B. (1972). Loss of a circadian adrenal corticosterone rhythm following suprachiasmatic lesions in the rat. *Brain Res.* **42**, 201–206.

Moore, R. Y., and Lenn, N. J. (1972). A retinohypothalamic projection in the rat. *J. Comp. Neurol.* **146**, 1–14.

Moore, R. Y., Speh, J. C., and Card, J. P. (1995). The retinohypothalamic tract originates from a distinct subset of retinal ganglion cells. *J. Comp. Neurol.* **352**, 351–366.

Morin, L. P., and Allen, C. N. (2006). The circadian visual system, 2005. *Brain Res. Rev.* **51**, 1–60.

Muller, B., and Grossniklaus, U. (2010). Model organisms—A historical perspective. *J. Proteomics* **73**, 2054–2063.

Nader, N., Chrousos, G. P., and Kino, T. (2009). Circadian rhythm transcription factor CLOCK regulates the transcriptional activity of the glucocorticoid receptor by acetylating its hinge region lysine cluster: Potential physiological implications. *FASEB J.* **23**, 1572–1583.

Nagel, R., Clijsters, L., and Agami, R. (2009). The miRNA-192/194 cluster regulates the *Period* gene family and the circadian clock. *FEBS J.* **276**, 5447–5455.

Nagoshi, E., Saini, C., Bauer, C., Laroche, T., Naef, F., and Schibler, U. (2004). Circadian gene expression in individual fibroblasts; cell-autonomous and self-sustained oscillators pass time to daughter cells. *Cell* **119**, 693–705.

Nagy, A., Gertsenstein, M., Vintersten, K., and Behringer, R. (eds.) (2003). Manipulating the Mouse Embryo: A Laboratory Manual. Cold Spring Harbor Laboratory Press, Cold Spring Harbor, New York.

Nakahata, Y., Kaluzova, M., Grimaldi, B., Sahar, S., Hirayama, J., Chen, D., Guarente, L. P., and Sassone-Corsi, P. (2008). The NAD^+-dependent deacetylase SIRT1 modulates CLOCK-mediated chromatin remodeling and circadian control. *Cell* **134**, 329–340.

Nakahata, Y., Sahar, S., Astarita, G., Kaluzova, M., and Sassone-Corsi, P. (2009). Circadian control of the NAD^+ salvage pathway by CLOCK-SIRT1. *Science* **324**, 654–657.

Nakajima, M., Imai, K., Ito, H., Nishiwaki, T., Murayama, Y., Iwasaki, H., Oyama, T., and Kondo, T. (2005). Reconstitution of circadian oscillation of cyanobacterial KaiC phosphorylation in vitro. *Science* **308**, 414–415.

Nakashima, A., Kawamoto, T., Honda, K. K., Ueshima, T., Noshiro, M., Iwata, T., Fujimoto, K., Kubo, H., Honma, S., Yorioka, N., et al. (2008). DEC1 modulates the circadian phase of clock gene expression. Mol. Cell. Biol. **28**, 4080–4092.

Nandi, D., Tahiliani, P., Kumar, A., and Chandu, D. (2006). The ubiquitin-proteasome system. J. Biosci. **31**, 137–155.

Nelson, R. J., and Zucker, I. (1981). Absence of extraocular photoreception in diurnal and nocturnal rodents exposed to direct sunlight. Comp. Biochem. Physiol. **69A**, 145–148.

Ohno, T., Onishi, Y., and Ishida, N. (2007). A novel E4BP4 element drives circadian expression of mPeriod2. Nucleic Acids Res. **35**, 648–655.

Ohsaki, K., Oishi, K., Kozono, Y., Nakayama, K., Nakayama, K. I., and Ishida, N. (2008). The role of β-TrCP1 and β-TrCP2 in circadian rhythm generation by mediating degradation of clock protein PER2. J. Biochem.,Tokyo **144**, 609–618.

Oishi, K., Miyazaki, K., Kadota, K., Kikuno, R., Nagase, T., Atsumi, G., Ohkura, N., Azama, T., Mesaki, M., Yukimasa, S., et al. (2003). Genome-wide expression analysis of mouse liver reveals CLOCK-regulated circadian output genes. J. Biol. Chem. **278**, 41519–41527.

Okamura, H., Miyake, S., Sumi, Y., Yamaguchi, S., Yasui, A., Muijtjens, M., Hoeijmakers, J. H., and van der Horst, G. T. (1999). Photic induction of mPer1 and mPer2 in Cry-deficient mice lacking a biological clock. Science **286**, 2531–2534.

O'Neill, J. S., and Reddy, A. B. (2011). Circadian clocks in human red blood cells. Nature **469**, 498–503.

O'Neill, J. S., van Ooijen, G., Dixon, L. E., Troein, C., Corellou, F., Bouget, F. Y., Reddy, A. B., and Millar, A. J. (2011). Circadian rhythms persist without transcription in a eukaryote. Nature **469**, 554–558.

Oster, H., Yasui, A., van der Horst, G. T., and Albrecht, U. (2002). Disruption of mCry2 restores circadian rhythmicity in mPer2 mutant mice. Genes Dev. **16**, 2633–2638.

Oster, H., Baeriswyl, S., Van Der Horst, G. T., and Albrecht, U. (2003). Loss of circadian rhythmicity in aging mPer1$^{-/-}$ mCry2$^{-/-}$ mutant mice. Genes Dev. **17**, 1366–1379.

Palkova, M., Sigmund, L., and Erkert, H. G. (1999). Effect of ambient temperature on the circadian activity rhythm in common marmosets, Callithrix j. jacchus (primates). Chronobiol. Int. **16**, 149–161.

Panda, S., Antoch, M. P., Miller, B. H., Su, A. I., Schook, A. B., Straume, M., Schultz, P. G., Kay, S. A., Takahashi, J. S., and Hogenesch, J. B. (2002a). Coordinated transcription of key pathways in the mouse by the circadian clock. Cell **109**, 307–320.

Panda, S., Sato, T. K., Castrucci, A. M., Rollag, M. D., DeGrip, W. J., Hogenesch, J. B., Provencio, I., and Kay, S. A. (2002b). Melanopsin (Opn4) requirement for normal light-induced circadian phase shifting. Science **298**, 2213–2216.

Panda, S., Provencio, I., Tu, D. C., Pires, S. S., Rollag, M. D., Castrucci, A. M., Pletcher, M. T., Sato, T. K., Wiltshire, T., Andahazy, M., et al. (2003). Melanopsin is required for non-image-forming photic responses in blind mice. Science **301**, 525–527.

Panda, S., Nayak, S. K., Campo, B., Walker, J. R., Hogenesch, J. B., and Jegla, T. (2005). Illumination of the melanopsin signaling pathway. Science **307**, 600–604.

Partch, C. L., Shields, K. F., Thompson, C. L., Selby, C. P., and Sancar, A. (2006). Posttranslational regulation of the mammalian circadian clock by cryptochrome and protein phosphatase 5. Proc. Natl. Acad. Sci. USA **103**, 10467–10472.

Pezuk, P., Mohawk, J. A., Yoshikawa, T., Sellix, M. T., and Menaker, M. (2010). Circadian organization is governed by extra-SCN pacemakers. J. Biol. Rhythms **25**, 432–441.

Pittendrigh, C. S. (1993). Temporal organization: Reflections of a Darwinian clock-watcher. Annu. Rev. Physiol. **55**, 16–54.

Preitner, N., Damiola, F., Lopez-Molina, L., Zakany, J., Duboule, D., Albrecht, U., and Schibler, U. (2002). The orphan nuclear receptor REV-ERBα controls circadian transcription within the positive limb of the mammalian circadian oscillator. *Cell* **110,** 251–260.

Price, J. L., Blau, J., Rothenfluh, A., Abodeely, M., Kloss, B., and Young, M. W. (1998). *double-time* is a novel *Drosophila* clock gene that regulates PERIOD protein accumulation. *Cell* **94,** 83–95.

Prolo, L. M., Takahashi, J. S., and Herzog, E. D. (2005). Circadian rhythm generation and entrainment in astrocytes. *J. Neurosci.* **25,** 404–408.

Provencio, I., Jiang, G., De Grip, W. J., Hayes, W. P., and Rollag, M. D. (1998). Melanopsin: An opsin in melanophores, brain, and eye. *Proc. Natl. Acad. Sci. USA* **95,** 340–345.

Qiu, X., Kumbalasiri, T., Carlson, S. M., Wong, K. Y., Krishna, V., Provencio, I., and Berson, D. M. (2005). Induction of photosensitivity by heterologous expression of melanopsin. *Nature* **433,** 745–749.

Quintero, J. E., Kuhlman, S. J., and McMahon, D. G. (2003). The biological clock nucleus: A multiphasic oscillator network regulated by light. *J. Neurosci.* **23,** 8070–8076.

Ralph, M. R., and Menaker, M. (1988). A mutation of the circadian system in golden hamsters. *Science* **241,** 1225–1227.

Ralph, M. R., Foster, R. G., Davis, F. C., and Menaker, M. (1990). Transplanted suprachiasmatic nucleus determines circadian period. *Science* **247,** 975–978.

Ramsey, K. M., Yoshino, J., Brace, C. S., Abrassart, D., Kobayashi, Y., Marcheva, B., Hong, H. K., Chong, J. L., Buhr, E. D., Lee, C., et al. (2009). Circadian clock feedback cycle through NAMPT-mediated NAD$^+$ biosynthesis. *Science* **324,** 651–654.

Rana, T. M. (2007). Illuminating the silence: Understanding the structure and function of small RNAs. *Nat. Rev. Mol. Cell Biol.* **8,** 23–36.

Reddy, A. B., Karp, N. A., Maywood, E. S., Sage, E. A., Deery, M., O'Neill, J. S., Wong, G. K., Chesham, J., Odell, M., Lilley, K. S., et al. (2006). Circadian orchestration of the hepatic proteome. *Curr. Biol.* **16,** 1107–1115.

Refinetti, R., and Menaker, M. (1992). The circadian rhythm of body temperature. *Physiol. Behav.* **51,** 613–637.

Reick, M., Garcia, J. A., Dudley, C., and McKnight, S. L. (2001). NPAS2: An analog of Clock operative in the mammalian forebrain. *Science* **293,** 506–509.

Reinke, H., Saini, C., Fleury-Olela, F., Dibner, C., Benjamin, I. J., and Schibler, U. (2008). Differential display of DNA-binding proteins reveals heat-shock factor 1 as a circadian transcription factor. *Genes Dev.* **22,** 331–345.

Reischl, S., Vanselow, K., Westermark, P. O., Thierfelder, N., Maier, B., Herzel, H., and Kramer, A. (2007). Beta-TrCP1-mediated degradation of PERIOD2 is essential for circadian dynamics. *J. Biol. Rhythms* **22,** 375–386.

Reppert, S. M. (1998). A clockwork explosion!. *Neuron* **21,** 1–4.

Reppert, S. M., and Weaver, D. R. (2002). Coordination of circadian timing in mammals. *Nature* **418,** 935–941.

Reyes, B. A., Pendergast, J. S., and Yamazaki, S. (2008). Mammalian peripheral circadian oscillators are temperature compensated. *J. Biol. Rhythms* **23,** 95–98.

Ripperger, J. A., and Merrow, M. (2011). Perfect timing: Epigenetic regulation of the circadian clock. *FEBS Lett.* **585,** 1406–1411.

Rivers, A., Gietzen, K. F., Vielhaber, E., and Virshup, D. M. (1998). Regulation of casein kinase I ε and casein kinase I δ by an *in vivo* futile phosphorylation cycle. *J. Biol. Chem.* **273,** 15980–15984.

Robles, M. S., Boyault, C., Knutti, D., Padmanabhan, K., and Weitz, C. J. (2010). Identification of RACK1 and protein kinase Cα as integral components of the mammalian circadian clock. *Science* **327,** 463–466.

Rossner, M. J., Oster, H., Wichert, S. P., Reinecke, L., Wehr, M. C., Reinecke, J., Eichele, G., Taneja, R., and Nave, K. A. (2008). Disturbed clockwork resetting in Sharp-1 and Sharp-2 single and double mutant mice. *PLoS One* **3**, e2762.

Ruby, N. F., and Heller, H. C. (1996). Temperature sensitivity of the suprachiasmatic nucleus of ground squirrels and rats in vitro. *J. Biol. Rhythms* **11**, 126–136.

Ruby, N. F., Burns, D. E., and Heller, H. C. (1999). Circadian rhythms in the suprachiasmatic nucleus are temperature-compensated and phase-shifted by heat pulses *in vitro*. *J. Neurosci.* **19**, 8630–8636.

Ruby, N. F., Brennan, T. J., Xie, X., Cao, V., Franken, P., Heller, H. C., and O'Hara, B. F. (2002). Role of melanopsin in circadian responses to light. *Science* **298**, 2211–2213.

Rutter, J., Reick, M., Wu, L. C., and McKnight, S. L. (2001). Regulation of *Clock* and NPAS2 DNA binding by the redox state of NAD cofactors. *Science* **293**, 510–514.

Rutter, J., Reick, M., and McKnight, S. L. (2002). Metabolism and the control of circadian rhythms. *Annu. Rev. Biochem.* **71**, 307–331.

Ryves, W. J., and Harwood, A. J. (2001). Lithium inhibits glycogen synthase kinase-3 by competition for magnesium. *Biochem. Biophys. Res. Commun.* **280**, 720–725.

Sahar, S., Zocchi, L., Kinoshita, C., Borrelli, E., and Sassone-Corsi, P. (2010). Regulation of BMAL1 protein stability and circadian function by GSK3β-mediated phosphorylation. *PLoS One* **5**, e8561.

Sanada, K., Okano, T., and Fukada, Y. (2002). Mitogen-activated protein kinase phosphorylates and negatively regulates basic helix-loop-helix-PAS transcription factor BMAL1. *J. Biol. Chem.* **277**, 267–271.

Sasaki, M., Yoshitane, H., Du, N. H., Okano, T., and Fukada, Y. (2009). Preferential inhibition of BMAL2-CLOCK activity by PER2 reemphasizes its negative role and a positive role of BMAL2 in the circadian transcription. *J. Biol. Chem.* **284**, 25149–25159.

Sato, T. K., Panda, S., Miraglia, L. J., Reyes, T. M., Rudic, R. D., McNamara, P., Naik, K. A., FitzGerald, G. A., Kay, S. A., and Hogenesch, J. B. (2004). A functional genomics strategy reveals Rora as a component of the mammalian circadian clock. *Neuron* **43**, 527–537.

Sato, T. K., Yamada, R. G., Ukai, H., Baggs, J. E., Miraglia, L. J., Kobayashi, T. J., Welsh, D. K., Kay, S. A., Ueda, H. R., and Hogenesch, J. B. (2006). Feedback repression is required for mammalian circadian clock function. *Nat. Genet.* **38**, 312–319.

Seggie, J., Werstiuk, E. S., and Grota, L. (1982). Effect of chronic lithium treatment on twenty four hour variation in plasma and red blood cell lithium and sodium concentrations, drinking behavior, body weight, kidney weight, and corticosterone levels. *Prog. Neuropsychopharmacol. Biol. Psychiatry* **6**, 455–458.

Shearman, L. P., Zylka, M. J., Weaver, D. R., Kolakowski, L. F., Jr., and Reppert, S. M. (1997). Two *period* homologs: Circadian expression and photic regulation in the suprachiasmatic nuclei. *Neuron* **19**, 1261–1269.

Shearman, L. P., Jin, X., Lee, C., Reppert, S. M., and Weaver, D. R. (2000a). Targeted disruption of the *mPer3* gene: Subtle effects on circadian clock function. *Mol. Cell. Biol.* **20**, 6269–6275.

Shearman, L. P., Sriram, S., Weaver, D. R., Maywood, E. S., Chaves, I., Zheng, B., Kume, K., Lee, C. C., van der Horst, G. T., Hastings, M. H., *et al.* (2000b). Interacting molecular loops in the mammalian circadian clock. *Science* **288**, 1013–1019.

Shi, S., Hida, A., McGuinness, O. P., Wasserman, D. H., Yamazaki, S., and Johnson, C. H. (2010). Circadian clock gene *Bmal1* is not essential; functional replacement with its paralog, *Bmal2*. *Curr. Biol.* **20**, 316–321.

Shigeyoshi, Y., Taguchi, K., Yamamoto, S., Takekida, S., Yan, L., Tei, H., Moriya, T., Shibata, S., Loros, J. J., Dunlap, J. C., *et al.* (1997). Light-induced resetting of a mammalian circadian clock is associated with rapid induction of the *mPer1* transcript. *Cell* **91**, 1043–1053.

Shim, H. S., Kim, H., Lee, J., Son, G. H., Cho, S., Oh, T. H., Kang, S. H., Seen, D. S., Lee, K. H., and Kim, K. (2007). Rapid activation of CLOCK by Ca2+−dependent protein kinase C mediates resetting of the mammalian circadian clock. *EMBO Rep.* **8**, 366–371.

Shirogane, T., Jin, J., Ang, X. L., and Harper, J. W. (2005). SCF$^{\beta\text{-TRCP}}$ controls clock-dependent transcription via casein kinase 1-dependent degradation of the mammalian period-1 (Per1) protein. *J. Biol. Chem.* **280**, 26863–26872.

Siepka, S. M., and Takahashi, J. S. (2005). Forward genetic screens to identify circadian rhythm mutants in mice. *Methods Enzymol.* **393**, 219–229.

Siepka, S. M., Yoo, S. H., Park, J., Song, W., Kumar, V., Hu, Y., Lee, C., and Takahashi, J. S. (2007). Circadian mutant *Overtime* reveals F-box protein FBXL3 regulation of *Cryptochrome* and *Period* gene expression. *Cell* **129**, 1011–1023.

Silver, L. M. (1995). Mouse Genetics: Concepts and Applications. Oxford University Press, New York, NY.

Silver, R., LeSauter, J., Tresco, P. A., and Lehman, M. N. (1996). A diffusible coupling signal from the transplanted suprachiasmatic nucleus controlling circadian locomotor rhythms. *Nature* **382**, 810–813.

Spengler, M. L., Kuropatwinski, K. K., Schumer, M., and Antoch, M. P. (2009). A serine cluster mediates BMAL1-dependent CLOCK phosphorylation and degradation. *Cell Cycle* **8**, 4138–4146.

Sprouse, J., Reynolds, L., Kleiman, R., Tate, B., Swanson, T. A., and Pickard, G. E. (2010). Chronic treatment with a selective inhibitor of casein kinase I δ/ε yields cumulative phase delays in circadian rhythms. *Psychopharmacology (Berl.)* **210**, 569–576.

Staiger, D., and Koster, T. (2011). Spotlight on post-transcriptional control in the circadian system. *Cell. Mol. Life Sci.* **68**, 71–83.

Stambolic, V., Ruel, L., and Woodgett, J. R. (1996). Lithium inhibits glycogen synthase kinase-3 activity and mimics wingless signalling in intact cells. *Curr. Biol.* **6**, 1664–1668.

Stephan, F. K., and Zucker, I. (1972). Circadian rhythms in drinking behavior and locomotor activity of rats are eliminated by hypothalamic lesions. *Proc. Natl. Acad. Sci. USA* **69**, 1583–1586.

Stokkan, K. A., Yamazaki, S., Tei, H., Sakaki, Y., and Menaker, M. (2001). Entrainment of the circadian clock in the liver by feeding. *Science* **291**, 490–493.

Storch, K. F., and Weitz, C. J. (2009). Daily rhythms of food-anticipatory behavioral activity do not require the known circadian clock. *Proc. Natl. Acad. Sci. USA* **106**, 6808–6813.

Storch, K. F., Lipan, O., Leykin, I., Viswanathan, N., Davis, F. C., Wong, W. H., and Weitz, C. J. (2002). Extensive and divergent circadian gene expression in liver and heart. *Nature* **417**, 78–83.

Strahl, B. D., and Allis, C. D. (2000). The language of covalent histone modifications. *Nature* **403**, 41–45.

Sweeney, B. M., and Hastings, J. W. (1960). Effects of temperature upon diurnal rhythms. *Cold Spring Harbor Symp. Quant. Biol.* **25**, 87–104.

Takahashi, J. S., Pinto, L. H., and Vitaterna, M. H. (1994). Forward and reverse genetic approaches to behavior in the mouse. *Science* **264**, 1724–1733.

Takahashi, J. S., Turek, F. W., and Moore, R. Y. (eds.) (2001). Handbook of Behavioral Neurobiology: Circadian Clocks, Vol. 2. Kluwer Acadamic/Plenum Publishers, New York.

Takahashi, J. S., Hong, H. K., Ko, C. H., and McDearmon, E. L. (2008). The genetics of mammalian circadian order and disorder: Implications for physiology and disease. *Nat. Rev. Genet.* **9**, 764–775.

Tamaru, T., Hirayama, J., Isojima, Y., Nagai, K., Norioka, S., Takamatsu, K., and Sassone-Corsi, P. (2009). CK2α phosphorylates BMAL1 to regulate the mammalian clock. *Nat. Struct. Mol. Biol.* **16**, 446–448.

Thresher, R. J., Vitaterna, M. H., Miyamoto, Y., Kazantsev, A., Hsu, D. S., Petit, C., Selby, C. P., Dawut, L., Smithies, O., Takahashi, J. S., et al. (1998). Role of mouse cryptochrome blue-light photoreceptor in circadian photoresponses. Science 282, 1490–1494.

Tischkau, S. A., Mitchell, J. W., Pace, L. A., Barnes, J. W., Barnes, J. A., and Gillette, M. U. (2004). Protein kinase G type II is required for night-to-day progression of the mammalian circadian clock. Neuron 43, 539–549.

Tomita, J., Nakajima, M., Kondo, T., and Iwasaki, H. (2005). No transcription-translation feedback in circadian rhythm of KaiC phosphorylation. Science 307, 251–254.

Tosini, G., and Menaker, M. (1996). Circadian rhythms in cultured mammalian retina. Science 272, 419–421.

Tosini, G., and Menaker, M. (1998). The tau mutation affects temperature compensation of hamster retinal circadian oscillators. Neuroreport 9, 1001–1005.

Travnickova-Bendova, Z., Cermakian, N., Reppert, S. M., and Sassone-Corsi, P. (2002). Bimodal regulation of mPeriod promoters by CREB-dependent signaling and CLOCK/BMAL1 activity. Proc. Natl. Acad. Sci. USA 99, 7728–7733.

Triqueneaux, G., Thenot, S., Kakizawa, T., Antoch, M. P., Safi, R., Takahashi, J. S., Delaunay, F., and Laudet, V. (2004). The orphan receptor Rev-erbα gene is a target of the circadian clock pacemaker. J. Mol. Endocrinol. 33, 585–608.

Tsuchiya, Y., Akashi, M., and Nishida, E. (2003). Temperature compensation and temperature resetting of circadian rhythms in mammalian cultured fibroblasts. Genes Cells 8, 713–720.

Tsuchiya, Y., Akashi, M., Matsuda, M., Goto, K., Miyata, Y., Node, K., and Nishida, E. (2009). Involvement of the protein kinase CK2 in the regulation of mammalian circadian rhythms. Sci. Signal. 2, ra26.

Ueda, H. R., Hayashi, S., Chen, W., Sano, M., Machida, M., Shigeyoshi, Y., Iino, M., and Hashimoto, S. (2005). System-level identification of transcriptional circuits underlying mammalian circadian clocks. Nat. Genet. 37, 187–192.

Ukai-Tadenuma, M., Kasukawa, T., and Ueda, H. R. (2008). Proof-by-synthesis of the transcriptional logic of mammalian circadian clocks. Nat. Cell Biol. 10, 1154–1163.

van der Horst, G. T., Muijtjens, M., Kobayashi, K., Takano, R., Kanno, S., Takao, M., de Wit, J., Verkerk, A., Eker, A. P., van Leenen, D., et al. (1999). Mammalian Cry1 and Cry2 are essential for maintenance of circadian rhythms. Nature 398, 627–630.

Vanselow, K., Vanselow, J. T., Westermark, P. O., Reischl, S., Maier, B., Korte, T., Herrmann, A., Herzel, H., Schlosser, A., and Kramer, A. (2006). Differential effects of PER2 phosphorylation: Molecular basis for the human familial advanced sleep phase syndrome (FASPS). Genes Dev. 20, 2660–2672.

Vaquero, A., Scher, M., Lee, D., Erdjument-Bromage, H., Tempst, P., and Reinberg, D. (2004). Human SirT1 interacts with histone H1 and promotes formation of facultative heterochromatin. Mol. Cell 16, 93–105.

Vielhaber, E., Eide, E., Rivers, A., Gao, Z. H., and Virshup, D. M. (2000). Nuclear entry of the circadian regulator mPER1 is controlled by mammalian casein kinase I ε. Mol. Cell. Biol. 20, 4888–4899.

Vitaterna, M. H., King, D. P., Chang, A. M., Kornhauser, J. M., Lowrey, P. L., McDonald, J. D., Dove, W. F., Pinto, L. H., Turek, F. W., and Takahashi, J. S. (1994). Mutagenesis and mapping of a mouse gene, Clock, essential for circadian behavior. Science 264, 719–725.

Vitaterna, M. H., Selby, C. P., Todo, T., Niwa, H., Thompson, C., Fruechte, E. M., Hitomi, K., Thresher, R. J., Ishikawa, T., Miyazaki, J., et al. (1999). Differential regulation of mammalian period genes and circadian rhythmicity by cryptochromes 1 and 2. Proc. Natl. Acad. Sci. USA 96, 12114–12119.

Vujovic, N., Davidson, A. J., and Menaker, M. (2008). Sympathetic input modulates, but does not determine, phase of peripheral circadian oscillators. *Am. J. Physiol. Regul. Integr. Comp. Physiol.* **295,** R355–R360.

Walton, K. M., Fisher, K., Rubitski, D., Marconi, M., Meng, Q. J., Sladek, M., Adams, J., Bass, M., Chandrasekaran, R., Butler, T., *et al.* (2009). Selective inhibition of casein kinase 1ε minimally alters circadian clock period. *J. Pharmacol. Exp. Ther.* **330,** 430–439.

Wang, Y., Osterbur, D. L., Megaw, P. L., Tosini, G., Fukuhara, C., Green, C. B., and Besharse, J. C. (2001). Rhythmic expression of Nocturnin mRNA in multiple tissues of the mouse. *BMC Dev. Biol.* **1,** 9.

Welsh, D. K., Logothetis, D. E., Meister, M., and Reppert, S. M. (1995). Individual neurons dissociated from rat suprachiasmatic nucleus express independently phased circadian firing rhythms. *Neuron* **14,** 697–706.

Welsh, D. K., Yoo, S. H., Liu, A. C., Takahashi, J. S., and Kay, S. A. (2004). Bioluminescence imaging of individual fibroblasts reveals persistent, independently phased circadian rhythms of clock gene expression. *Curr. Biol.* **14,** 2289–2295.

Welsh, D. K., Takahashi, J. S., and Kay, S. A. (2010). Suprachiasmatic nucleus: Cell autonomy and network properties. *Annu. Rev. Physiol.* **72,** 551–577.

Wilkinson, K. D. (1999). Ubiquitin-dependent signaling: The role of ubiquitination in the response of cells to their environment. *J. Nutr.* **129,** 1933–1936.

Wilkinson, K. A., and Henley, J. M. (2010). Mechanisms, regulation and consequences of protein SUMOylation. *Biochem. J.* **428,** 133–145.

Woo, K. C., Kim, T. D., Lee, K. H., Kim, D. Y., Kim, W., Lee, K. Y., and Kim, K. T. (2009). Mouse period 2 mRNA circadian oscillation is modulated by PTB-mediated rhythmic mRNA degradation. *Nucleic Acids Res.* **37,** 26–37.

Woo, K. C., Ha, D. C., Lee, K. H., Kim, D. Y., Kim, T. D., and Kim, K. T. (2010). Circadian amplitude of cryptochrome 1 is modulated by mRNA stability regulation via cytoplasmic hnRNP D oscillation. *Mol. Cell. Biol.* **30,** 197–205.

Xu, Y., Padiath, Q. S., Shapiro, R. E., Jones, C. R., Wu, S. C., Saigoh, N., Saigoh, K., Ptacek, L. J., and Fu, Y. H. (2005). Functional consequences of a CKIδ mutation causing familial advanced sleep phase syndrome. *Nature* **434,** 640–644.

Xu, Y., Toh, K. L., Jones, C. R., Shin, J. Y., Fu, Y. H., and Ptacek, L. J. (2007). Modeling of a human circadian mutation yields insights into clock regulation by PER2. *Cell* **128,** 59–70.

Yagita, K., Tamanini, F., van Der Horst, G. T., and Okamura, H. (2001). Molecular mechanisms of the biological clock in cultured fibroblasts. *Science* **292,** 278–281.

Yamada, Y. R., and Forger, D. B. (2010). Multiscale complexity in the mammalian circadian clock. *Curr. Opin. Genet. Dev.* **20,** 626–633.

Yamaguchi, S., Isejima, H., Matsuo, T., Okura, R., Yagita, K., Kobayashi, M., and Okamura, H. (2003). Synchronization of cellular clocks in the suprachiasmatic nucleus. *Science* **302,** 1408–1412.

Yamazaki, S., Numano, R., Abe, M., Hida, A., Takahashi, R., Ueda, M., Block, G. D., Sakaki, Y., Menaker, M., and Tei, H. (2000). Resetting central and peripheral circadian oscillators in transgenic rats. *Science* **288,** 682–685.

Yan, L., and Silver, R. (2004). Resetting the brain clock: Time course and localization of mPER1 and mPER2 protein expression in suprachiasmatic nuclei during phase shifts. *Eur. J. Neurosci.* **19,** 1105–1109.

Yang, X., Downes, M., Yu, R. T., Bookout, A. L., He, W., Straume, M., Mangelsdorf, D. J., and Evans, R. M. (2006). Nuclear receptor expression links the circadian clock to metabolism. *Cell* **126,** 801–810.

Yin, L., Wang, J., Klein, P. S., and Lazar, M. A. (2006). Nuclear receptor Rev-erbα is a critical lithium-sensitive component of the circadian clock. *Science* **311**, 1002–1005.

Yoo, S. H., Yamazaki, S., Lowrey, P. L., Shimomura, K., Ko, C. H., Buhr, E. D., Siepka, S. M., Hong, H. K., Oh, W. J., Yoo, O. J., *et al.* (2004). PERIOD2::LUCIFERASE real-time reporting of circadian dynamics reveals persistent circadian oscillations in mouse peripheral tissues. *Proc. Natl. Acad. Sci. USA* **101**, 5339–5346.

Yoo, S. H., Ko, C. H., Lowrey, P. L., Buhr, E. D., Song, E. J., Chang, S., Yoo, O. J., Yamazaki, S., Lee, C., and Takahashi, J. S. (2005). A noncanonical E-box enhancer drives mouse *Period2* circadian oscillations *in vivo*. *Proc. Natl. Acad. Sci. USA* **102**, 2608–2613.

Yoshitane, H., Takao, T., Satomi, Y., Du, N. H., Okano, T., and Fukada, Y. (2009). Roles of CLOCK phosphorylation in suppression of E-box-dependent transcription. *Mol. Cell. Biol.* **29**, 3675–3686.

Young, M. W., and Kay, S. A. (2001). Time zones: A comparative genetics of circadian clocks. *Nat. Rev. Genet.* **2**, 702–715.

Yu, W., Zheng, H., Price, J. L., and Hardin, P. E. (2009). DOUBLETIME plays a noncatalytic role to mediate CLOCK phosphorylation and repress CLOCK-dependent transcription within the *Drosophila* circadian clock. *Mol. Cell. Biol.* **29**, 1452–1458.

Zhang, X., Odom, D. T., Koo, S. H., Conkright, M. D., Canettieri, G., Best, J., Chen, H., Jenner, R., Herbolsheimer, E., Jacobsen, E., *et al.* (2005). Genome-wide analysis of cAMP-response element binding protein occupancy, phosphorylation, and target gene activation in human tissues. *Proc. Natl. Acad. Sci. USA* **102**, 4459–4464.

Zhao, W. N., Malinin, N., Yang, F. C., Staknis, D., Gekakis, N., Maier, B., Reischl, S., Kramer, A., and Weitz, C. J. (2007). CIPC is a mammalian circadian clock protein without invertebrate homologues. *Nat. Cell Biol.* **9**, 268–275.

Zheng, B., Larkin, D. W., Albrecht, U., Sun, Z. S., Sage, M., Eichele, G., Lee, C. C., and Bradley, A. (1999). The *mPer2* gene encodes a functional component of the mammalian circadian clock. *Nature* **400**, 169–173.

Zheng, B., Albrecht, U., Kaasik, K., Sage, M., Lu, W., Vaishnav, S., Li, Q., Sun, Z. S., Eichele, G., Bradley, A., *et al.* (2001). Nonredundant roles of the *mPer1* and *mPer2* genes in the mammalian circadian clock. *Cell* **105**, 683–694.

7

The Genetics of the Human Circadian Clock

Luoying Zhang,* Chris R. Jones,† Louis J. Ptacek,* and Ying-Hui Fu*

*Department of Neurology, University of California, San Francisco, California, USA
†Department of Neurology, University of Utah, Salt Lake City, Utah, USA

ABSTRACT

Daily rhythms of behavioral and physiological processes are believed to arise from endogenous circadian clocks. Unlike model organisms, genetic studies of human behavioral traits present extra challenges due to many factors such as the heterogeneous genetic background and environmental influences. Identifying molecular components of the human circadian clock were not possible until

Advances in Genetics, Vol. 74
0065-2660/11 $35.00
DOI: 10.1016/B978-0-12-387690-4.00007-6

the recognition of Mendelian circadian traits in human subjects in recent years. Characterizing these rare Mendelian traits therefore established the foundation for identification of the genetic components for human circadian and sleep mechanisms. This line of investigation has proven fruitful and provided new insights into these pathways. Genetic association studies have also offered many possible genetic contributions to these mechanisms. Studies of these genes/proteins in conjunction with modeling human mutations in model organisms afford the opportunity to unravel the molecular mechanisms which in time will lead to pharmacological interventions that may not only help modify these behavioral traits but also may prove effective for treating other sleep-related disorders.

Humans, like most organisms, exhibit daily rhythms in behavior and physiology. Jürgen Aschoff, one of the pioneers in the field of human circadian rhythms, demonstrated that these rhythms persist in (relatively) constant conditions with an approximately 24-h (i.e., circadian) periodicity (Dunlap *et al.*, 2004). In one of his early studies, human subjects lived in a converted former bomb shelter underground without time cues or direct social interactions and could select their own sleep–wake schedules. The subjects exhibited an approximately 25-h rest–activity cycle as well as circadian oscillations of other physiological variables, suggesting the presence of an endogenous circadian clock. However, the mechanism for this endogenous clock remained largely unclear until some 30 years later when a culmination of data from animal studies suggested a transcriptional–translational feedback loop with a circadian period length (Hastings *et al.*, 2008).

I. HUMAN CIRCADIAN PHENOTYPES

The molecular mechanism of the "master" human circadian clock in the suprachiasmatic nuclei (SCN) of the anterior hypothalamus was initially inferred from animal studies (Hastings *et al.*, 2008), subsequently supported by the discovery of *in vitro* human peripheral tissue circadian rhythms (McNamara *et al.*, 2001), and more recently explored by examining the circadian cell biology of transgenic mice carrying a causative mutation from a human with a profoundly altered circadian rhythm (Xu *et al.*, 2007). Human circadian conditions are classified according to the altered phase alignment between the internal clock and the external day–night cycle. As this altered phase relationship (or, in the case of free-running circadian sleep disorder, desynchrony) will almost always affect the manifest sleep–wake cycle, the American Academy of Sleep Medicine categorized these conditions as circadian rhythm sleep disorders (CRSDs) if the condition is enduring and associated with a sleep–wake complaint. Some people with persistently and profoundly altered sleep–wake schedules (and thus potentially profoundly altered circadian clocks) do not perceive the condition as a problem, usually because they succeed at scheduling their social schedule around their biological sleep schedule.

For patient care purposes, there are four major types of chronic CRSDs that are all stubbornly resistant to enforced rest–activity schedules and thus likely to be intrinsic biological propensities with a genetic substrate: advanced (advanced sleep phase disorder, ASPD), delayed (delayed sleep phase disorder, DSPD), free-running (free-running sleep disorder, FRSD), and irregular (irregular sleep–wake disorder, ISWD; American Academy of Sleep Medicine, 2005). In ASPD, both sleep onset and final awakening time occur earlier than the desired solar clock time. In DSPD, sleep onset and final awakening are later than desired. The sleep–wake cycle in ASPD, and to a lesser extent in DSPD, maintains a relatively stable phase relationship with solar time. In FRSD, sleep onset and final awakening are typically delayed by 1–2 h every day and thus not entrained to the day–night cycle. ISWD is usually associated with diffuse brain pathology and characterized by at least three short-sleep episodes across the solar day and night. The first primate animal model of ISWD was produced in squirrel monkeys with surgical SCN lesions (Edgar *et al.*, 1993).

In humans, it is not possible to perform mutagenesis screens as in model organisms. Therefore, searching for spontaneous dominant mutations in people with definite CRSDs or strikingly unusual sleep–wake schedules that cluster in families or small kindreds is one alternative strategy to correlate genotype with circadian phenotype. Another approach is to search for statistically significant associations in larger sample sizes of people between more common or subtle circadian rhythm phenotypes and different polymorphisms in genes known or suspected to influence circadian function. In either case, phenotyping the circadian rhythm of human research participants is also performed primarily by two approaches. The first critical property often used to characterize any circadian "clock" is its endogenous period of oscillation (tau). Historically, this was measured in the absence of any obvious internal or external stimuli that reset the phase of the oscillation or masked its overt expression (e.g., in continuous dim light); the so-called "free-running period" (Wever, 1970). However, in the absence of daily time cues under such constant laboratory conditions, complex temporal patterns of whole-organism physiology and behavior, such as internal desynchrony, that confound the estimation of human tau can arise (Czeisler and Dijk, 2001). "Forced desynchrony" protocols were then devised to circumvent these limitations of free-running experiments by distributing rest and activity times uniformly across all circadian clock times (Carskadon and Dement, 1975; Czeisler *et al.*, 1990; Johnson *et al.*, 1992). Unfortunately, the prolonged and strict conditions of forced desynchrony protocols are generally only attempted in physically and psychologically robust individuals and are formidably expensive. Other complex properties of the circadian system in an intact mammal have led to discussion of whether tau itself is actually invariant (Campbell, 2000; Daan, 2001; Scheer *et al.*, 2007). Nonetheless, free-running estimates of tau can be both readily obtained and revealing when animal models bearing a human transgene suspected of altering the human clock are found to have an altered tau compared to wild-type (WT) littermates (Xu *et al.*, 2005, 2007).

When the organism under study is a human, and forced desynchrony measurements of tau are not feasible, then more emphasis is placed on investigating the second critical property of any circadian system by estimating the (usually relatively stable) temporal relationship ("phase angle") between some measureable marker of the phase of the biological clock and a relevant marker of the phase of the external day–night cycle (the "phase angle of entrainment"). Though an advanced or delayed phase angle of entrainment between the end of the sleep period and solar dawn (e.g., ASPD or DSPD, respectively) may be associated with a shorter or longer tau, respectively, the circadian system (including the "clock") is sufficiently complicated that this is not necessarily the case (Carskadon *et al.*, 1999; Dijk *et al.*, 2000). Phase angles of entrainment can be measured under constant laboratory conditions in far less time than a forced desynchrony protocol using the "constant routine" (Mills *et al.*, 1978) but are still more time consuming and expensive as well as less widely available than assessments in the community setting. The latter exposes the research subject to their habitual, but variable, entrainment signals from the environment though this source of noise can be mitigated to a degree by averaging over many days. As a general rule then, human circadian studies outside the laboratory are usually based largely on measurements of the phase angle of entrainment of different measurable outputs of the circadian system relative to solar time. Some of those outputs are more tightly coupled to the SCN clock than others (see below).

One reliable marker of circadian phase is the dim light melatonin onset (DLMO), which can be measured from blood or saliva samples obtained in the home or during an evening in the laboratory if light exposure, posture, activity, eating, and wakefulness are controlled (Lewy and Sack, 1989; Sletten *et al.*, 2010). As in any human research, care must be taken not to extrapolate the accuracy and precision of measurement techniques from one population (e.g., the young) to all populations (e.g., the elderly; Gooneratne *et al.*, 2003). However, the saliva DLMO does appear more promising in several nonelderly populations (Leibenluft *et al.*, 1996; Martin and Eastman, 2002; Nagtegaal *et al.*, 1998). Another commonly used phase marker is core body temperature, which is easier to monitor in the laboratory setting than the DLMO, but perhaps more easily masked by other factors (Reinberg, 1975).

The melatonin and temperature rhythms are generally considered to more closely reflect the molecular-genetic phase of the SCN circadian clock than the rest–activity schedule and sleep–wake oscillation (which are influenced by multiple other factors including the sleep homeostat; Edgar *et al.*, 1993) but are usually more expensive and time consuming than the latter. With the development of small, wrist-worn movement detectors called actigraphs, rest–activity rhythm measurement has become quite convenient for the research participant and reliably records wrist acceleration in the ambulatory setting (Ancoli-Israel et al., 2003). Actigraphy recordings correlate well with melatonin and core body temperature rhythms in some subjects, probably when sleeping on their preferred intrinsic schedule, but like other assessments of overt behavior are subject to masking effects and must be considered as influenced only partially by the circadian clock in combination with multiple other factors. In order to assess circadian phase in a larger number of human subjects, questionnaires such as the Horne–Ostberg (Horne and Ostberg, 1976) and Munich Chronotype (Roenneberg et al., 2003) were developed to assess sleep–wake preferences. Chronotypes identified by the Horne–Ostberg questionnaire correlate with body temperature rhythms (Horne and Ostberg, 1976) and circadian period (Duffy et al., 2001) in selected, but not all, populations. More recently, a novel multichannel ambulatory recorder interpreted with the help of a multiple regression model has been proposed (Kolodyazhniy et al., 2011).

II. INSIGHTS GAINED FROM FAMILIAL ADVANCED SLEEP PHASE SYNDROME

FASPS (Familial Advanced Sleep Phase Syndrome) was first identified as a highly penetrant autosomal dominant trait in three large Utah families in which affected individuals exhibit very early sleep onset and offset (Jones et al., 1999). When the Horne–Ostberg questionnaire, which measures "morning lark" or "night owl" tendency, was performed on clinically affected and unaffected family members, individuals thought to be phase-advanced did score significantly higher than unaffected relatives. FASPS appears to be early onset: the youngest affected individual was 8 years old, and most FASPS subjects knew they were obligate "morning larks" by 30 years of age, which is distinctly different from ASPS caused by aging (Campbell et al., 1989; Haimov and Lavie, 1997). In one of the families, FASPS subjects demonstrated a 4-h phase advance of the time of sleep onset, sleep offset, first slow wave sleep, and first rapid eye movement (REM) sleep compared to that of the controls, although sleep quality and quantity were not significantly different between the two groups. Narcolepsy, obstructive sleep apnea, "restless legs" syndrome, and depression were ruled out as possible causes of early sleep onset. Consistent with the sleep–wake cycle, DLMO and core body temperature rhythms were also advanced by approximately 4 h in FASPS subjects. Sleep–wake and temperature rhythms of one FASPS subject were monitored in a time isolation facility and showed a very short

circadian period of 23.3 h (Fig. 7.1), which is substantially shorter than that
of control subjects (24.2 h) and is consistent with an advanced sleep–wake
cycle phase.

A. Identifying and characterizing a causative mutation in PERIOD 2

To determine the genetic mechanism underlying FASPS, linkage analysis was
performed for one large family, which mapped the allele to chromosome 2qter
(Toh *et al.*, 2001). Further physical mapping was carried out and localized the
mutation to a homolog of the *Drosophila period* gene, human *PERIOD 2* (h*PER2*;
Table 7.1). A base change at position 2106 (A–G) of the h*PER2* cDNA was
identified, which results in substitution of a serine at amino acid 662 with a
glycine (S662G). This change cosegregates with the ASPS phenotype except for
one branch of the family with a few "affected" individuals who do not carry this
mutation and are believed to be FASPS phenocopies. These phenocopies demonstrate the challenge and complexity of human behavioral genetics, as these
strong "morning lark" individuals evaluated by Horne–Ostberg questionnaire do
not actually carry the FASPS mutation segregating in the family. At the same
time, this also emphasizes the power of large families in human genetic analysis.

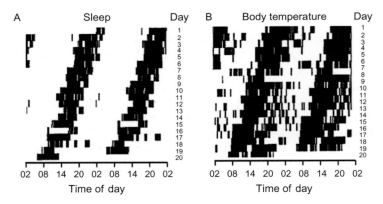

Figure 7.1. Free-running period of sleep/wake and body temperature cycles in a FASPS patient.
Sleep/wake (A) and body temperature (B) rhythms of a 69-year-old female monitored in
time isolation for 18 days. The data are double plotted. (A) Filled bars indicate periods
of sleep derived from polygraphically recorded sleep scored using "standard" criteria. (B)
Filled bars indicate periods when body temperature is below the daily mean. The free-
running period of both variables is 23.3 h based on chi-squared periodogram (adapted
from Jones *et al.*, 1999).

Table 7.1. Known and Putative Human Circadian Genes and Their Roles in Regulation of Circadian Rhythms and Sleep

Gene	Circadian rhythm and sleep phenotype	Comment
hPER1	Morningness	
hPER2	ASPD and morningness	Causative mutation for ASPD
hPER3	DSPD and morningness	Predisposition for DSPD
hCLOCK	Eveningness	Not reproducible according to some studies
hCKIdelta	ASPD	Causative mutation for ASPD
hCKIepsilon	DSPD and FRSD	Protection against DSPD and FRSD
hDEC2	Short sleep	Causative mutation for short sleep
hCOP9S3	ISWD	Patients have Smith–Magenis syndrome

ASPD, advanced sleep phase disorder; DSPD, delayed sleep phase disorder; FRSD, free-running sleep disorder; ISWD, irregular sleep–wake disorder.

Functional characterization was carried out to establish whether the S662G mutation causes FASPS. The *in vitro* study using PER2 truncation mutants demonstrates that S662 is located within the casein kinase I (CKI) epsilon binding region and the S662G mutation causes hypophosphorylation by CKIepsilon (Toh *et al.*, 2001). Sequence analysis of hPER2 reveals four additional serine residues that are C-terminal to S662 and each with two amino acids in between (i.e., S665, S668, S671, and S674), consistent with the CKI recognition consensus motif (Toh *et al.*, 2001). Moreover, mutating S662 to aspartate (S662D), which mimics a phosphoserine, restores CKI-dependent phosphorylation (Toh *et al.*, 2001), suggesting that S662 is a phosphorylation site on PER2.

To characterize the functional consequences of the S662G mutation *in vivo*, transgenic mice carrying wild-type hPER2, as well as hPER2 with S662G or S662D mutations, were generated using a human bacterial artificial chromosome (BAC) which carries the *cis*-acting genomic regulatory elements that can faithfully recapitulate endogenous PER2 expression (Xu *et al.*, 2007). The behavior analysis showed that the S662G transgenic mice exhibited ~2 h shorter free-running period, whereas the S662D mice exhibited 0.5 h lengthening of period compared to wild-type mice. Under 12 h light:12 h dark (12L:12D) conditions, the S662G mice showed ~4 h phase advance of locomotor activity rhythms which is almost identical to that of human FASPS subjects that carry this mutation. The S662G mutation does not appear to significantly affect PER2 degradation or nuclear localization. Instead, the mutation affects *PER2* transcript levels. In the transgenic mice, both hPER2 and the endogenous mouse *Per2* (mPer2) mRNA levels peaked earlier for S662G and later for S662D compared to wild-type mice, corresponding to the shorter and longer behavioral periods, respectively. Moreover, the mRNA levels were lower in S662G mice and higher in S662D mice relative to wild-type mice. Because both mutant hPER2 and the endogenous wild-type mPer2 transcript levels are reduced in the S662G mice; this argues for reduced transcriptional activity rather than reduced PER2 mRNA stability.

B. Identifying and characterizing a causative mutation in CKI delta

As the number of genes identified for the molecular clock from model organisms and human genome sequence data increased rapidly in the 1990s and early 2000s, candidate gene approach for identifying causative mutations in FASPS became possible. Exon sequencing was performed on circadian genes for individuals that belong to a moderate-sized family with FASPS (Xu et al., 2005). This led to the identification of a second mutation that causes FASPS, which is a threonine-to-alanine alteration at amino acid 44 of CKIdelta (CKIdelta-T44A; Table 7.1), a residue that is conserved with other mammalian CKIs and Drosophila CKI (dDBT). In vitro kinase assay demonstrated that this mutation results in decreased phosphorylation of both exogenous substrates (phosvitin and alpha-casein) and circadian substrates (PER1-3). To examine the effects of this mutation on circadian rhythms in vivo, BAC transgenic mice carrying either the wild-type (hCKIdelta-WT) or the mutant (hCKIdelta-T44A) human CKI delta were generated. The behavioral period under free-running condition was significantly shorter in the mutant transgenic mice compared to wild-type mice, consistent with the phase-advanced phenotype of the human patients that carry this mutation. Neither CKIdelta$^{+/-}$ nor hCKIdelta-WT mice exhibit altered period, suggesting that the period is not affected by CKIdelta gene dosage. Thus, the shorter period observed in hCKIdelta-T44A transgenic mice is due to the T44A mutation and not altered CKIdelta gene dosage. Interestingly, expression of hCKIdelta-T44A in Drosophila circadian neurons resulted in longer period compared to expression of wild-type hCKIdelta (hCKIdelta-WT). This may reflect differences in the regulatory mechanism of the mammalian clock versus invertebrate clock.

C. Characterization of the interaction between PERIOD 2 and CKI

The hPER2-S662G and hCKIdelta-T44A mutations indicate that phosphorylation of PER2 by CKI is critical for circadian timing in humans. Indeed, biochemical characterization supports the idea that the four serine residues C-terminal to S662 are phosphorylated by CKI (Xu et al., 2007). In vitro phosphorylation assays using PER2 peptides that encompass residues from 660 to 674 demonstrated that

PER2 peptide with a phosphate covalently linked to S662 was phosphorylated at the other residues by CKI, whereas PER2 peptide without a phosphate at S662 was not phosphorylated by CKI. Moreover, a quantitative assay showed that an additional four moles of phosphate were incorporated per mole of the PER2 peptide, corresponding to the four serine residues C-terminal to S662. Subsequent phosphoamino acid analysis revealed that the threonine and tyrosine residues on the peptide were not phosphorylated, implying that phosphorylation occurred at the serine residues. Taken together, these results suggest that phosphorylation at S662 of PER2 serves as a priming event that is critical for a cascade of phosphorylations downstream of S662 by CKI.

To characterize the functional relevance of the interaction between PER2 and CKI *in vivo*, hPER2 transgenic mice were crossed with both hCKIdelta-WT transgenic and CKIdelta knockout mice (Xu et al., 2007). As described earlier in this chapter, hPER2-S662G transgenic mice exhibit a short period of ~22 h, whereas neither hCKIdelta-WT transgenic nor CKIdelta$^{+/-}$ exhibit altered circadian period. However, in mice carrying both hPER2-S662G and hCKIdelta-WT transgenes, the period is shorter than hPER2-S662G single transgenic animals by over 1 h. Consistently, expressing hPER2-S662G in CKIdelta$^{+/-}$ background slightly lengthens the period compared to expressing hPER2-S662G in wild-type background. However, hPER2-S662D transgenic mice which show long period on wild-type background exhibit even longer period in CKIdelta$^{+/-}$ background and a shorter period in hCKIdelta-WT background. Therefore, decreasing CKIdelta dosage lengthens period for hPER2 transgenic mice whether it is S662G or S662D. Similarly, increasing CKIdelta dosage shortens period for these mice regardless of the mutation at S662.

Collectively, these results led to the proposal of the following model regarding how CKI acts on PER2 to regulate circadian period: CKI phosphorylates the serine residues downstream of S662 on PER2 after S662 is phosphorylated by a priming kinase. Phosphorylation in this region of PER2 increases PER2 mRNA and thus protein, while CKI likely phosphorylates some other site(s) that result in degradation of PER2. The net effect of these two opposing processes determines the level of PER2 and in turn sets the circadian period (Fig. 7.2A). In wild-type background, the balance of these opposing effects can be maintained, thus decreasing or increasing CKIdelta gene dosage does not change the period (Fig. 7.2B and C). In the case of S662G mutation, the S662 residue can no longer be phosphorylated by the priming kinase, leading to hypophosphorylation of the downstream residues by CKI. Therefore the net effect of CKI on PER2 results in reduced PER2 levels and a shorter period (Fig. 7.2D). Decreasing CKIdelta gene dosage partially suppresses the period shortening effect by reducing phosphorylation-mediated PER2 degradation (Fig. 7.2E), whereas increasing CKIdelta gene dosage further shortens the period by enhancing phosphorylation-mediated PER2 degradation (Fig. 7.2F).

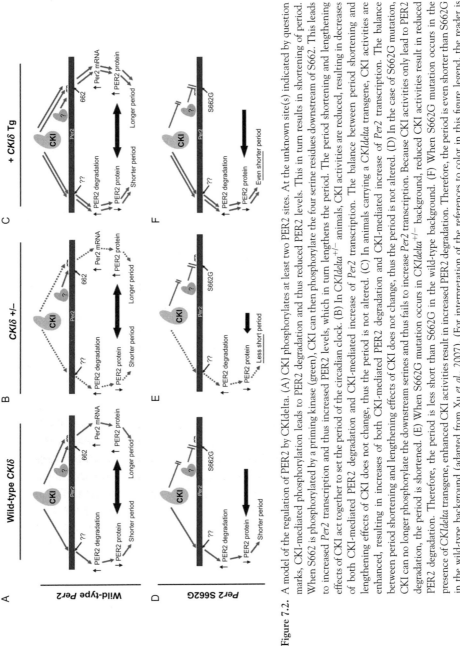

Figure 7.2. A model of the regulation of PER2 by CKIdelta. (A) CKI phosphorylates at least two PER2 sites. At the unknown site(s) indicated by question marks, CKI-mediated phosphorylation leads to PER2 degradation and thus reduced PER2 levels. When S662 is phosphorylated by a priming kinase (green), CKI can then phosphorylate the four serine residues downstream of S662. This leads to increased *Per2* transcription and thus increased PER2 levels, which in turn lengthens the period. The period shortening and lengthening effects of CKI act together to set the period of the circadian clock. (B) In *CKIdelta+/−* animals, CKI activities are reduced, resulting in decreases of both CKI-mediated PER2 degradation and CKI-mediated increase of *Per2* transcription. The balance between period shortening and lengthening effects of CKI does not change, thus the period is not altered. (C) In animals carrying a *CKIdelta* transgene, CKI activities are enhanced, resulting in increases of both CKI-mediated PER2 degradation and CKI-mediated increase of *Per2* transcription. The balance between period shortening and lengthening effects of CKI does not change, thus the period is not altered. (D) In the case of S662G mutation, CKI can no longer phosphorylate the downstream serines and thus fails to increase *Per2* transcription. Because CKI activities only lead to PER2 degradation, the period is shortened. (E) When S662G mutation occurs in *CKIdelta+/−* background, reduced CKI activities result in reduced PER2 degradation. Therefore, the period is less short than S662G in the wild-type background. (F) When S662G mutation occurs in the presence of *CKIdelta* transgene, enhanced CKI activities result in increased PER2 degradation. Therefore, the period is even shorter than S662G in the wild-type background (adapted from Xu et al., 2007). (For interpretation of the references to color in this figure legend, the reader is referred to the Web version of this chapter.)

D. Identification of a mutation in DEC2 that leads to short-sleep phenotype

In the process of identifying mutations underlying FASPS, a small family with two individuals that have lifelong shorter daily sleep times emerged (He *et al.*, 2009; Fig. 7.3). These individuals demonstrate sleep-offset times that are much earlier than nonaffected family members and general controls, similar to the early sleep-offset times of FASPS patients. However, the sleep-onset time in these two individuals is similar to conventional sleepers. This results in about 2 h shorter total sleep time per 24-h day in the affected individuals compared to nonaffected family members.

Candidate gene sequencing revealed a point mutation in human *DEC2* (h*DEC2*) that cosegregated with this short-sleep phenotype (He *et al.*, 2009; Table 7.1). Mouse *Dec2* (m*Dec2*) is a basic helix-loop-helix (bHLH) transcription factor that represses the transcriptional activities of CLOCK/BMAL1 (Honma *et al.*, 2002). The mutation in h*DEC2* causes a proline-to-arginine alteration at amino acid position 384 (P384R), which is located in a highly conserved region of the protein (He *et al.*, 2009). This mutation results in attenuated DEC2 repressive activity of CLOCK/BMAL1-mediated transcriptional activity in cell culture.

To validate that the P384R mutation is indeed causing the short-sleep phenotype, BAC transgenic mice were generated to carry wild-type h*DEC2* (h*DEC2-WT*) or h*DEC2-P384R*. h*DEC2-P384R* mice do not exhibit an altered circadian period, but the duration of the activity period (alpha) is 1.2 h longer relative to h*DEC2-WT* transgenic, wild-type littermates, and *Dec2* knockout

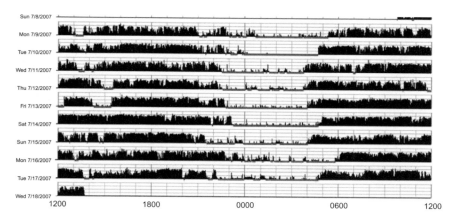

Figure 7.3. Activity recording of a *DEC2-P384R* mutation carrier. Filled bars indicate periods of activity by wrist actigraphy. Extended periods of activity can be observed (adapted from He *et al.*, 2009).

mice. This corresponds to shorter sleep duration (i.e., inactive period) in humans. Moreover, when hDEC2-P384R is expressed in *Dec2* knockout background, alpha is further lengthened (~2.5 h).

To assay the effects of the *DEC2-P384R* mutation on sleep, electroencephalography (EEG) and electromyography (EMG) were performed on mutant transgenic mice and their littermates as control. hDEC2-P384R mice were awake for significantly longer period of time during the light phase compared to wild type, accompanied by significant reduction of both non-rapid eye movement (NREM) and REM sleep. Analysis of sleep architecture demonstrated decreased wake duration and an increase in the number of wake episodes in hDEC2-P384R mice relative to wild-type mice. In addition, these animals also showed significantly more NREM episodes during the light phase, but each episode was shorter in duration. These results indicate that sleep (in particular NREM sleep) is more fragmented in hDEC2-P384R mice than that of wild-type mice. To better understand the role of DEC2 in sleep regulation, hDEC2-P384R mice and wild-type littermates were subjected to acute sleep deprivation. hDEC2-P384R mice exhibited significantly less rebound in both NREM and REM sleep, as well as a slower recovery of acute sleep loss. hDEC2-P384R mice also showed lower NREM delta power density change after sleep deprivation compared to wild type, which indicates that the depth of the rebound of NREM sleep following acute sleep deprivation is affected in hDEC2-P384R animals. Consistent with the mammalian data, expressing mDec2-P384R in the sleep/rest center of *Drosophila* brain led to significantly less sleep-like behavior with decreased sleep bout duration and increased sleep bout number compared to flies expressing mDec2-WT. Collectively, these results demonstrate DEC2 as an important player in the regulation of sleep homeostasis.

III. INSIGHTS GAINED FROM ASSOCIATION STUDIES

Our understanding of the other CRSDs, including DSPD, FRSD, ISWD, and diurnal preference primarily thus far come from association studies (Table 7.1). Structural polymorphism in hPER3 (Dijk *et al.*, 2000) has been shown to be associated with DSPD and extreme diurnal preference, although there are some discrepancies in that certain polymorphisms are associated with DSPD (eveningness) in some populations but morningness in others (Archer *et al.*, 2003, 2010; Ebisawa *et al.*, 2001; Johansson *et al.*, 2003; Pereira *et al.*, 2005). However, a missense variation in hCKIepsilon may play a protective role in DSPD and FRSD (Takano *et al.*, 2004). In addition, polymorphisms in hPER1, hPER2, and hCLOCK have all been implicated in contributing to diurnal preferences (Carpen *et al.*, 2005, 2006; Katzenberg *et al.*, 1998), although there has been some controversies regarding the reported hCLOCK polymorphism as well, with some studies failing to reproduce the association between hCLOCK polymorphism and evening preference (Iwase *et al.*, 2002; Pedrazzoli *et al.*, 2007; Robilliard *et al.*, 2002). Patients with Smith–Magenis syndrome (SMS) exhibit disrupted sleep patterns similar to individuals with ISWD, and the majority of SMS patients also exhibit inverted circadian rhythm of melatonin (Boudreau *et al.*, 2009; Potocki *et al.*, 2000). SMS patients demonstrate haploinsufficiency for subunit 3 of the COP9 signalsome (hCOP9S3), which is conserved from plants to human (Potocki *et al.*, 2000). This may contribute to the alteration in circadian rhythms and sleep. Interestingly, COP9 has recently been shown to be required for light-mediated clock resetting in *Drosophila* (Knowles *et al.*, 2009). One possible explanation for the discrepant results among different association studies is that the linkage disequilibrium of the genetic variant that is responsible for the phenotype and the variant being investigated is different in different populations. This could contribute to different results from studies carried out in different populations, whereas heterogeneity within a population may contribute to different results from studies performed in the same population. Other contributors to the discrepant results include less than optimal phenotyping, as well as complex gene–gene and gene–environment interactions that could influence the phenotype. A more detailed discussion regarding this issue is reviewed in Allebrandt and Roenneberg (2008).

All in all, these discrepant results demonstrate the great difficulties and challenges faced by researchers studying the genetics of human circadian rhythms. At the same time, they underscore the powerfulness of Mendelian circadian variants with extreme phenotypes (such as FASPS mutations described earlier in this chapter) in probing the molecular mechanism of the human circadian clock. On the other hand, insights gained from association studies may reveal more regarding genetic adaptation to the environment and the complex gene network that function together to regulate circadian rhythms.

IV. FUTURE DIRECTIONS

Based on years of research on the genetics of the human circadian clock, one important lesson we learned is that most insights are gained from studies of extreme phenotypes. Although these phenotypes are often quite rare, they are most likely to reveal novel components of the core clock and/or characterize the regulation and function of domains/motifs of known clock genes. Findings from these extreme phenotypes have significantly advanced our understanding of the molecular mechanism underlying human circadian rhythms and sleep. As more and more FASPD and FDSPD families are characterized, and genome sequencing becoming cheaper and more plausible, we will be able to identify more players in the regulation of circadian rhythms and sleep in human with high efficiency and precision. Some of these may be conserved clock components characterized in model organisms, whereas others may be novel members that are unique to the human clock. Learning about the similarities and differences between the human clock and that of model organisms can assist us in understanding not only the regulatory mechanism but also the underlying function of circadian rhythm and sleep. Moreover, most of the screens to identify circadian genes in model organisms focus on mutations that alter circadian period under free-running conditions, whereas in human subjects, the focus is on mutations that lead to altered phase in the presence of entrainment signals. Therefore, the search for human circadian genes may reveal novel components involved in clock entrainment.

As a fundamental property of life, it is not surprising that disrupted circadian rhythms have been implicated as a contributor to various disorders and diseases. Besides sleep disorders which have been extensively discussed here, psychiatric and neurodegenerative diseases may also be influenced by altered circadian rhythms and sleep (Wulff *et al.*, 2010). Clock genes have been shown to be involved in diseases and disorders including, but not limited to, cancer (Fu *et al.*, 2002; Gery *et al.*, 2006), epilepsy (Gachon *et al.*, 2004), hypertension (Doi *et al.*, 2010), and metabolic disorders such as diabetes and obesity (Marcheva *et al.*, 2009). Therefore, identifying human circadian genes and characterizing the regulatory mechanism of circadian rhythms and sleep will facilitate the development of treatment for a myriad of disorders/diseases and ultimately improve human health.

References

Allebrandt, K. V., and Roenneberg, T. (2008). The search for circadian clock components in humans: New perspectives for association studies. *Braz. J. Med. Biol. Res.* **41,** 716–721.

American Academy of Sleep Medicine (2005). Diagnostic and Coding Manual. American Academy of Sleep Medicine, Westchester, IL.

Ancoli-Israel, S., Cole, R., Alessi, C., Chambers, M., Moorcroft, W., and Pollak, C. P. (2003). The role of actigraphy in the study of sleep and circadian rhythms. *Sleep* **26,** 342–392.

Archer, S. N., Robilliard, D. L., Skene, D. J., Smits, M., Williams, A., Arendt, J., and von Schantz, M. (2003). A length polymorphism in the circadian clock gene Per3 is linked to delayed sleep phase syndrome and extreme diurnal preference. *Sleep* **26,** 413–415.

Archer, S. N., Carpen, J. D., Gibson, M., Lim, G. H., Johnston, J. D., Skene, D. J., and von Schantz, M. (2010). Polymorphism in the PER3 promoter associates with diurnal preference and delayed sleep phase disorder. *Sleep* **33,** 695–701.

Boudreau, E. A., Johnson, K. P., Jackman, A. R., Blancato, J., Huizing, M., Bendavid, C., Jones, M., Chandrasekharappa, S. C., Lewy, A. J., Smith, A. C., and Magenis, R. E. (2009). Review of disrupted sleep patterns in Smith-Magenis syndrome and normal melatonin secretion in a patient with an atypical interstitial 17p11.2 deletion. *Am. J. Med. Genet. A* **149A,** 1382–1391.

Campbell, S. (2000). Is there an intrinsic period of the circadian clock? *Science* **288,** 1174–1175.

Campbell, S. S., Gillin, J. C., Kripke, D. F., Erikson, P., and Clopton, P. (1989). Gender differences in the circadian temperature rhythms of healthy elderly subjects: Relationships to sleep quality. *Sleep* **12,** 529–536.

Carpen, J. D., Archer, S. N., Skene, D. J., Smits, M., and von Schantz, M. (2005). A single-nucleotide polymorphism in the 5'-untranslated region of the hPER2 gene is associated with diurnal preference. *J. Sleep Res.* **14**, 293–297.

Carpen, J. D., von Schantz, M., Smits, M., Skene, D. J., and Archer, S. N. (2006). A silent polymorphism in the PER1 gene associates with extreme diurnal preference in humans. *J. Hum. Genet.* **51**, 1122–1125.

Carskadon, M. A., and Dement, W. C. (1975). Sleep studies on a 90-minute day. *Electroencephalogr. Clin. Neurophysiol.* **39**, 145–155.

Carskadon, M. A., Labyak, S. E., Acebo, C., and Seifer, R. (1999). Intrinsic circadian period of adolescent humans measured in conditions of forced desynchrony. *Neurosci. Lett.* **260**, 129–132.

Czeisler, C. A., and Dijk, D. (2001). Human Circadian Physiology and Sleep-Wake Regulation. Kluwer Academic/Plenum Publishers, New York.

Czeisler, C. A., Allan, J., and Kronauer, R. E. (1990). A Method for Assaying the Effects of Therapeutic Agents on the Period of the Endogenous Circadian Pacemaker in Man. Oxford University Press, New York.

Daan, S., and Aschoff, J. (eds.), (2001). The Entrainment of Circadian Systems, pp. 7–43. Kluwer Academic/Plenum, New York. Chapter 1, Vol. 12.

Dijk, D. J., Duffy, J. F., and Czeisler, C. A. (2000). Contribution of circadian physiology and sleep homeostasis to age-related changes in human sleep. *Chronobiol. Int.* **17**, 285–311.

Doi, M., Takahashi, Y., Komatsu, R., Yamazaki, F., Yamada, H., Haraguchi, S., Emoto, N., Okuno, Y., Tsujimoto, G., Kanematsu, A., Ogawa, O., Todo, T., *et al.* (2010). Salt-sensitive hypertension in circadian clock-deficient Cry-null mice involves dysregulated adrenal Hsd3b6. *Nat. Med.* **16**, 67–74.

Duffy, J. F., Rimmer, D. W., and Czeisler, C. A. (2001). Association of intrinsic circadian period with morningness-eveningness, usual wake time, and circadian phase. *Behav. Neurosci.* **115**, 895–899.

Dunlap, J. C. L., Loros, J. J., and DeCoursey, P. J. (eds.), (2004). Chronobiology: Biological Timekeeping, pp. 1–406. Sinauer Associate, Sunderland.

Ebisawa, T., Uchiyama, M., Kajimura, N., Mishima, K., Kamei, Y., Katoh, M., Watanabe, T., Sekimoto, M., Shibui, K., Kim, K., Kudo, Y., Ozeki, Y., *et al.* (2001). Association of structural polymorphisms in the human period3 gene with delayed sleep phase syndrome. *EMBO Rep.* **2**, 342–346.

Edgar, D. M., Dement, W. C., and Fuller, C. A. (1993). Effect of SCN lesions on sleep in squirrel monkeys: Evidence for opponent processes in sleep-wake regulation. *J. Neurosci.* **13**, 1065–1079.

Fu, L., Pelicano, H., Liu, J., Huang, P., and Lee, C. (2002). The circadian gene Period2 plays an important role in tumor suppression and DNA damage response in vivo. *Cell* **111**, 41–50.

Gachon, F., Fonjallaz, P., Damiola, F., Gos, P., Kodama, T., Zakany, J., Duboule, D., Petit, B., Tafti, M., and Schibler, U. (2004). The loss of circadian PAR bZip transcription factors results in epilepsy. *Genes Dev.* **18**, 1397–1412.

Gery, S., Komatsu, N., Baldjyan, L., Yu, A., Koo, D., and Koeffler, H. P. (2006). The circadian gene per1 plays an important role in cell growth and DNA damage control in human cancer cells. *Mol. Cell* **22**, 375–382.

Gooneratne, N. S., Metlay, J. P., Guo, W., Pack, F. M., Kapoor, S., and Pack, A. I. (2003). The validity and feasibility of saliva melatonin assessment in the elderly. *J. Pineal Res.* **34**, 88–94.

Haimov, I., and Lavie, P. (1997). Circadian characteristics of sleep propensity function in healthy elderly: A comparison with young adults. *Sleep* **20**, 294–300.

Hastings, M. H., Maywood, E. S., and Reddy, A. B. (2008). Two decades of circadian time. *J. Neuroendocrinol.* **20**, 812–819.

He, Y., Jones, C. R., Fujiki, N., Xu, Y., Guo, B., Holder, J. L., Jr., Rossner, M. J., Nishino, S., and Fu, Y. H. (2009). The transcriptional repressor DEC2 regulates sleep length in mammals. *Science* **325**, 866–870.

Honma, S., Kawamoto, T., Takagi, Y., Fujimoto, K., Sato, F., Noshiro, M., Kato, Y., and Honma, K. (2002). Dec1 and Dec2 are regulators of the mammalian molecular clock. *Nature* **419**, 841–844.

Horne, J. A., and Ostberg, O. (1976). A self-assessment questionnaire to determine morningness-eveningness in human circadian rhythms. *Int. J. Chronobiol.* **4**, 97–110.

Iwase, T., Kajimura, N., Uchiyama, M., Ebisawa, T., Yoshimura, K., Kamei, Y., Shibui, K., Kim, K., Kudo, Y., Katoh, M., Watanabe, T., Nakajima, T., *et al.* (2002). Mutation screening of the human Clock gene in circadian rhythm sleep disorders. *Psychiatry Res.* **109**, 121–128.

Johansson, C., Willeit, M., Levitan, R., Partonen, T., Smedh, C., Del Favero, J., Bel Kacem, S., Praschak-Rieder, N., Neumeister, A., Masellis, M., Basile, V., Zill, P., *et al.* (2003). The serotonin transporter promoter repeat length polymorphism, seasonal affective disorder and seasonality. *Psychol. Med.* **33**, 785–792.

Johnson, M. P., Duffy, J. F., Dijk, D. J., Ronda, J. M., Dyal, C. M., and Czeisler, C. A. (1992). Short-term memory, alertness and performance: A reappraisal of their relationship to body temperature. *J. Sleep Res.* **1**, 24–29.

Jones, C. R., Campbell, S. S., Zone, S. E., Cooper, F., DeSano, A., Murphy, P. J., Jones, B., Czajkowski, L., and Ptacek, L. J. (1999). Familial advanced sleep-phase syndrome: A short-period circadian rhythm variant in humans. [see comments]*Nat. Med.* **5**, 1062–1065.

Katzenberg, D., Young, T., Finn, L., Lin, L., King, D. P., Takahashi, J. S., and Mignot, E. (1998). A CLOCK polymorphism associated with human diurnal preference. *Sleep* **21**, 569–576.

Knowles, A., Koh, K., Wu, J. T., Chien, C. T., Chamovitz, D. A., and Blau, J. (2009). The COP9 signalosome is required for light-dependent timeless degradation and Drosophila clock resetting. *J. Neurosci.* **29**, 1152–1162.

Kolodyazhniy, V., Spati, J., Frey, S., Gotz, T., Wirz-Justice, A., Krauchi, K., Cajochen, C., and Wilhelm, F. H. (2011). Estimation of human circadian phase via a multi-channel ambulatory monitoring system and a multiple regression model. *J. Biol. Rhythms* **26**, 55–67.

Leibenluft, E., Feldman-Naim, S., Turner, E. H., Schwartz, P. J., and Wehr, T. A. (1996). Salivary and plasma measures of dim light melatonin onset (DLMO) in patients with rapid cycling bipolar disorder. *Biol. Psychiatry* **40**, 731–735.

Lewy, A. J., and Sack, R. L. (1989). The dim light melatonin onset as a marker for circadian phase position. *Chronobiol. Int.* **6**, 93–102.

Marcheva, B., Ramsey, K. M., Affinati, A., and Bass, J. (2009). Clock genes and metabolic disease. *J. Appl. Physiol.* **107**, 1638–1646.

Martin, S. K., and Eastman, C. I. (2002). Sleep logs of young adults with self-selected sleep times predict the dim light melatonin onset. *Chronobiol. Int.* **19**, 695–707.

McNamara, P., Seo, S. B., Rudic, R. D., Sehgal, A., Chakravarti, D., and FitzGerald, G. A. (2001). Regulation of CLOCK and MOP4 by nuclear hormone receptors in the vasculature: A humoral mechanism to reset a peripheral clock. *Cell* **105**, 877–889.

Mills, J. N., Minors, D. S., and Waterhouse, J. M. (1978). Adaptation to abrupt time shifts of the oscillator(s) controlling human circadian rhythms. *J. Physiol.* **285**, 455–470.

Nagtegaal, E., Peeters, T., Swart, W., Smits, M., Kerkhof, G., and van der Meer, G. (1998). Correlation between concentrations of melatonin in saliva and serum in patients with delayed sleep phase syndrome. *Ther. Drug Monit.* **20**, 181–183.

Pedrazzoli, M., Louzada, F. M., Pereira, D. S., Benedito-Silva, A. A., Lopez, A. R., Martynhak, B. J., Korczak, A. L., Koike Bdel, V., Barbosa, A. A., D'Almeida, V., and Tufik, S. (2007). Clock polymorphisms and circadian rhythms phenotypes in a sample of the Brazilian population. *Chronobiol. Int.* **24**, 1–8.

Pereira, D. S., Tufik, S., Louzada, F. M., Benedito-Silva, A. A., Lopez, A. R., Lemos, N. A., Korczak, A. L., D'Almeida, V., and Pedrazzoli, M. (2005). Association of the length polymorphism in the human Per3 gene with the delayed sleep-phase syndrome: Does latitude have an influence upon it? *Sleep* **28**, 29–32.

Potocki, L., Glaze, D., Tan, D. X., Park, S. S., Kashork, C. D., Shaffer, L. G., Reiter, R. J., and Lupski, J. R. (2000). Circadian rhythm abnormalities of melatonin in Smith-Magenis syndrome. *J. Med. Genet.* **37**, 428–433.

Reinberg, A. (1975). Circadian changes in the temperature of human beings. *Bibl. Radiol.* 128–139.

Robilliard, D. L., Archer, S. N., Arendt, J., Lockley, S. W., Hack, L. M., English, J., Leger, D., Smits, M. G., Williams, A., Skene, D. J., and Von Schantz, M. (2002). The 3111 Clock gene polymorphism is not associated with sleep and circadian rhythmicity in phenotypically characterized human subjects. *J. Sleep Res.* **11**, 305–312.

Roenneberg, T., Wirz-Justice, A., and Merrow, M. (2003). Life between clocks: Daily temporal patterns of human chronotypes. *J. Biol. Rhythms* **18**, 80–90.

Scheer, F. A., Wright, K. P., Jr., Kronauer, R. E., and Czeisler, C. A. (2007). Plasticity of the intrinsic period of the human circadian timing system. *PLoS One* **2**, e721.

Sletten, T. L., Vincenzi, S., Redman, J. R., Lockley, S. W., and Rajaratnam, S. M. (2010). Timing of sleep and its relationship with the endogenous melatonin rhythm. *Front. Neurol.* **1**, 137.

Takano, A., Uchiyama, M., Kajimura, N., Mishima, K., Inoue, Y., Kamei, Y., Kitajima, T., Shibui, K., Katoh, M., Watanabe, T., Hashimotodani, Y., Nakajima, T., *et al.* (2004). A missense variation in human casein kinase I epsilon gene that induces functional alteration and shows an inverse association with circadian rhythm sleep disorders. *Neuropsychopharmacology* **29**, 1901–1909.

Toh, K. L., Jones, C. R., He, Y., Eide, E. J., Hinz, W. A., Virshup, D. M., Ptacek, L. J., and Fu, Y. H. (2001). An hPer2 phosphorylation site mutation in familial advanced sleep phase syndrome. *Science* **291**, 1040–1043.

Wever, R. (1970). The effects of electric fields on circadian rhythmicity in men. *Life Sci. Space Res.* **8**, 177–187.

Wulff, K., Gatti, S., Wettstein, J. G., and Foster, R. G. (2010). Sleep and circadian rhythm disruption in psychiatric and neurodegenerative disease. *Nat. Rev. Neurosci.* **11**, 589–599.

Xu, Y., Pdaiath, Q., Shapiro, R., Jones, C. R., Wu, S. M., Saigoh, N., Saitgoh, K., Ptacek, L., and Fu, Y. (2005). Functional consequences of a CKId mutation causing familial advanced sleep phase syndrome. *Nature* **434**, 640–644.

Xu, Y., Toh, K. L., Jones, C. R., Shin, J. Y., Fu, Y. H., and Ptacek, L. J. (2007). Modeling of a human circadian mutation yields insights into clock regulation by PER2. *Cell* **128**, 59–70.

Index

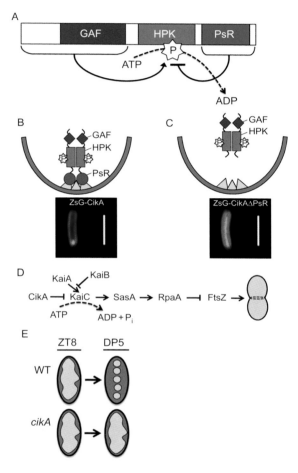

Chapter 2, Figure 2.3. (See Page 32 of this volume).

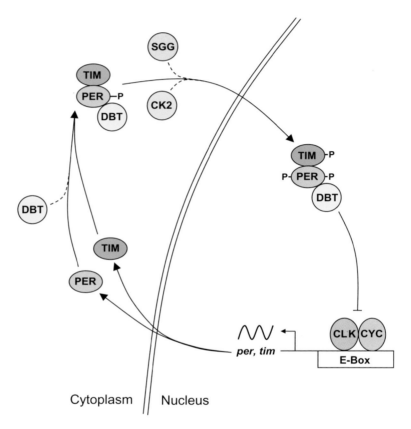

Chapter 5, Figure 5.1. (See Page 146 of this volume).

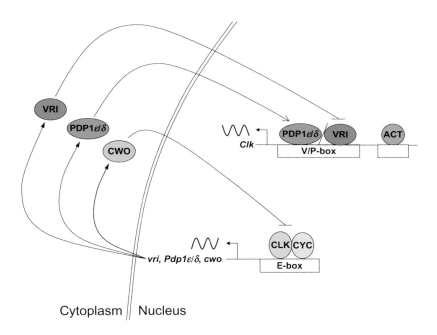

Chapter 5, Figure 5.2. (See Page 148 of this volume).

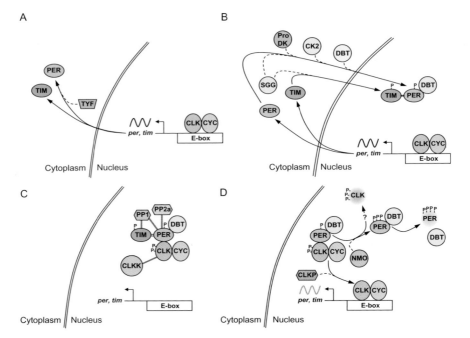

Chapter 5, Figure 5.3. (See Page 151 of this volume).

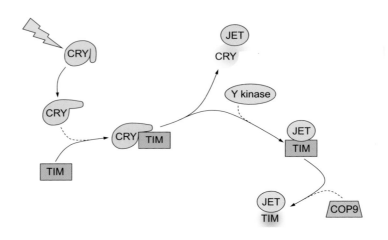

Chapter 5, Figure 5.4. (See Page 157 of this volume).